BIOCHEMISTRY AND MOLECULAR GENETICS OF CANCER METASTASIS

DEVELOPMENTS IN ONCOLOGY

F.J. Cleton and J.W.I.M. Simons, eds.: Genetic Origins of Tumour Cells. 90-247-2272-1.
J. Aisner and P. Chang, eds.: Cancer Treatment and Research. 90-247-2358 2.
B.W. Ongerboer de Visser, D.A. Bosch and W.M.H. van Woerkom-Eykenboom, eds.: Neuro-oncology: Clinical and Experimental Aspects. 90-247-2421-X.
K. Hellmann, P. Hilgard and S. Eccles, eds.: Metastasis: Clinical and Experimental Aspects. 90-247-2424-4.
H.F. Seigler, ed.: Clinical Management of Melanoma. 90-247-2584-4.
P. Correa and W. Haenszel, eds.: Epidemiology of Cancer of the Digestive Tract. 90-247-2601-8.
L.A. Liotta and I.R. Hart, eds.: Tumour Invasion and Metastasis. 90-247-2611-5.
J. Banoczy, ed.: Oral Leukoplakia. 90-247-2655-7.
C. Tijssen, M. Halprin and L. Endtz, eds.: Familial Brain Tumours. 90-247-2691-3.
F.M. Muggia, C.W. Young and S.K. Carter, eds.: Anthracycline Antibiotics in Cancer. 90-247-2711-1.
B.W. Hancock, ed.: Assessment of Tumour Response. 90-247-2712-X.
D.E. Peterson, ed.: Oral Complications of Cancer Chemotherapy. 0-89838-563-6.
R. Mastrangelo, D.G. Poplack and R. Riccardi, eds.: Central Nervous System Leukemia. Prevention and Treatment. 0-89838-570-9.
A. Polliack, ed.: Human Leukemias. Cytochemical and Ultrastructural Techniques in Diagnosis and Research. 0-89838-585-7.
W. Davis, C. Maltoni and S. Tanneberger, eds.: The Control of Tumor Growth and its Biological Bases. 0-89838-603-9.
A.P.M. Heintz, C. Th. Griffiths and J.B. Trimbos, eds.: Surgery in Gynecological Oncology. 0-89838-604-7.
M.P. Hacker, E.B. Double and I. Krakoff, eds.: Platinum Coordination Complexes in Cancer Chemotherapy. 0-89838-619-5.
M.J. van Zwieten. The Rat as Animal Model in Breast Cancer Research: A Histopathological Study of Radiation- and Hormone-Induced Rat Mammary Tumors. 0-89838-624-1.
B. Lowenberg and A. Hogenbeck, eds.: Minimal Residual Disease in Acute Leukemia. 0-89838-630-6.
I. van der Waal and G.B. Snow, eds.: Oral Oncology. 0-89838-631-4.
B.W. Hancock and A.M. Ward, eds.: Immunological Aspects of Cancer. 0-89838-664-0.
K.V. Honn and B.F. Sloane, eds.: Hemostatic Mechanisms and Metastasis. 0-89838-667-5.
K.R. Harrap, W. Davis and A.N. Calvert, eds.: Cancer Chemotherapy and Selective Drug Development. 0-89838-673-X.
V.D. Velde, J.H. Cornelis and P.H. Sugarbaker, eds.: Liver Metastasis. 0-89838-648-5.
D.J. Ruiter, K. Welvaart and S. Ferrone, eds.: Cutaneous Melanoma and Precursor Lesions. 0-89838-689-6.
S.B. Howell, ed.: Intra-Arterial and Intracavitary Cancer Chemotherapy. 0-89838-691-8.
D.L. Kisner and J.F. Smyth, eds.: Interferon Alpha-2: Pre-Clinical and Clinical Evaluation. 0-89838-701-9.
P. Furmanski, J.C. Hager and M.A. Rich, eds.: RNA Tumor Viruses, Oncogenes, Human Cancer and Aids: On the Frontiers of Understanding. 0-89838-703-5.
J.E. Talmadge, I.J. Fidler and R.K. Oldham: Screening for Biological Response Modifiers: Methods and Rationale. 0-89838-712-4.
J.C. Bottino, R.W. Opfell and F.M. Muggia, eds.: Liver Cancer. 0-89838-713-2.
P.K. Pattengale, R.J. Lukes and C.R. Taylor, eds.: Lymphoproliferative Diseases: Pathogenesis, Diagnosis, Therapy. 0-89838-725-6.
F. Cavalli, G. Bonadonna and M. Rozencweig, eds.: Malignant Lymphomas and Hodgkin's Disease. 0-89838-727-2.
L. Baker, F. Valeriote and V. Ratanatharathorn, eds.: Biology and Therapy of Acute Leukemia. 0-89838-728-0.
J. Russo, ed.: Immunocytochemistry in Tumor Diagnosis. 0-89838-737-X.
R.L. Ceriani, ed.: Monoclonal Antibodies and Breast Cancer. 0-89838-739-6.
D.E. Peterson, G.E. Elias and S.T. Sonis, eds.: Head and Neck Management of the Cancer Patient. 0-89838-747-7.
D.M. Green: Diagnosis and management of Malignant Solid Tumors in Infants and Children. 0-89838-750-7.
K.A. Foon and A.C. Morgan, Jr., eds.: Monoclonal Antibody Therapy of Human Cancer. 0-89838-754-X.
J.G. McVie, et al, eds., Clinical and Experimental Pathology of Lung Cancer. 0-89838-764-7.
K.V. Honn, W.E. Powers and B.F. Sloane, eds.: Mechanisms of Cancer Metastasis. 0-89838-765-5.

BIOCHEMISTRY AND MOLECULAR GENETICS OF CANCER METASTASIS

Proceedings of the Symposium on Biochemistry and Molecular Genetics of Cancer Metastasis
Bethesda, Maryland — March 18–20, 1985

edited by

Karoly Lapis
Semmelweis Medical University
Budapest, Hungary

Lance A. Liotta
National Institutes of Health
Bethesda, Maryland

Alan S. Rabson
National Institutes of Health
Bethesda, Maryland

Martinus Nijhoff Publishing
a member of the Kluwer Academic Publishers Group
Boston / Dordrecht / Lancaster

Distributors for North America:
Kluwer Academic Publishers
190 Old Derby Street
Hingham, Massachusetts 02043, USA

Distributors for the UK and Ireland:
Kluwer Academic Publishers
MTP Press Limited
Falcon House, Queen Square
Lancaster LA1 1RN, UNITED KINGDOM

Distributors for all other countries:
Kluwer Academic Publishers Group
Distribution Centre
Post Office Box 322
3300 AH Dordrecht, THE NETHERLANDS

Library of Congress Cataloging-in-Publication Data

Symposium on Biochemistry and Molecular Genetics of
Cancer Metastasis (1985 : Bethesda, Md.)
Biochemistry and molecular genetics of cancer
metastasis.

(Developments in oncology)
Includes bibliographies and index.
1. Metastasis—Congresses. 2. Pathology, Molecular—
Congresses. 3. Cancer—Genetic aspects—Congresses.
4. Cancer—Immunological aspects—Congresses.
5. Molecular genetics—Congresses. I. Lapis, Károly.
II. Liotta, L. A. (Lance A.) III. Rabson, Alan S.,
1926- . IV. Title. V. Series. [DNLM: 1. Neoplasm
Metastasis—congresses. W1 DE998N / QZ 202 S9795b 1985]
RC269.S39 1985 616.99'407 85-30980
ISBN 0-89838-785-X

Printed in the United States of America

CONTENTS

5-2058

2/24/86

CONTRIBUTING AUTHORS

Altevogt, Dr. P.
Institut für Immunologie und Genetik
Deutsches Krebsforschungszentrum
D-6900 Heidelberg
Federal Republic of Germany

Balaban, Dr. G.
Department of Human Genetics
University of Pennsylvania School of Medicine
Philadelphia, PA 19104

Basara, Dr. M.
Department of Laboratory Medicine and Pathology
University of Minnesota
Minneapolis, MN 55455

Blick, Dr. M.
Department of Clinical Immunology and Biological Therapy
The University of Texas
M.D. Anderson Hospital and Tumor Institute
Houston, TX 77030

Brys, Dr. L.
Instituut Voor Moleculaire Biologie
Vrije Universiteit Brussel
B 1640 St.-Genesius-Rode
Paardenstraat 65
Brussels, Belgium

Buck, Dr. C. A.
The Wistar Institute
36th & Spruce Streets
Philadelphia, PA 19104

Carter, Dr. S. K.
Vice President, Anti-Cancer Research
Bristol-Myers Company
Pharmaceutical Research and Developmental Division
345 Park Avenue
New York, NY 10154

Chambers, Dr. A. F.
The London Regional Cancer Centre
and Department of Radiation Oncology
University of Western Ontario
London, Ontario, N6A 4G5, Canada

Damsky, Dr. C. H.
Departments of Oral Biology and Anatomy
Schools of Dentistry and Medicine
HSW 681
University of California
San Francisco, CA 94341

De Baetselier, Dr. P.
Instituut Voor Moleculaire Biologie
Vrije Universiteit Brussel
B 1640 St.-Genesius-Rode
Paardenstraat 65
Brussels, Belgium

Dennis, Dr. J.
Department of Pathology
Richardson Laboratory
Queen's University
Kingston, Canada

Dus, Dr. D.
Department of Tumor Immunology
The Ludwik Hirszfeld Institute of Immunology
and Experimental Therapy
Polish Academy of Sciences
Wroclaw, Poland

Eisenbach, Dr. L.
Department of Cell Biology
The Weizmann Institute of Science
Rehovot, Israel

Feldman, Dr. M.
Department of Cell Biology
The Weizmann Institute of Science
Rehovot, Israel

Fidler, Dr. Isaiah J.
Department of Cell Biology
University of Texas
M.D. Anderson Hospital and Tumor Institute
6723 Bertner Avenue
Houston, TX 77030

Franke, Dr. W. W.
Institute of Cell and Tumor Biology
German Cancer Research Center
D-6900 Heidelberg
Federal Republic of Germany

Frazier, Dr. M.
Department of Medicine
Baylor College of Medicine
Houston, TX 77030

Frost, Dr. P.
Department of Cell Biology
University of Texas System Cancer Center
M.D. Anderson Hospital and Tumor Institute
Houston, TX 77030

Furcht, Dr. L. T.
Stone Professor of Pathology
Department of Laboratory Medicine and Pathology
University of Minnesota Medical School
Box 609 Mayo Memorial Building
420 Delaware Street S.E.
Minneapolis, MN 55455-0315

Gallick, Dr. G.
Department of Tumor Biology
University of Texas
M.D. Anderson Hospital and Tumor Institute
6723 Bertner Avenue
Houston, TX 77030

Gorelik, Dr. E.
Biological Therapeutics Branch
Biological Response Modifiers Program
Division of Cancer Treatment
National Cancer Institute
Frederick, MD 21701

Greig, Dr. R.
Assistant Director
Department of Tumor Biology L-109
Smith Kline & French Laboratories
1500 Spring Garden Street
P.O. Box 7929
Philadelphia, PA 19101

Gruber, Dr. P.
The Wistar Institute
36th & Spruce Streets
Philadelphia, PA 19104

Gutterman, Dr. J. U.
Clinical Immunology
M.D. Anderson Hospital and Tumor Institute
Houston, TX 77030

Hämmerling, Dr. G.
Institute for Immunology and Genetics
German Cancer Research Center
D-6900 Heidelberg
Federal Republic of Germany

Helman, Dr. L. J.
Pediatric Branch
National Cancer Institute
National Institutes of Health
Bethesda, MD 20205

Herberman, Dr. R. B.
Biological Therapeutics Branch
Biological Response Modifier Program, NCI
Frederick Cancer Research Facility
P.O Box B
Frederick, MD 21701

Horwitz, Dr. A. F.
Department of Biochemistry and Biophysics
University of Pennsylvania
Philadelphia, PA 19104

Irimura, Dr. T.
Department of Tumor Biology
The University of Texas
M.D. Anderson Hospital and Tumor Institute
Houston, TX 77030

Israel, Dr. M. A.
Pediatric Branch
National Cancer Institute
National Institutes of Health
Bethesda, MD 20205

Katzav, Dr. S.
Department of Cell Biology
The Weizmann Institute of Science
Rehovot, Israel

Kerbel, Dr. R. S.
Mount Sinai Research Institute
Division of Cancer Research
600 University Avenue
Toronto, Ontario, Canada, M5G 1X5

Knudsen, Dr. K. A.
Lankenau Medical Research Center
Philadelphia, PA 19151

Kopper, Dr. L.
I. Institute of Pathology and Experimental Cancer Research
Semmelweis Medical University
1085 Budapest, Hungary

Koprowski, Dr. H.
The Wistar Institute of Anatomy and Biology
36th Street at Spruce
Philadelphia, PA 19104

Kurzrock, Dr. R.
Clinical Immunology
M.D. Anderson Hospital and Tumor Institute
Houston, TX 77030

La Biche, Dr. R.
Department of Tumor Biology
The University of Texas
M.D. Anderson Hospital and Tumor Institute
Houston, TX 77030

Lapis, Dr. K.
I. Institute of Pathology and Experimental Cancer Research
Semmelweis Medical University
1085 Budapest, Hungary

Linehan, Dr. W. M.
Surgery Branch
National Cancer Institute
National Institutes of Health
Bethesda, MD 20205

Liotta, Dr. L. A.
Laboratory of Pathology
National Cancer Institute
National Institutes of Health
Bethesda, MD 20205

Liteplo, Dr. R. G.
Department of Pathology
Queen's University
Kingston, Ontario, Canada

McCarthy, Dr. J.
Department of Laboratory Medicine and Pathology
University of Minnesota
Minneapolis, MN 55455

Moll, Dr. R.
Institute of Pathology
University of Mainz Medical School
D-6500 Mainz
Federal Republic of Germany

Mulé, Dr. J. J.
Surgery Branch
Division of Cancer Treatment
National Cancer Institute
Bethesda, MD 20205

Nicolson, Dr. G. L.
Professor and Chairman
Department of Tumor Biology (108)
University of Texas System Cancer Center
M.D. Anderson Hospital & Tumor Institute
Houston, TX 77030

Norden-Skubitz, Dr. A.
Department of Laboratory Medicine and Pathology
University of Minnesota
Minneapolis, MN 55455

Nowell, Dr. P. C.
Department of Pathology and Laboratory Medicine
University of Pennsylvania
School of Medicine
Philadelphia, PA 19104

Opolski, Dr. A.
Department of Tumor Immunology
The Ludwik Hirszfeld Institute of Immunology
and Experimental Therapy
Polish Academy of Sciences
Wroclaw, Poland

Ortaldo, Dr. J. R.
Biological Therapeutics Branch
Biological Response Modifiers Program
Division of Cancer Treatment
National Cancer Institute
Frederick, MD 21701

Pal, Dr. K.
I. Institute of Pathology and Experimental Cancer Research
Semmelweis Medical University
1085 Budapest, Hungary

Palm, Dr. S.
Department of Laboratory Medicine and Pathology
University of Minnesota
Minneapolis, MN 55455

Pauli, Dr. B. U.
Department of Pathology
Rush-Presbyterian-St. Luke's Medical Center
1753 West Congress Parkway
Chicago, IL 60612

Radzikowski, Dr. C.
Department of Tumor Immunology
The Ludwik Hirszfeld Institute of Immunology
and Experimental Therapy
Polish Academy of Sciences
Wroclaw, Poland

Reading, Dr. C. L.
Department of Tumor Biology
The University of Texas
M.D. Anderson Hospital and Tumor Institute
Houston, TX 77030

Remels, Dr. L.
Instituut Voor Moleculaire Biologie
Vrije Universiteit Brussel
B 1640 St.-Genesium-Rode
Paardenstraat 65
Brussels, Belgium

Roos, Dr. E.
Division of Cell Biology
The Netherlands Cancer Institute
121 Plesmanlaan
1066 CX Amsterdam, The Netherlands

Rosenberg, Dr. S. A.
Surgery Branch
Division of Cancer Treatment
National Cancer Institute
Bethesda, MD 20205

Rotter, Dr. V.
Department of Biophysics
The Weizmann Institute of Science
Rehovot 76100, Israel

Salup, Dr. R. R.
Biological Therapeutics Branch
Biological Response Modifiers Program
Division of Cancer Treatment
National Cancer Institute
Frederick, MD 21701

Sas, Dr. D.
Department of Laboratory Medicine and Pathology
University of Minnesota
Minneapolis, MN 55455

Schirrmacher, Dr. V.
Institut für Immunologie und Genetik
Deutsches Krebsforschungszentrum
D-6900 Heidelberg
Federal Republic of Germany

Segal, Dr. S.
Department of Microbiology
Ben Gurion University
Beer-Sheva, Israel

Sobel, Dr. M. E.
Laboratory of Pathology
National Cancer Institute
National Institutes of Health
Bethesda, MD 20205

Strzadala, Dr. L.
Department of Tumor Immunology
The Ludwik Hirszefld Institute of Immunology
and Experimental Therapy
Polish Academy of Sciences
Wroclaw, Poland

Talmadge, Dr. J. E.
Preclinical Screening Laboratory
Program Resources, Inc.
National Cancer Institute
Fredrick Cancer Research Facility
Frederick, MD 21701

Thiele, Dr. C. J.
Pediatric Branch
National Cancer Institute
National Institutes of Health
Bethesda, MD 20205

Thorgeirsson, Dr. U. P.
Laboratory of Pathology
National Cancer Institute
National Institutes of Health
Bethesda, MD 20205

Timar, Dr. F.
I. Institute of Pathology and Experimental Cancer Research
Semmelweis Medical University
1085 Budapest, Hungary

Timar, Dr. J.
I. Institute of Pathology and Experimental Cancer Research
Semmelweis Medical University
1085 Budapest, Hungary

Tryggvason, Dr. K.
Department of Biochemistry and Department of Pathology
University of Medicine and Dentistry, New Jersey
Rutgers Medical School
Piscataway, NJ 08854

Turpeenniemi-Hujanen, Dr. T.
Laboratory of Pathology
National Cancer Institute
National Institutes of Health
Bethesda, MD 20205

Waller, Dr. C. A.
Institut für Immunologie und Genetik
Deutsches Krebsforschungszentrum
D-6900 Heidelberg
Federal Republic of Germany

Weinstein, Dr. R. S.
Department of Pathology
Rush-Presbyterian-St. Luke's Medical Center
1753 West Congress Parkway
Chicago, IL 60612

Wheelock, Dr. M. J.
The Wistar Institute
36th & Spruce Streets
Philadelphia, PA 19104

Wiltrout, Dr. R. H.
Biological Therapeutics Branch
Biological Response Modifiers Program
Division of Cancer Treatment
National Cancer Institute
Frederick, MD 21701

Wolf, Dr. D.
Department of Biophysics
The Weizmann Institute of Science
Rehovot 76100, Israel

PREFACE

The success rate for treatment of primary neoplasms has improved significantly due to improved surgical, radiotherapy, and chemotherapy methods, and by supportive patient care. In contrast, the treatment of cancer metastases, the cause of most cancer deaths, has not been very successful. Approximately 50% or more of patients with primary malignant neoplasms already have established metastases. Consequently, the most important problem in cancer treatment is the destruction or prevention of metastases.

Metastases research has obvious clinical importance. Yet it has only been recently that investigators have attempted to study the mechanisms involved in this process. This is in part due to the complexity of metastases formation. A metastatic colony is the result of a complicated series of steps involving multiple tumor host interactions. It is expected that multiple biochemical factors and gene products derived both from the host and the tumor cell may be required for the metastasizing tumor cell to invade, survive host defenses, travel in the circulation, arrest and adhere in the target organ, invade out, and grow as a metastatic colony. Some of these factors have recently been identified by investigators who have focused on individual steps in the metastatic process and have employed new technologies in immunology, biochemistry and molecular biology. The purpose of this volume is to capture some of the excitement in the field of metastases based on such new discoveries. The ultimate hope is that these discoveries will be translated into widely adopted clinical methods for the prevention, detection, and eradication of metastases.

Karoly Lapis
Lance A. Liotta
Alan Rabson

Part One

Biochemistry and Molecular Biology Aspects

1

BIOCHEMICAL AND MOLECULAR BIOLOGY APPROACHES TO STUDY CANCER METASTASES

LANCE A. LIOTTA, M.D., Ph.D.

Laboratory of Pathology, National Cancer Institute, National Institutes of Health, Bethesda, MD 20205

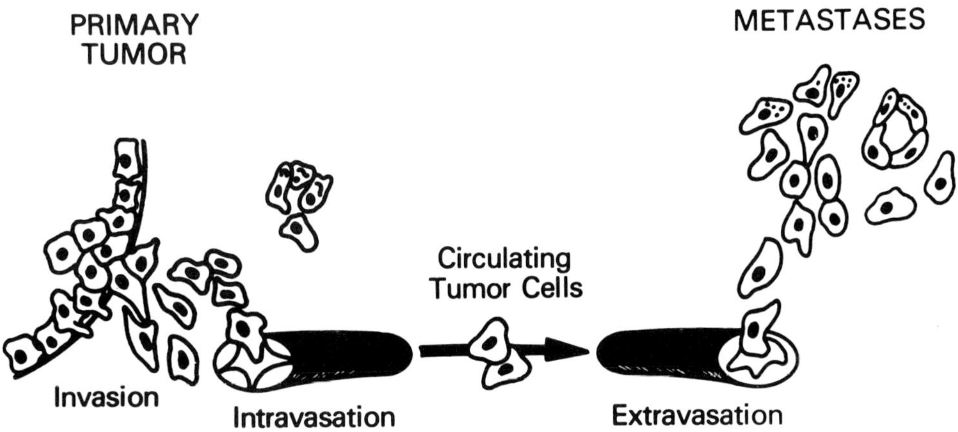

Fig. 1. Diagram of the metastatic cascade. Tumor cells invade at the primary tumor site and enter the interstitial stroma. They thereby gain access to blood vessels for further dissemination. Tumor cells invade the vascular wall and are dislodged into the circulation in single cells and clumps. Circulating tumor cells arrest in the precapillary venules of the target organ by adherence or mechanical wedging. They must then exit the circulation to initiate a secondary tumor colony called a metastasis.

Metastases are the major cause of treatment failure for cancer patients. There is a great need to develop new clinical methods to a) predict the aggressiveness of a patient's individual tumor and b) localize and destroy established metastases growing in internal body sites. Therefore, a major challenge to cancer scientists is to identify biochemical factors (specific gene products) which are augmented in metastatic tumor cells and are functionally associated with their malignant propensity. Measurement of the level of such factors may provide a predictive index of the aggressiveness of the patient's tumor. Antibodies to the same metastatic factors could be applied to radioscintography to localize clinically silent metastases. Finally, pharmacologic agents which bind to and/or block the metastatic factors may be ultimately useful in the therapy of metastases.

A metastasis is the end result of a complex series of steps (Fig. 1) through which tumor cells from the primary mass invade host tissue, enter the circulation, evade host defenses, arrest in the target organ, and invade and grow out in that organ to initiate a metastatic colony. It is apparent that this complex process must involve multiple gene products. Metastatic propensity is assumed to be distinctly separate from tumorigenicity alone. Defined "oncogenes" have been identified which can induce the tumorigenic state. However, the role of oncogenes in the metastatic process has been unclear. A number of laboratories are now attempting to identify "metastases genes" which illicit or augment the metastatic phenotype. The following major approaches are being used to identify metastases genes.

A. Traditional Molecular Biology: Cloning of genes which code for proteins known to be involved in at least one step in the metastatic process (1-4).

B. Somatic Cell Hybridization: Identification of augmented or suppressed gene products in hybrid tumor cells formed from normal cells x metastatic tumor cells or metastatic x nonmetastatic tumor cells (8,9).

C. DNA Transfection: Transfection of high molecular weight DNA fragments or isolated genes (including known oncogenes) into normal or nonmetastatic tumor cells in order to induce the metastatic phenotype (10-13).

D. Differential cDNA Screening: Selection of specific genes which are differentially expressed in metastatic versus nonmetastatic tumor cells (14,15).

Examples of recent progress using some of the above approaches will now be briefly reviewed.

A variety of biochemical factors have been associated with defined phases of invasion or metastases. Cell surface receptors for collagen, glycoproteins such as laminin, or proteoglycans are all postulated to facilitate tumor cell attachment, which is thought to be the first step of invasion (1). Attachment may trigger the release of degradative enzymes which facilitate the penetration of the tumor cells through tissue barriers. Important enzymes include collagenases, serine proteases, and cathepsins. Once the tumor cell has entered the target tissue, specific growth factors may be required for colony formation. Throughout all steps of metastases, evasion of host immune defenses may involve a variety of tumor gene products.

Basement membranes are dense continuous layers of extracellular matrix which separate organ parenchymal cells from the interstitial stroma. They are composed of a type IV collagen backbone, noncollagenous glycoproteins such as laminin and heparin sulfate proteoglycans. Metastasizing tumor cells must penetrate basement membranes at multiple stages in the metastatic process. Histologic and biochemical analysis indicates that such penetration takes place in three steps: step 1: attachment to the basement membrane surface. This may be mediated in large part by tumor cell surface receptors which bind to specific components of the basement membrane such as the glycoprotein laminin, step 2: local dissolution of the basement membrane mediated by tumor cell proteases such as type IV collagenase, and step 3: locomotion of the tumor cell through the basement membrane and the interstitial stroma. The laminin receptor is a 67,000 KDa cell surface protein isolated in our laboratory and thought to play a role in tumor cell attachment to basement membranes (2-4). The unoccupied receptor is present in higher amounts in more aggressive human breast carcinomas. Laminin ligand fragments which block the receptor inhibit or abolish metastases formation by circulating tumor cells (5,6). Monoclonal antibodies to the human breast cancer laminin receptor have been used to isolate a putative cDNA clone for the receptor (4, and M. Sobel et al., manuscript in preparation). The monoclonal antibody (LRmAb) was used to screen a lambda gt11 expression vector cDNA library (Meloy Labs.). When the lambda phage infects bacteria it forms plaques. Since the inserted cDNA expresses the protein of its sequence, the cDNA for a specific protein inserted in the phage can then be identified using antibodies to that protein to screen the plaques. 1.6 million plaques were screened and 6 plaques were selected which exhibited specific immunoreactivity with LRmAb. All six plaques bred true. The recombinant phage was purified from the plaques and the DNA was analyzed by restriction endonuclease mapping. All six clones were found to have an overlapping region containing the same internal restriction sites. This region codes for the presumed laminin receptor binding domain since LRmAb inhibits binding of laminin to the receptor. The cDNA clones were used to prepare a nick translated probe for the laminin receptor. The probe was hybridized to RNA from human tumor cells previously studied for their ability to specifically bind labeled laminin. The probe was specific for a single message of the size sufficient to code for the laminin receptor protein. Furthermore, the level of message correlated with the laminin binding activity of each tumor cell type. The probe should be useful to study the expression of the laminin receptor during tumor progression. Furthermore, knowledge of the laminin receptor amino acid sequence should provide strategies for designing synthetic compounds to block the receptor.

Another example of a cloned gene for a protein involved in metastases is the gene for plasminogen activator (PA) (7). This protease is thought to augment tumor invasion through the generation of plasmin which can degrade extracellular matrix proteins and activate other latent proteases. Analysis of the product of the urokinase PA gene has revealed that the enzyme is secreted as an inactive single-chain pro-urokinase zymogen. It is activated by proteolysis which removes lys residue to generate a two chain active molecule. The catalytic activity is located on the amino terminal portion (residues 1-135) of the chain. The regulation of PA function in tissues may be linked to the activation of the pro-urokinase, and this has a number of implications for tumor invasion.

Somatic cell hybridization is another major approach to study genes involved in metastases. The results from this type of approach must be analyzed carefully since hybrid cells may be unstable and the exact karyotypic features of each hybrid clone may be different. However, it is possible to reliably interpret the data derived from an individual hybridization system in regard to the correlation of a specific gene product with a biologic phenotype. Specifically, the metastatic propensity of individual hybrid cell lines has been compared with their secretion of proteases. In one such study, fusion of tumor cell lines of high and low metastatic propensity resulted in maintenance or enhancement of the high metastatic phenotype (8). When the metastatic cells were fused with normal cells, both the metastatic phenotype and the collagenase IV activity was suppressed. The secretion of basement membrane degrading type IV collagenase in these hybrids correlated well with their metastatic behavior (Table 1). Thus for this system, type IV collagenase may be genetically linked with the expression of the metastatic phenotype. Another proteolytic enzyme, PA, has also been studied in tumor cell hybrids. Ramshaw et al. (9) demonstrated that nonmetastatic hybrids expressed low levels of PA. However, as the hybrids reverted to the metastatic phenotype

Table 1. Collagenase IV activity and *in vivo* behavior of tumor cell hybrids.

Cell line	Collagenase IV activity[a] ng/10^6 cells	Lung metastases formation[b] Nude mice	B6C3F$_1$ mice
B16-F10^{RR}	128 (41)	238 (2-316) (T+)*	200 (100>300) (T+)
UV-2237^{RR}	72 (22)	60 (1-158) (T+)	0 (T+)
K-1735 clone 16	19 (10)	0 (T+)	0 (T+)
C3H-F	3 (1)	0 (T-)	0 (T-)
PEC$_A$	18 (10)	0 (T-)	0 (T-)
PEC$_{NA}$	15 (6)	ND	ND
Cell hybrid			
2237^{RR} X F10	201 (31)	149 (0->400)	>400
F10^{RR} X clone 16-1	203 (64)	46 (9-300)	109 (9-294)
F10^{RR} X clone 16-3	223 (56)	300 (41>300)	259 (22>300)
F10^{RR} X C3H	22 (13)	0 (0-1)	0
F10^{RR} X PEC	32 (13)	0	0
F10^{RR} X PEC$_w$	38 (9)	0	0
2237^{RR} X PEC$_1$	2 (1)	0	0 (0-1)
2237^{RR} X PEC$_q$	6 (5)	0	5 (0-18)

[a]Means from 3-10 expts. The standard errors are given in parentheses.
[b]Median number of lung nodules from 10 animals per group. The ranges are given in parentheses.
[c]PEC - activated (PEC$_a$) or nonactivated (PEC$_{NA}$) B6C3F$_1$ peritoneal macrophages.
[d]C3H-F - mouse embryo fibroblasts.
*Tumorigenicity (T+ or T-). All cell hybrids were highly tumorigenic.

during culture *in vitro*, the PA level was still found to be low. Thus in the latter system, high levels of PA were apparently not required for metastases.

A series of recent reports have demonstrated that transfection of tumor DNA into suitable nontumorigenic and nonmetastatic recipient cells can induce those cells to become fully metastatic. Studies from our laboratory (10) as well as those of Bernstein and Weinberg (11) and Koestler et al. (12) have shown that NIH-3T3 cells transfected with human tumor DNA containing the ras^H oncogene are metastatic when injected intravenously or subcutaneously into nude mice. The ras^H oncogene, by itself, transfected into NIH-3T3 cells can induce these cells to be fully metastatic in nude mice. Further studies by Muschel et al. (13) have shown that the activated ras^H but not the proto-oncogene (normal cellular counterpart of the ras oncogene) can induce NIH-3T3

Table 2. Metastatic behavior of early passage diploid fibroblasts transformed by the c-ras^{H}-oncogene in Nu/Nu mice: Pozzatti et al. (submitted) and Muschel et al. (13). The untransfected fibroblasts were devoid of metastatic propensity.

Cell clones	No. animals injected intravenously	No. animals with metastases	Average no. lung nodules/ mouse $\pm$ standard deviation	Spontaneous metastases assayed within 4 weeks
FH06T1-1 (Chinese hamster lung fibroblasts)	4	4	11 $\pm$ 6	not tested
Ren 1 (rat embryo fibroblasts)	9	9	117 $\pm$ 40	numerous in multiple organs
Ren 2 (rat embryo fibroblasts)	9	9	>200	moderate in number
Ren 3 (rat embryo fibroblasts)	4	4	>200	rare in number

cells to be fully metastatic. A plasmid construct of the proto-oncogene which induces large amounts of the normal p21 protein will induce NIH-3T3 cells to be fully tumorigenic but not metastatic in nude mice. Tumor growth potential alone was not sufficient to induce the metastatic behavior in nude mice. Diploid fibroblasts transfected with the T24 ras^{H} also produced extensive spontaneous metastases (Table 2). Thus, induction of the complete metastatic phenotype by ras^{H} does not require the use of aneuploid or otherwise "unusual" recipient cells such as NIH-3T3 cells.

More than one hypothesis can be set forth to explain how a single transforming gene can induce the complex metastatic process in the appropriate cell type. The oncogene, in an additive fashion, could confer growth potential on cells which are already expressing all the necessary gene products for metastases. A second explanation is that the integrated oncogene induces a genetic instability. This instabilty then results in the production of metastatic variants which are selected _in vivo_. A third possibility is that transformation by the ras oncogene induces a cascade of cellular gene products normally expressed by migrating cells during embryogenesis or tissue remodeling. Evaluation of these alternative hypotheses will be the subject of future research.

We can conclude that ras^H oncogene transformation may provide a useful model system to identify metastatic genes. Through transfection or cDNA library screening, genes can now be identified which are altered in their expression in the metastatic ras^H transformants compared to the benign counterparts. It remains unknown as to whether the ras oncogene plays a necessary role in human tumors. To date, the data is conflicting. Some investigators have found increased expression of ras in invasive metastatic tumors compared to benign counterparts (16). Other investigators have found no difference in ras content in metastases compared to the primary tumor (17,18). Unfortunately, most of these studies could not distinguish the activated ras from the proto-oncogene form.

The molecular genetics approach may be valuable to study the tumor cell host interactions. Bernstein and Weinberg (11) have found that ras^H transformed 3T3 cells which are fully metastatic in nude mice are only tumorigenic in an immunocompetent strain of mice. This is not unexpected since host immune factors are well known to modulate the metastatic process. This group went further to transfect a human tumor DNA segment which could restore the metastatic capacity in this alternate strain of mice. This gene may therefore play an important role, not in the intrinsic aggressiveness of tumor cells, but in their interaction with the host. Genes which regulate major histocompatibility antigens on the tumor cell surface have previously been shown to affect the metastatic process (19). Whether the Bernstein and Weinberg gene is related to MHC antigen expression, remains to be determined.

It is unlikely that one or two genes alone can be responsible for expression of the complete metastatic phenotype. Instead, it would seem more likely that a few key genetic changes (which may include activation of an oncogene) can induce a cascade of genetic expression for multiple gene products necessary for all the steps in metastases. This is supported by recent studies of differential messenger RNA levels in metastatic versus nonmetastatic tumor cells derived from the same parent (14,15). The studies revealed some genes which were expressed in greater amounts in the metastatic compared to the nonmetastatic tumor cells and other genes preferentially expressed in the nonmetastatic tumor cells. It should be possible to clone these genes and study their products. The hope is that analysis of the products of such genes and their regulation may provide strategies for tumor diagnosis and metastases therapy.

REFERENCES

1. Liotta, L.A. In: Important Advances in Oncology 1985 (Eds. V. DeVita, S. Hellman, and S. Rosenberg), J.B. Lippincott, 1985, pp. 28-41.
2. Terranova, V., Rao, C., Kalebic, T., Margulies, I., and Liotta, L.A. Proc. Natl. Acad. Sci. USA 80: 444-448, 1983.
3. Liotta, L.A., Horan Hand, P., Rao, C., Bryant, G., Barsky, S., and Schlom, J. Exp. Cell Res. 156: 117-126, 1985.
4. Wewer, U., Steeg, P., Claysmith, A., Liotta, L., and Sobel, M. Proceedings of the International Symposium on Biology and Chemistry of Basement Membranes. Mishima, Japan, June, 1985.
5. Horan Hand, P., Thor, A., Schlom, J., Rao, C.N., and Liotta, L. Cancer Res. 45: 2713-2719, 1985.
6. Barsky, S.H., Rao, C.N., Williams, J.E., and Liotta, L.A. J. Clin. Invest. 74: 843-848, 1984.
7. Blasi, F. Fogarty International Center Conference, NIH, March, 1985.
8. Turpeenniemi-Hujanen, T., Thorgeirsson, U.P., Hart, I.R., Grant, S., and Liotta, L.A. J. Natl. Cancer Inst. 75: 99-102, 1985.
9. Ramshaw, I.A., Carlsen, S., Wang, H.C., and Badenoch-Jones, P. Int. J. Cancer 32: 471-478, 1983.
10. Thorgeirsson, U., Hujanen, T., Williams, J., Westin, E., Heilman, C., Talmadge, J., and Liotta, L.A. Mol. Cell. Biol. 5: 259-262, 1985.
11. Bernstein, S. and Weinberg, R. Proc. Natl. Acad. Sci. USA 82: 1726-1730, 1985.
12. Koestler, T.P., Sweet, R., Yokoyama, S., Corwin, S., Greig, R., and Poste, G. Proc. 75th Annual Meeting of the American Association for Cancer Research. Abstr. No. 234, p. 49, 1984.
13. Muschel, R., Lowy, D., and Liotta, L.A. Am. J. Pathol. (in press)
14. Steeg, P., Kalebic, T., Claysmith, A., Liotta, L., and Sobel, M. Federation of American Society for Experimental Biology, Proceedings, 44: No. 5, Abstr. 5413, 1985.
15. Nicolson, G.L., Rotter, V., Wolf, D., Irimura, T., Reading, C.L., Blick, M., La Biche, R., and Frazier, M. In: Biochemistry and Molecular Genetics of Cancer Metastases (Eds. L. Liotta, K. Lapis, and A.S. Rabson), Martinus Nijhoff, The Hague, 1985 (in press)
16. Vousden, K. and Marshall, C.J. EMBO J. 3: 913-917, 1984.
17. Gallick, G., Kurzrock, R., Kloetzer, W., Arlinghaus, R., and Gutterman, J. Proc. Natl. Acad. Sci. USA 82: 1795-1799, 1985.
18. Kris, R.M., Avivi, A., Bar-Eli, M., Alon, Y., Carmi, P., Schlessinger, J., and Raz, A. Int. J. Cancer 35: 227-231, 1985.
19. Gorelik, E., Fogel, M., DeBaetselier, P., Katzav, S., Feldman, M., and Segal, S. In: Tumor Invasion and Metastases (Eds. L. Liotta and I. Hart), Martinus Nijhoff, The Hague, 1982, pp. 133-146.

2

USE OF THE CHICK EMBRYO IN STUDYING THE MOLECULAR GENETICS OF METASTASIS

A.F. CHAMBERS

The London Regional Cancer Centre and Department of Radiation Oncology, University of Western Ontario, London, Ontario, N6A 4G5, Canada

INTRODUCTION

Metastasis is an apparently complex process in which cells are required to perform a number of functions as they move from a primary tumor to establish new tumors at distant sites. One of the current challenges to our understanding of this process is to determine how molecular alterations (either genetic or epigenetic) in tumor cells contribute to the apparently separate functions performed by metastatic cells. The difficulty with this approach has been to find assays for metastatic properties that are appropriate to molecular studies. We have developed an assay in the embryonic chick that lends itself readily to molecular genetic analyses of metastatic properties of cells (1). In this review, this assay and the advantages that it offers to molecular studies will be discussed. We have used this assay to begin an examination of genes that can contribute to the metastatic phenotype; some of these studies will also be described here.

ASSAYS FOR METASTATIC PROPERTIES

Metastasis has been studied using a variety of assays, outlined in Fig. 1. Metastasis involves growth of a primary tumor, spread of cells from this tumor to a distant site, and growth of a secondary tumor. Natural metastasis, as it occurs in a host in which a primary tumor arose spontaneously, is obviously difficult to study experimentally; therefore, a variety of experimental assays have been developed.

"Spontaneous" assays attempt to model the whole metastatic process, and involve the implantation of tumor cells in a host animal, the development of a local tumor, and detection of any tumors that arise at distant sites. The primary tumor either can be left *in situ* or can be removed by, for example, surgery or irradiation. This assay cannot be assumed to duplicate natural

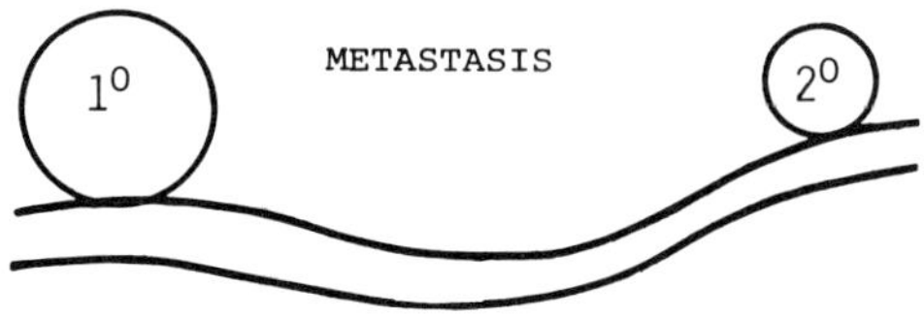

HOW TO MEASURE METASTASIS EXPERIMENTALLY

1. NATURAL METASTASIS
2. "SPONTANEOUS" METASTASIS ASSAYS
3. "EXPERIMENTAL" METASTASIS ASSAYS
4. *IN VITRO* ASSAYS

Fig. 1. Schematic model of metastasis, and approaches that have been taken to study this process.

metastasis, since the immune status of the experimental host may not be identical to that of an animal in which a tumor arose spontaneously. Because it is a time-consuming, multi-step assay that does not lend itself readily to quantitative experimentation, "experimental" metastasis assays (2,3) are often used instead, as models of the latter portion of hematogenous metastasis. In these assays, cells are injected i.v. into a host, circumventing the inital steps of metastasis. More readily quantifiable, these assays have proven to be extremely useful in describing the cell biology of metastatic cells, although most of the assays used present difficulties for molecular studies. Finally, *in vitro* assays of metastatic properties have been used to elucidate specific steps in metastasis (4-6), although our knowledge to date has been insufficient to devise assays covering all steps in metastasis.

Design of experimental metastasis assays involves two critical decisions (Fig. 2): 1) what host animal will be used, and 2) how will metastatic spread be detected. Molecular biological approaches to metastasis put constraints on both of these decisions. First, molecular approaches often involve the testing of genetically-mixed cells (*e.g.*, hybrids, interspecies DNA transfectants, or virally-transformed cells), which are likely to express foreign antigens. These antigens may result in the immune destruction of the

"EXPERIMENTAL" METASTASIS ASSAYS

1. CHOICE OF HOST ANIMAL
   a. Immune-competent eg. mouse, rat.
   b. Immune-deficient eg. nude mouse, chick embryo.

2. CHOICE OF DETECTION PROCEDURE
   a. Tumor formation.
   b. Quantitation of growth.

Fig. 2. Decisions to be made in designing an experimental metastasis assay.

cells, useful only when one wishes to study the role of the immune system in metastasis. For other studies, immune-compromised hosts are required. Second, assays in which detection of tumors is the endpoint require sufficient time for tumors to grow to detectable sizes. Genetically-manipulated cells often express unstable phenotypes and and genotypes (7-11), offering advantages to more rapid detection procedures. Furthermore, metastatic tumors may arise from a sub-population of cells (see 12-14 for review), and during long assay periods, spontaneous generation of new variants and selection in the host may well occur (15-19), making it difficult to know that the cells tested are the ones producing a result. There are advantages to being able to determine properties of whole populations of test cells immediately after their injection into the host. We have devised an experimental assay (1) that optimizes these two choices for a molecular genetic approach to metastasis, namely the ouabain-detection assay in the embryonic chick.

EXPERIMENTAL METASTASIS ASSAY IN THE CHICK EMBRYO

The embryonic chick is used because its immature immune system has been shown to permit growth of a wide variety of heterologous tumor cells following i.v. injection (20-24). The major limitation in the past had been the relatively short assay time permitted (7-10 days) when using tumor formation as the endpoint, since most tumor types did not grow rapidly enough to form detectable tumors during this period and detection of any tumors (e.g. histologically) was laborious and hard to quantify (21,24). This problem was solved with our observation that chick cells are considerably more sensitive to the cytotoxic drug ouabain than are rodent cells, resulting in our development of an assay to quantitate growth of rodent cells in chick liver following i.v. injection (Fig. 3) (1). In this assay, rodent cells are injected into chorioallantoic membrane (CAM) veins of 11-day chick embryos. Cell growth or death in a target organ (e.g., liver) is determined by dissecting out the liver, dissociating it into a suspension of single cells, and plating these cells in vitro in a concentration of ouabain sufficient to kill chicken cells while permitting growth of any rodent cells present. In this way, numbers of rodent cells per liver can be determined. If livers are dissected at various times following injection, a detailed kinetic description of the fate of the injected cells can be obtained.

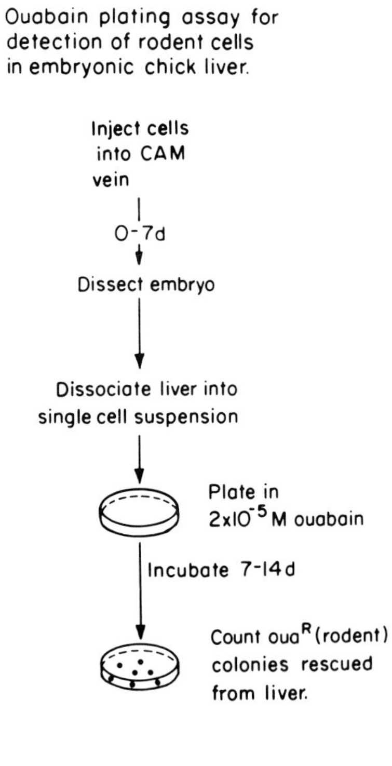

Fig. 3. Experimental metastasis assay and detection scheme in the chick embryo. This assay is described in more detail in Ref. 1.

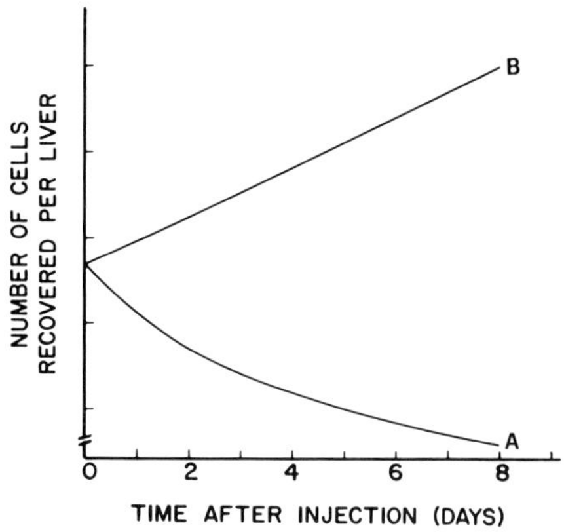

Fig. 4. General pattern of cell kinetics in embryonic liver after injection into CAM veins. Summarized from data presented in Refs. 1, 25-27 and unpublished observations.

Two general kinetic patterns have been observed (Fig. 4) (1, 25-27). The number of cells present in liver immediately following injection is relatively constant for both normal and transformed cells (~1-5% of the original inoculum) and is assumed to represent trapping by the liver. Subsequent behavior, however, depends on the type of cells injected. Normal cells and "normal" (immortalized) tissue culture cells decline in number in liver after their inital trapping (Pattern A). On the other hand, some transformed cells increase in number in liver (Pattern B). This growth following i.v. injection can be detected even when the growth rate is too slow to permit observation of micrometastases at the end of the 7-10 day assay period. This assay thus provides a more sensitive and rapid detection scheme than does detection of visible tumors, and permits description of subtle perturbations in growth rate.

Our goal has been to determine how cells that follow these two different metastatic growth responses (Patterns A vs. B) differ at the molecular genetic level. Two approaches can be taken. The first involves transfer of genomic DNA from metastatic cells (Pattern B) into non-metastatic cells (Pattern A), in hopes that dominant gene(s) will convert the recipient cells to a metastatic phenotype. This strategy has proven successful (25). The second approach is to transfer known genes into non-metastatic cells, to determine what effect (if any) these genes have on the metastatic phenotype of the cells. Some of our work using the latter approach will be described in detail here.

EFFECT OF THE VIRAL SRC ONCOGENE ON EXPERIMENTAL METASTATIC ABILITY

Temperature-sensitive mutants have provided powerful tools for the study of a variety of genetic loci. Because chick embryos do not regulate their body temperature and can develop over a range of temperatures (28), they are ideal hosts for studying the effects of temperature-sensitive mutations on metastatic properties. Little has been known about the effects of viral oncogenes on metastatic properties, in part because infection of cells with oncogenic viruses often leads to expression of antigenic cell surface molecules. We have examined the experimental metastatic ability of rat cells infected with an avian sarcoma virus that is temperature-sensitive for *in vitro* measures of *src* oncogene function (26).

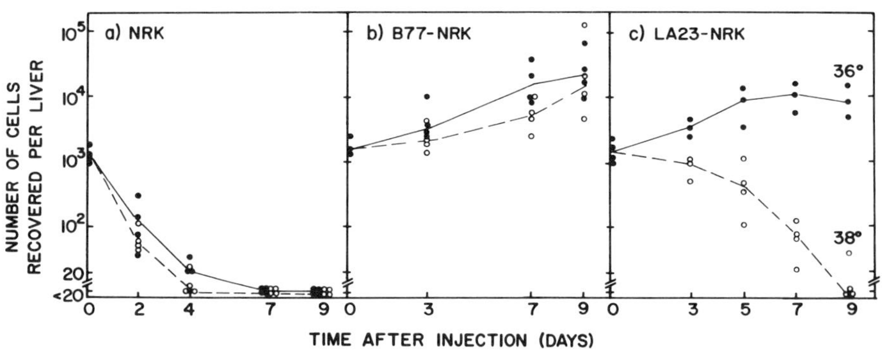

Fig. 5. Kinetics of growth in embryonic liver after injection of (a) NRK, (b) B77-NRK and (c) LA23-NRK cells (5×10^4 per embryo) into CAM veins of 11-day chick embryos. After injection, embryos were maintained at 36°C (●) or 38°C (○). Each point represents the number of viable rodent cells present in the liver of one embryo; lines connect median points. Figure reprinted with permission from Ref. 26.

Cells used in these studies were: NRK cells (Normal Rats Kidney cells, a "normal" cell line [29]), B77-NRK cells (NRK cells infected with a wild-type avian sarcoma virus, B77) and LA23-NRK (NRK cells infected with tsLA23, a mutant avian sarcoma virus that is temperature-sensitive for src gene kinase function and for src-related in vitro transformation properties [30-34]). As expected, normal NRK cells, when injected i.v. into chick embryos, did not grow (Fig. 5a). Cells transformed with the wild-type virus (B77-NRK cells) grew well after i.v. injection (Fig. 5b). Neither of these responses was temperature-dependent. On the other hand, LA23-NRK cells grew in chick liver only at 36°C (the permissive temperature, at which cells were transformed in vitro). At the non-permissive temperature (38°C, at which cells behaved normally in vitro), these cells followed a pattern similar to that of normal NRK cells and failed to grow (Fig. 5c).

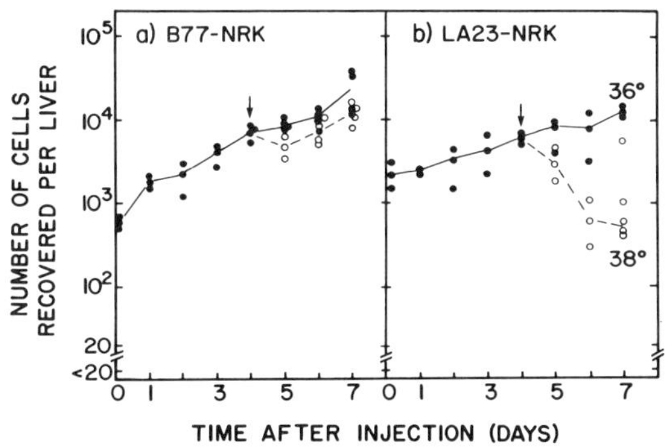

Fig. 6. Kinetics of growth in embryonic liver in temperature shift experiments. Experiments were performed as described in Fig. 5, except embryos were maintained at 36° (●) for four days after injection, at which point half of the embryos were shifted to 38°C (○). Reprinted with permission from Ref. 26.

To determined whether src function was required for maintenance of growth in liver once established, a temperature-shift experiment was performed (Fig. 6). LA23-NRK cells were injected and growth in liver was permitted for 4 days at 36°, at which point some of the embryos were transferred to 38°. A change in growth response, to that of normal cells, was detectable within one day of the temperature shift (Fig. 6b), showing clearly that src function contributes to both establishment and maintenance of growth in chick liver following i.v. injection.

The results presented in Fig. 6 demonstrate the quantitative sensitivity of this assay. The rapid response to changes in growth conditions obviously could not be detected in assays using tumor-formation as the endpoint. Such detailed day-to-day monitoring of cell growth is a useful feature of this assay in the study of the molecular biology of metastasis.

EFFECT OF THE HUMAN H-ras ONCOGENE ON EXPERIMENTAL METASTATIC ABILITY

Members of the ras oncogene family have been shown to transform NIH3T3 cells in vitro (35-37), and in some cases these cells have been shown to form tumors in nude mice (38). The relationship between morphological transformation in vitro and aspects of tumor growth and progression in vivo, however, is not well understood. Since oncogenic viruses may be able to act as mutagens, long-term assays increase opportunities for further genetic changes and selection, and may yield misleading conclusions about the immediate effects of oncogene transformation. The chick embryo assay, by measuring the initial responses of injected cells, avoids this difficulty. We have used this assay to examine metastatic properties of NIH3T3 cells recently transformed with the cloned human T24 bladder oncogene (H-ras)(27). To determine the relationship between in vitro transformation and metastatic proliferative ability, we also examined the metastatic properties of spontaneously-arising, morphologically transformed NIH3T3 cells.

H-ras-transformed NIH3T3 cells grew in embryonic chick liver following i.v. injection, whereas control cells (NIH3T3 cells and NIH3T3 cells transfected with NIH3T3 DNA) did not (Fig. 7)(27). These results were from two independent pools of ~100 ras-transformed in vitro foci; similar results were obtained when individual clones of ras-transformed cells were tested. Molecular analysis of the ras-transformed cells showed that they contained on average ~1-2 ras gene copies per cell, and that reinjection of cells obtained from chick livers did not select for increased growth ability or for increased gene copy number (data not shown). It thus appears that a single copy of the activated H-ras oncogene per NIH3T3 cell is sufficient to confer experimental metastatic ability on these cells.

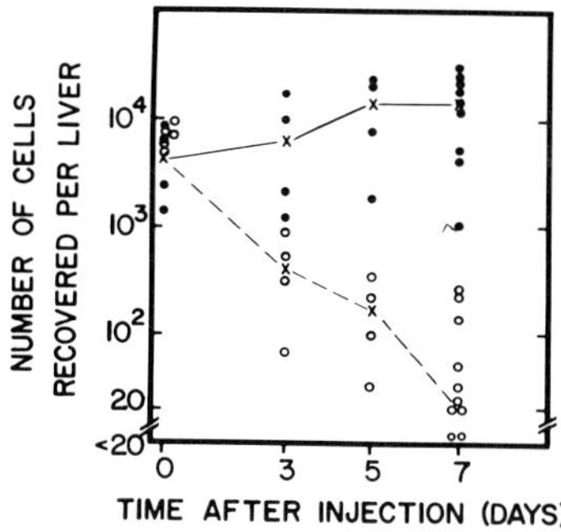

Fig. 7. Kinetics of growth in embryonic liver after injection of NIH3T3 cells morphologically transformed with cloned H-ras-T24 DNA in pBR322 (●) or control NIH3T3 cells transfected with NIH3T3 DNA (○) (5 x 10^5 cells per embryo) into 11-day chick CAM veins. Each point represents the number of viable rodent cells present in the liver of one embryo; lines connect median points.

To determine whether the in vivo phenotype detected in this assay was identical to in vitro morphological transformation, we examined the in vivo properties of ten clones of spontaneously-arising, morphologically transformed NIH3T3 cells. All ten clones failed to grow in the chick embryo following i.v. injection (Fig. 8), with behavior indistinguishable from that of normal

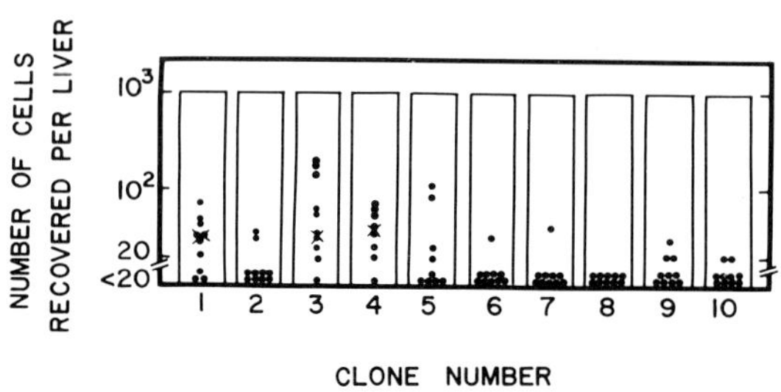

Fig. 8. Number of rodent cells recovered from embryonic chick livers 7 days after the injection of cells from ten clones of spontaneously-arising morphologically transformed NIH3T3 cells (5 x 10^5 cells per embryo) into 11-day chick CAM veins. Each point represents the number of rodent cells present in the liver of one embryo.

NIH3T3 control cells (*cf.* Fig. 7, 7-day point). It appears that the *in vitro* morphological transformation assay detects only a part of the malignant properties of the *ras*-transformed cells, and that *ras*-transformation changes NIH3T3 cells in some way beyond morphological transformation in conferring experimental metastatic ability to these cells.

CONCLUSIONS AND IMPLICATIONS

The assay used in these studies, in which cells are injected into chick embryo CAM veins and growth is detected in embryonic organs, models the latter half of hematogenous metastatic spread. This assay offers a number of advantages in molecular genetic studies of metastatic properties. First, molecular approaches often involve the generation of antigenic and immunogenic cells, necessitating the use of immune-deficient hosts such as the chick embryo. Second, the ability of the chick embryo to develop over a range of temperatures allows a powerful genetic tool, that of temperature-sensitive mutation, to be applied to the study of metastatic properties. Third, this assay is very sensitive to small changes in growth ability, to the point of being able to detect a response within one day of a change in growth conditions, as described for the temperature-sensitive *src*-transformed cells. Fourth, there are advantages to being able to determine the fate of injected cells immediately after injection, rather than having to wait for detectable tumors to form. Genetically-manipulated cells often have unstable genotypes (7-11), and subsequent changes in the cells must be considered in long-term, tumor-formation assays. This consideration plays a relatively smaller role in the rapid assay used here. Finally, because this assay permits detection of metastatic properties in immunogenic cells, the possibility must be considered that some of the variability seen when using metastasis assays in immune-competent hosts (see 12-14 for review) rests in heterogeneity of antigenicity in populations of otherwise metastasis-competent cells. It has been reported that experimental metastatic variants arise spontaneously at rapid rates in clonal populations of several tumor types (15,16,18,19). When using an immune-competent host, one cannot determine whether an increase in metastatic ability rests with enhanced ability to avoid detection by the host immune system or with other metastasis-related abilities. The availability of a quantitative metastasis assay in an immune-deficient host will permit an examination of the role of modulation of immunogenicity in generation of metastatic ability.

Using this assay, we have shown that members of two classes of oncogenes (H-ras and v-src) with two distinct molecular activities (GTP-binding and kinase activity, respectively [39]) are capable of conferring experimental metastatic ability when introduced into non-metastatic cell lines and tested in an immune-deficient host. The mechanism(s) by which these genes could cause changes in metastatic ability is as yet unclear, although growth control functions can be implicated. The molecular mechanisms by which these oncogenes are able to induce in vitro transformation are under active study by many groups; these studies must now be extended to the mechansims of conferring metastatic capabilities. Work to date has implicated a variety of growth-control functions in oncogene activity (39,40). For example, it has been shown that microinjection of the ras-encoded p21 protein into quiescent NIH3T3 and NRK cells results in transient cell proliferation and changes in morphology (41,42). In addition, the GTP-binding function of p21 raises the possibility that the ras gene product may function like GTP-binding hormone coupling factors (39). The plasma membrane location and kinase functions of src-encoded $pp60^{src}$ have suggested a role for this oncogene product in hormone-like signal transductions, although the transformation-related kinase substrate(s) remain to be determined (39).

It should be noted that the studies described here use cell lines, which are immortalized and are thus thought to be part way along a path of "tumor progression" (43). Use of a variety of cells at different stages of progression in studies on the molecular genetics of metastasis will help us to determine ways in which cells can acquire metastatic properties, the kinds and numbers of genes involved, and the role that any of these genes may play in natural metastasis. The observation that some oncogenes can not only transform cells by in vitro criteria and render them tumorigenic but can also confer on them aspects of metastatic ability raises questions about the nature of and interrelationships among these phenotypes.

ACKNOWLEDGMENTS

The contributions of G.P. Bondy, V. Ling, M. Naik, and S. Wilson to the research described here is gratefully acknowledged, as is the secretarial assistance of K. McIntyre in preparation of this manuscript. This research is supported by the National Cancer Institute of Canada and the Ontario Cancer Treatment and Research Foundation.

REFERENCES

1. Chambers, A.F., Shafir, R., and Ling, V., Cancer Res. 42: 4018-4025, 1982.
2. Hill, R.P., and Bush, R.S. Int. J. Radiat. Biol. Relat. Stud. Phys. Chem. Med. 15: 435-444, 1969.
3. Fidler, I.J., Nature New Biol. 242: 148-149, 1973.
4. Cifone, M.A., and Fidler, I.J. Proc. Natl. Acad. Sci. USA 77: 1039-1043, 1980.
5. Seppa, S.T., Seppa, H.E., Liotta, L.A., Glaser, B.M., and Martin, G.R., Invasion Met. 3: 139-150, 1983.
6. Thorgeirsson, U.P., Liotta, L.A., Kalebic, T., Margulies, I.M., Thomas, K. and Rios-Candelore, M., J. Natl. Cancer Inst. 69: 1049-1054, 1982.
7. Croce, C.M., Biochim. Biophys. Acta. 605: 411-430, 1980.
8. Calos, M.P., Lebkowski, J.S., and Botchan, M.R., Proc. Natl. Acad. Sci. USA 80: 3015-3019, 1983.
9. Kopchick, J.J., and Stacey, D.W., Mol. Cell. Biol. 4: 240-246, 1984.
10. Larizza, L., and Schirrmacher, V., Cancer Met. Rev. 3: 193-222, 1984.
11. Schafer, R., Griegel, S., Dubbert, M.A., and Willecke, K., EMBO J. 3: 659-663, 1984.
12. Fidler, I.J., and Hart, I.R., Science 217: 998-1003, 1982.
13. Heppner, G.H., and Miller, B.E., Cancer Met. Rev. 2: 5-23, 1983.
14. Nicolson, G.L., Cancer Met. Rev. 3: 25-42, 1984.
15. Harris, J.F., Chambers, A.F., Hill, R.P., and Ling, V., Proc. Natl. Acad. Sci. USA 79: 5547-5551, 1982.
16. Lagarde, A.F., Invasion Met. 3: 52-64, 1983.
17. Chambers, A.F., Harris, J.F., Ling, V., and Hill, R.P., Invasion Met. 4: 225-237, 1984.
18. Hill, R.P., Chambers, A.F., Ling, V., and Harris, J.F., Science 224: 998-1001, 1984.
19. Ling, V., Chambers, A.F., Harris, J.F., and Hill, R.P., J. Cell. Physio. Supplement 3: 99-103, 1984.
20. Clarkson, B., Katz, A., Dann, M.A., and Karnofsky, D.A., J. Natl. Cancer Inst. 32: 471-495, 1964.
21. Leighton, J., Prog. Exp. Tumor Res. 4: 98-125, 1964.
22. Easty, G.C., Easty, D.M., and Tchao, R., Eur. J. Cancer 5: 287-295, 1969.
23. Easty, G.C., Easty, D.M., and Tchao, R., Eur. J. Cancer 5: 297-305, 1969.
24. Locker, J., Goldblatt, P.J., and Leighton, J., Cancer Res. 30: 1632-1644, 1970.
25. Chambers, A.F., and Ling, V., Cancer Res. 4: 3970-3975, 1984.
26. Chambers, A.F., and Wilson, S., Mol. Cell. Biol., in press, 1985.
27. Bondy, G.P., Wilson, S. and Chambers, A.F., Manuscript submitted.
28. Whittow, G.C., In: Avian Physiology (Ed. P.D. Sturkie), New York, Springer-Verlag, 1976, pp. 147-173.
29. Duc-Nguyen, H., Rosenblum, E.N., and Zeigel, R.F., J. Bacteriol. 92: 1133-1140, 1966.
30. Wyke, J.A., Virology 52: 587-590, 1973.
31. Wyke, J.A., and Linial, M., Virology 53: 152-161, 1973.
32. Chen, Y.C., Hayman, M.J., and Vogt, P.K., Cell 11: 513-521, 1977.
33. Durkin, J.P., Boynton, A.L., and Whitfield, J.F., Biochem. Biophys. Res. Comm. 103: 233-239, 1981.
34. Durkin, J.P., Whitfield, J.F. and MacManus, J.P. J. Cell. Physiol. 115: 313-319, 1983.
35. Der, C.J., Krontiris, T.G., and Cooper, G.M., Proc. Natl. Acad. Sci. USA 79: 3637-3640, 1982.

36. Goldfarb, M., Shimizu, K., Perucho, M., and Wigler, M., Nature 296: 404-409, 1982.
37. Shih, C., and Weinberg, R.A., Cell 29: 161-169, 1982.
38. Chang, E.H., Furth, M.E., Scolnick, E.M., and Lowy, D.R., Nature 297: 479, 1982.
39. Hunter, T., J. Natl. Cancer Inst. 73: 773-786, 1984.
40. Heldin, C.-H., and Westermark, B., Cell 37: 9-20, 1984.
41. Feramisco, J.R., Gross, M., Kamata, T., Rosenberg, M., and Sweet, R.W., Cell 38: 109-117, 1984.
42. Stacey, D.W., and Kung, H.-F., Nature 310: 508-511, 1984.
43. Land, H., Parada, L.F., and Weinberg, R.A., Science 222: 771-778, 1983.

3

INTEGRAL MEMBRANE ADHESION GLYCOPROTEINS: WHAT IS THEIR FATE DURING METASTASIS?

C. H. DAMSKY,[1,3,4] K. A. KNUDSEN,[1,5] A.F. HORWITZ,[2] M. J. WHEELOCK,[1]

P. GRUBER,[1] C. A. BUCK [1]

1. The Wistar Institute, 36th & Spruce Streets, Philadelphia, PA 19104

2. Department of Biochemistry and Biophysics, University of Pennsylvania Philadelphia, PA 19104

3. Dr. Damsky's present address: Departments of Stomatology and Anatomy, Schools of Dentistry and Medicine, University of California, San Francisco, CA 94143

4. To whom reprint requests should be sent.

5. Dr. Knudsen's present address: Lankenau Medical Research Center Philadelphia, PA 19151

Data from a large number of laboratories over several decades have shown that adhesive interactions among cells and between cells and extracellular matrix (ECM) are essential to early cell lineage segregation, cell recognition, cell migration and maintenance of normal tissue architecture in the adult. Adhesive interactions are also a fundamental feature of the complex metastatic process. Part of the job of understanding both normal and aberrant adhesive behavior is very basic: identification of the molecules involved at adhesion sites. This job is becoming increasingly complicated for several reasons. We have learned that any one cell type has the ability to adhere to more than one type of extracellular matrix ligand (1, 2) and that cells can adhere to one another not by one, but by several kinds of adhesive complexes (reviewed in 3). Thus, the cells of mature tissues appear to express a repertoire of adhesive molecules. In several well-documented situations, it is clear that the repertoires of different tissue types are distinct. For example, nervous tissue expresses calcium dependent and independent cell-cell adhesion molecules that are distinguishable from analagous molecules expressed by epithelial or connective tissue cells (4, 5, 6, 7). Studies on the adhesive mechanisms of cell populations during development tell us that the repertoire of adhesion molecules expressed by different cell types is plastic: that there can be periods of expression and non-expression of particular adhesion molecules (8), and there can be modifications of their chemical structure and surface distribution during the course of development (9). Thus, the adhesive options for cells are multiple and the overall adhesion strategies adopted by particular cell types are distinct and subject to modulation. The major implication of these features for metastasis is that tumor cells might alter their adhesive behavior by altering the level of expression, the surface distribution, turnover, or post translational modifications of the repertoire of cell adhesion molecules existing in adult tissue. On the other hand, they might change their repertoire by expressing some developmentally regulated adhesion antigens at an inappropriate time, or by terminating the expression of other antigens more characteristic of the mature tissue of origin. In one well-studied example, characterization of the 68kd cell surface laminin receptor in normal and tumor tissue has revealed a difference in its cell surface distribution which might contribute to the higher affinity of malignant cells for the laminin in the basal lamina (10). In a second example, the neural cell adhesion molecule (N-CAM) which is normally expressed in neuroectoderm from 6d chick embryos, is greatly reduced following transformation of these cells with Rous sarcoma virus (11). Very little is known about how other cell surface adhesion molecules behave during metastatic progression. Our goal in this article is to describe the properties of two different cell surface adhesion antigens studied by our laboratories and to stimulate interest in them among those who have the appropriate models for tumor metastasis with which to study the relevance of these molecules to the metastatic process.

Our approach has been to use tissue culture systems which exhibit cell-cell and/or cell-matrix interactions and to use, initially, heterospecific antisera to perturb the adhesive phenomena of interest in a manner similar to that originally described by Gerisch (12). Figure 1 shows selective disruption of the cell-substratum and cell-cell adhesion of a cultured murine mammary tumor epithelial cell monolayer (MMTE cells) by antisera against serum-free medium conditioned by two different mammary tumor cell lines: the MMTE cells used as the target (anti SFMI) and the MCF-7 human mammary carcinoma cell line (anti SFMII). We have been successful in identifying and purifying the cell surface molecules capable of blocking the effects of each of these antisera (13, 14, 15: Reviewed in 3, 16).

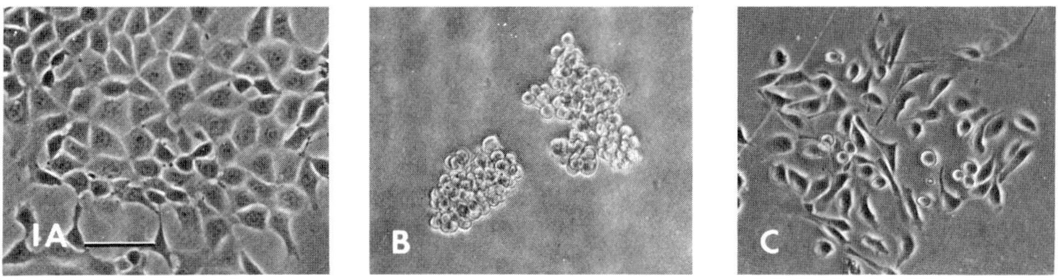

Fig. 1. a) Control monolayer of MMTE cells. b) MMTE cells treated with a 1:100 dilution of anti SFM I. c) MMTE cells treated with a 1:64 dilution of anti SFM II. Bar = 100 um

An 80kd glycopeptide was purified from SFM conditioned medium of human MCF-7 cells that is able to block the anti SFMII induced disruption of cell-cell interactions in MMTE. A specific antiserum against this component (anti GP80) has the same cell-cell adhesion disruptive activity as the polyspecific anti SFMII. Anti GP80 and a monoclonal antibody made against the 80kd component, recognize a single 80kd glycopeptide in unfractionated SFM and a 120kd glycoprotein in unfractionated nonionic detergent extracts of human epithelial cells (15 and Wheelock et al. in preparation). This suggests that the parent molecule on the cell surface is a 120kd glycoprotein which is turned over and shed into the medium as an 80kd glycopeptide. A summary of other properties of this molecule, which we call Cell-CAM 120/80, are given in Table 1. These characteristics are shared by the calcium dependent cell adhesion molecule found on liver (17) and on mouse teratocarcinoma cells (18, 19, 20). One characteristic in particular which might be relevant to the regulation of function of this molecule is its extreme protease sensitivity. It breaks down easily to components with M_r of 105K and 92K during NP-40 extraction unless protease inhibitors are included. Diisopropylfluorophosphate (DFP) is the most effective, outperforming even an elaborate cocktail of inhibitors. The extent of breakdown seems to depend on the cell line being examined. For example, the 120kd form from the MCF-7 line, originally isolated from a metastatic lesion, appears to breakdown more easily during NP40 extraction than that from JAR human choriocarcinoma cells. Whether the degree to which the molecule is broken down in different cell lines correlates consistently with metastatic potential is unknown. It may be that _in vivo_, the general increase in the level of proteases in the environment of a metastasizing tumor affects the integrity of cell adhesion molecules helping to compromise the ability of the tumor cells to remain associated with the primary tumor site.

In both cultured cells and embryonic epithelial tissue, Cell-CAM 120/80 is restricted to the cell-cell borders of neighboring cells (Fig. 2). Boller et al. (22) have shown by immunoelectron microscopy that antibody against uvormorulin isolated from mouse teratocarcinoma cells (20) is localized primarily at the intermediate junction (zonula adherens) of mature intestinal epithelial tissue. Whether the distribution changes in tumor cells or is more diffuse in embryonic than in mature tissue remains to be determined. Cell-CAM 120/80 is restricted to epithelial cells in adult tissues and is not found in endothelium, nerve or connective tissue.

Table I
Characteristics of Cell-CAM 120/80

120kd cell surface associated glycoprotein
Solubilized only in the presence of detergent; pI 4.8
Shed into conditioned medium as an 80kd glycopeptide
Protease sensitive: break down products at 105, 92, 80kd
Degraded when trypsinized in the absence of Ca^{2+}
Present on all cells of preimplantation mouse embryo
Restricted to epithelial tissue in adults
Localized to cell-cell borders in epithelial tissue and cultured cells

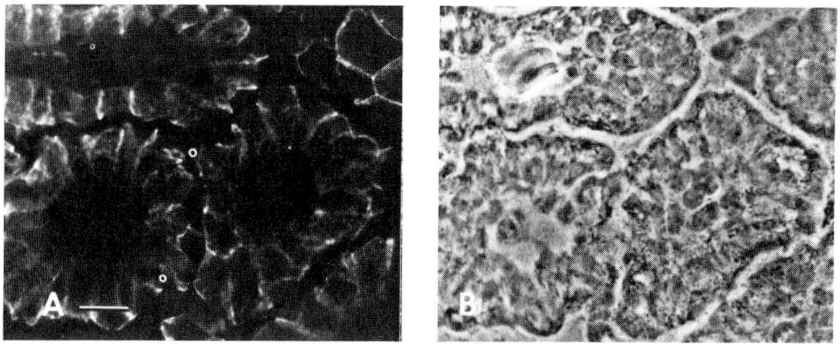

Fig. 2. a) Cross section of villi of 17d. mouse embryo small intestine stained with anti GP80 and goat rhodamine-anti-rabbit. Staining is confined to lateral border of epithelial cells. Connective tissue cells in central lamina propria are unstained. b) Phase micrograph of same field. Bar = 10 um

The relevance of Cell-CAM 120/80 to cell-cell adhesion in adult epithelia *in vivo* and in primary epithelial cultures *in vitro* was difficult to establish because anti GP80 does not disrupt cell-cell adhesion in these systems although it strongly stains the cell-cell borders indicating that it is present. We then investigated the role of Cell-CAM 120/80 in the earliest cell adhesion events in mouse development and determined that anti GP80 inhibits compaction of the 8-16 cell mouse embryo (15,16) and the segregation of a layer of primative endoderm cells around immunosurgically isolated inner cell masses (22). Thus, Cell-CAM 120/80 is expressed very early and is present on all cells of the preimplanatation mouse embryo. Since Cell-CAM 120/80 is restricted to epithelial tissue in the adult, the regulation of its expression during development becomes a question of great interest. Localization studies in collaboration with Dr. Ivan Damjanov, Hahnemann University, Philadelphia, PA., indicate that the first cell population to lose Cell-CAM 120/80 is the parietal endoderm which migrates out from the primitive endoderm to colonize the roof of the blastocoelic cavity (Damjanov et al. in preparation). These cells migrate individually rather than in sheets, and it stands to reason that they would no longer need their cell-cell adhesion molecules. The earliest mesodermal cells also lose Cell-CAM 120/80 as they come to lie between the embryonic endoderm and ectoderm (Damjanov et al. in preparation).

The ability of anti GP80 to perturb cell-cell adhesion in very early embryos and in the MMTE cultures initially used as target cells for our antisera, but not in other epithelial preparations which clearly contain the antigen, emphasizes the point raised earlier that cells have multiple cell-cell adhesion mechanisms. Our results suggest that anti GP80 can only disrupt adhesion if Cell-CAM 120/80 is the dominant adhesion mechanism present in the tissue being tested. This appears to be the case in certain stages of early development, but not in many mature epithelia. Testing of carcinomas of different metastatic potential with antibodies against Cell-CAM 120/80 and against other cell-cell adhesion components (e.g., anti-desmosomal antibodies: 23) may allow one to parse the adhesion strategy of epithelial tissues and to determine whether there are systematic or diagnostic changes in the adhesive capabilities of metastatic cells.

Regulation of cell-matrix interactions is also critical to both normal development and metastatic progression. The identification of cell surface molecules which can interact with components of the extracellular matrix (ECM) is an important step in building a complete picture of the possible range of interactions that might be involved in these processes. Receptors for laminin (24, 25) and for various types of collagens (26, 27) have recently been identified as has a membrane intercalated heparan sulfate proteoglycan (28,29). In general, these molecules are found in both normal and at least some tumor cells. Differences in their post translational modification, turnover, abundance or cell-surface distribution pattern, however, could be of significance for the development of the malignant phenotype.

The complex of cell-matrix adhesion glycoproteins which constitutes the subject of the remainder of this paper was first identified in transformed BHK fibroblasts (30). Following biochemical purification, a group of 3 glycoproteins of 120-160kd was found to inhibit the crude adhesion-disrupting antiserum used in our early studies (Fig. 3a, d, e; 14). A more specific antiserum was made against these glycoproteins (anti-GP140). Subsequently, similar material was purified from MMTE cells (Fig. 3b; 13). Recently, an adhesion disrupting monoclonal antibody, CSAT Mab, made against surface membrane vesicles of chick myoblasts has been isolated by Nicola Neff, A. F. Horwitz, and colleagues at the University of Pennsylvania which, by immunoaffinity purification, identifies a triplet of glycoproteins in the same molecular weight range (see Fig. 3c; 31, 32).

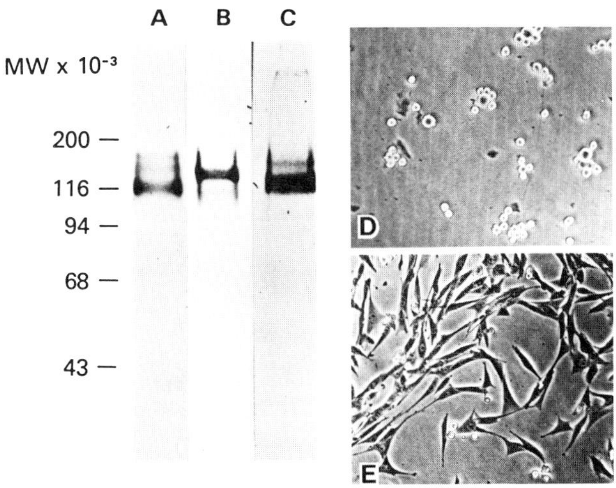

Fig. 3. a) SDS-PAGE run under reducing conditions showing the polypeptide composition of cleanest material purified from NP-40 extracts of transformed baby hamster kidney fibroblasts that can block the cell-substratum adhesion-disruption effects of anti-GP140. b) Similar material isolated from extracts of MMTE cells. Purified material in A and B were labeled with ^{125}I prior to analysis. c) Material from NP40 extracts of ^{35}S methionine labeled chick embryo fibroblasts purified by immunoaffinity chromatography using immobilized CSAT monoclonal antibody. d) Morphology of hamster fibroblasts detached by anti-GP140. e) Hamster fibroblasts treated with a mixture of anti-GP140 and material purified from either hamster fibroblasts (see A) or MMTE Cells (B). Details of purification protocols and antibody blocking assays can be found in (13,14,32). Bar = 100 um

This group of glycoproteins is ubiquitous in adherent tissue cells, being present on connective tissue cells, endothelium, epithelium (33) muscle and nerve (31). It is absent from circulating blood cells such as lymphocytes and polymorphonuclear (PMN) leukocytes. It is also absent from blood cells which have been stimulated to adhere such as PMN's migrating in response to chemotactic peptide, and activated peritoneal macrophages.

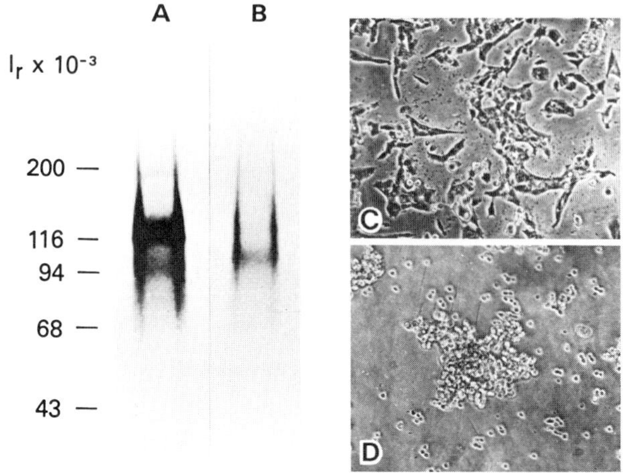

Fig. 4. a) Cleanest material purified from NP40 extracts of an adherent hamster melanoma cell line which blocks disruption of their cell-substratum adhesion caused by anti-GP140. b) Material purified by the same protocol from an NP40 extract of a nonadherent melanoma cell line. The same amount of starting cell protein was used in both cases. The purified fractions were labeled with ^{125}I before analysis by SDS-PAGE under reducing conditions. c) Morphology of the adherent melanoma cell line. d) Morphology of the nonadherent melanoma cell line. Bar = 100 um (see Ref. 34)

Its association with adherent tissue cells has been demonstrated in a particularly striking manner in a study carried out with adherent and nonadherent hamster melanoma cells (34). Results showed that, while the adherent line contained an amount of the 140kd glycoprotein complex comparable to other control and transformed cultured cell lines, the nonadherent line showed a 20-fold reduction of these components as determined by antibody absorption and antibody blocking activity. Purification of antibody blocking activity from ^{125}I surface labeled adherent and nonadherent cell extracts revealed that the nonadherent cell line contained significantly less material in the 120-140kd range than the adherent line (Fig 4). What is particularly interesting about this system is that the nonadherent melanoma line can be induced to adhere following growth in the presence of bromodeoxyuridine (BrdU). These now adherent cells re-express antibody blocking activity and produce the 120-140 kd glycoproteins (34).

Studies such as these can be extended to human melanomas where primary tumors and metastatic lesions can often be obtained from the same patient. Cell lines from such pairs of related tumors have been established at the Wistar Institute by Meenhard Herlyn, Hilary Koprowski and associates (35). These can have significantly different adhesive properties and should provide an excellent source for correlative studies of a cell's metastatic potential, its adhesive properties and the extent of the 120-140kd cell-substratum adhesion glycoproteins.

An additional feature of note about these glycoproteins is that they, like the Cell-CAM 120/80 cell-cell adhesion glycoprotein, are expressed very early in development. They are expressed at least as early as the blastocyst stage in the mouse, since anti-GP140, produced against the glycoproteins purified from hamster fibroblasts, inhibits mouse blastocyst attachment and spreading *in vitro* (22).

In attempting to determine the mechanism by which the 140kd glycoproteins participate in cell matrix adhesion, we have asked three kinds of questions: Are these glycoproteins localized in places consistent with their direct participation in adhesion? Do they interact directly with known extracellular matrix molecules? Since these 120-160kd glycoproteins seem to exist as a complex, can we gain any insight into the functions of the individual components? For these studies, we have relied heavily on the monoclonal antibody CSAT (2, 31, 32, 36). This antibody detaches the myoblasts present in 48 hr. primary chick myoblast cultures plated on gelatin in the presence of serum, leaving the fibroblasts behind. Fibroblast adhesion is also perturbed by CSAT Mab, but this is more difficult to detect unless these cells are provided with defined matrices in the absence of the adhesive factors present in serum.

Following affinity chromatography on immobilized CSAT Mab, the immunopurified antigen appears as in Fig. 3c when analyzed by SDS-PAGE under reducing conditions. When analyzed under nonreducing conditions, better resolution of the 3 bands is obtained, largely because the central band runs more quickly under nonreducing conditons and assumes the lowest apparent molecular weight of the 3 when analyzed under these conditions (32). The components are referred to as bands 1, 2, and 3 based on their positions in nonreduced SDS polyacrylamide gels.

Localization studies have been carried out using both the CSAT Mab and an antiserum (anti GP3) raised in rabbits against the lowest molecular weight band (Band 3) excised from nonreduced gels (36). The CSAT ag co-aligns with portions of actin containing microfilament bundles (MFB) and has a characteristic relationship with their termini as delineated by the cytoskeleton-associated proteins vinculin (37, 38), and talin (39) which have been shown to co-localize with MFB termini at focal contacts. Fig. 5 shows that the CSAT ag is enriched at the periphery of MFB rather than directly beneath them. Its co-localization with fibronectin, however, is very precise (Fig. 6; 36), suggesting that CSAT ag may function along with Fn primarily at sites of close contacts rather than at vinculin-rich focal contacts. Similar results have been observed by Chen et al. (40), using the monoclonal antibody JG22 that probably recognizes the same antigen complex as CSAT Mab (41). All three bands of the CSAT ag are present on transformed fibroblasts. However, localization studies show that it has a diffuse distribution when these cells are plated under typical culture conditions (glass coverslips in the presence of serum; Fig. 7a). The distribution pattern of CSAT ag on these cells can be influenced by the presence of a defined matrix. In the presence of either a Fn or a LM coated substratum and in the absence of serum, and the presence of cycloheximide (CHI), CSAT ag can assume a distribution pattern typical of normal fibroblasts (Fig. 7b). These results suggest clearly that the CSAT ag is located in places where it can participate directly in adhesion. Not only is it in close proximity to cytoskeletal and extracellular matrix molecules already considered to be involved in adhesion, but its distribution can be modulated by components of the extracellular matrix.

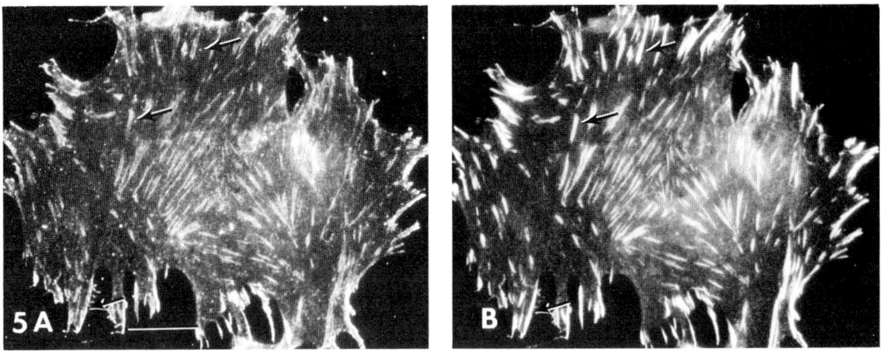

Fig. 5. a) Chick tendon fibroblast fixed and permeabilized with methanol at -20°C and stained with CSAT Mab and fluorescein-goat anti-mouse. b) Same cell stained with rabbit anti-talin and rhodamine-goat anti-rabbit. Bar = 1 um.

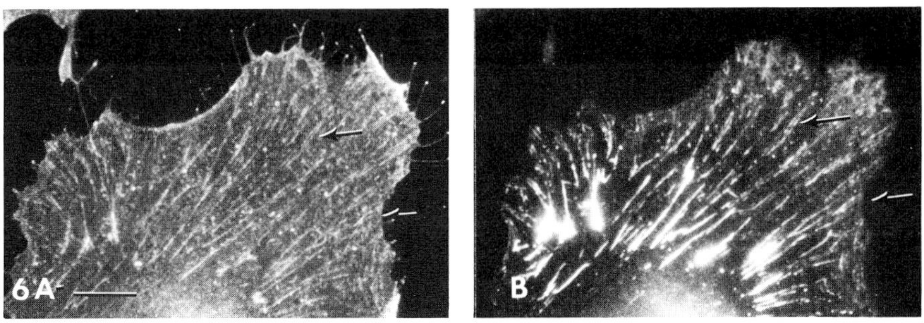

Fig. 6. a) Chick tendon fibroblast fixed with paraformaldehyde, permeabilized with acetone and stained with anti-GP3 and rhodamine goat anti-rabbit. b) Same cell stained with monoclonal anti-Fn and fluorescein goat anti-mouse. Bar = 10um

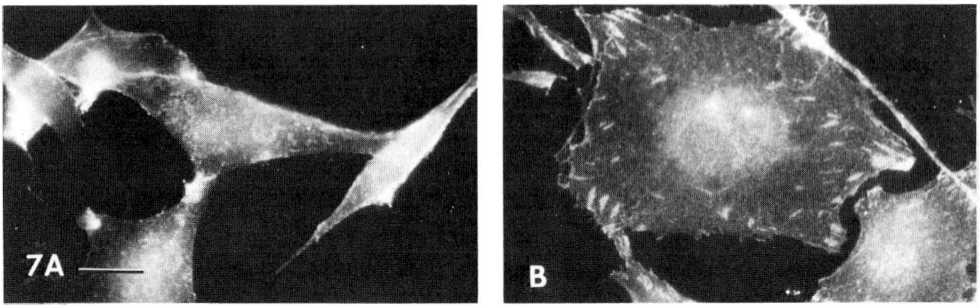

Fig. 7 a) Transformed tendon fibroblast plated on glass with serum containing medium and stained with CSAT Mab and fluorescein goat anti-mouse following permeabilization with methanol. b) Transformed tendon fibroblast plated on Fn-coated glass in serum-free medium and fixed and stained with CSAT Mab as described for 7a. Bar = 10 um. Fluorescence protocols for figures 5-7 are as described in (36).

The use of adhesion assays in conjunction with defined matrices can add functional information on the role of CSAT ag to the spatial information described above. Horwitz and associates have shown that when different cell types are plated in different environments, their responses to CSAT Mab reveal a hierarchy of adhesive strategies (see 2, 42 and Table II). Myoblasts can be detached by CSAT Mab from complex or defined matrices in the presence or absence of serum suggesting that the CSAT ag participates in the dominant adhesion mechanism for this cell type under all culture conditions. Tendon and cardiac fibroblasts cannot be detached from Fn or LM coated matrices in the presence of serum, apparently being able to compensate by taking advantage of spreading factors in serum. Tendon fibroblasts can be detached from either laminin or fibronectin coated substrata if plated in the absence of serum. However, cardiac fibroblasts can be detached from LM but not from Fn coated substrates even in serum-free medium (42 and Horwitz et al. in preparation). Thus, the cardiac fibroblasts have the most complex adhesion strategy of the cell types tested, being able to utilize serum spreading factors and a CSAT resistant Fn adhesion mechanism as well as laminin. Both the adhesion assays and the localization data are consistent with the possibility that the CSAT ag is involved in the mechanisms by which cells attach to both Fn and Lm.

The antigen could function in two different ways. It could participate in the organization of ECM ligand-cell surface receptor complexes, serving to stabilize them and promote their association with the cytoskeleton. This hypothesis would permit distinct receptors for every adhesive ligand but would allow the CSAT ag to serve more than one kind of receptor-ligand complex. The second possibility is that the CSAT ag acts as a receptor for both laminin and fibronectin. It needn't be the only receptor for either one of these ligands. Clearly, it wouldn't be the only receptor for laminin since a 68kd laminin receptor has already been described in several cell types by several investigators. Thus, this second possibility requires not only that there be multiple receptors for at least some adhesive ligands, but that one surface receptor can serve more than one ECM ligand. The localization and adhesion assays described above do not distinguish between these possibilities. However, if it could be demonstrated that the CSAT ag complex binds directly and specifically to Fn and/or LM, either in solution or using a ligand affinity column, this would constitute strong evidence that the CSAT ag functions as a ligand receptor. Pytela et al. (43) have demonstrated the binding of 140kd glycoproteins from an NP40 extract of mammalian fibroblasts by immobilized Fn. These glycoproteins are eluted from the Fn column with the cell binding fragment of Fn. It is likely, although it has not been confirmed, that these 140kd glycoproteins are the mammalian counterparts of the chick CSAT ag complex.

Table II
Conditions Under Which CSAT Mab Inhibits Attachment of Various Cell Types

	Undefined Matrix + Serum	Fn	LM
Myoblasts	+	+	+
Tendon fibroblasts	-	+	+
Cardiac fibroblasts	-	-	+

A final series of experiments addresses the role of the individual bands within the CSAT ag (Damsky et al. in preparation). We have taken an approach of selective proteolysis (44, 45). Tendon fibroblasts preincubated with cycloheximide and then treated vigorously with trypsin in the presence of calcium, are still able to spread on Fn in the absence of serum and the presence of CHI (Fig. 8a). This interaction is inhibited by CSAT Mab (Fig. 8b). However, if the reducing agent dithiothreitol (DTT) is included with the trypsin, adhesion to Fn is destroyed (Fig. 8c). Analysis of the CSAT ag affinity purified by CSAT Mab from cells treated under these conditions reveals that band 1 is removed by treatment with trypsin alone (Fig. 9T). Examination of trypsinized cells stained for CSAT ag and vinculin shows a diffuse pattern for each, similar to that found for transformed fibroblasts (see Fig. 7a). Thus, band 1 may play some role in the organization of transmembrane adhesive complexes when cells contact an appropriate matrix. Cells treated with trypsin-DTT show that band 3 has been degraded (Fig. 9 T-D). This correlates with a loss of adhesion to Fn suggesting the importance of an intact band 3 for adhesion to Fn. Studies with murine sarcoma cells have implicated a glycoprotein with similar properties in the adhesion of these cells to Fn (45).

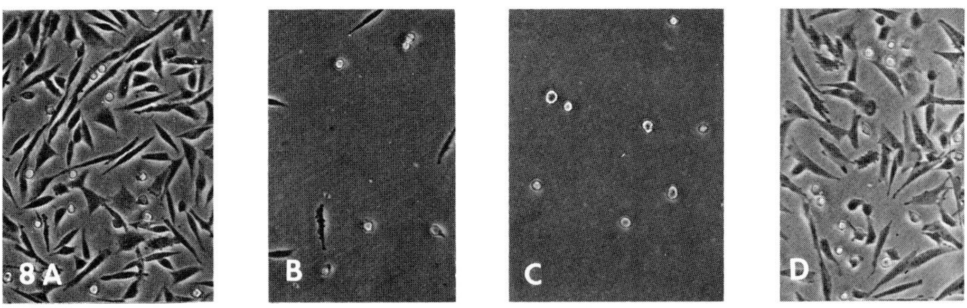

Fig. 8. a) Chick tendon fibroblasts harvested with 1mg/1ml trypsin in SFM for 20 min. at 37°C and plated for 2 hrs. on Fn-coated substrate in SFM and CHI. b) Cells treated as in A but plated in the presence of 50 ug/ml CSAT Mab. c) Chick tendon fibroblasts treated with 1mg/1ml trypsin and 50mM dithiothreitol (DTT) for 20 min. at 37°C and plated as in A. d) Cells treated with 5mM EDTA and 50mM DTT and plated as above. In each case, after 2 hrs. of plating, unattached cells were removed by standardized pipetting before photographs were taken. Bar = 100 um. See (44) for details of protocols.

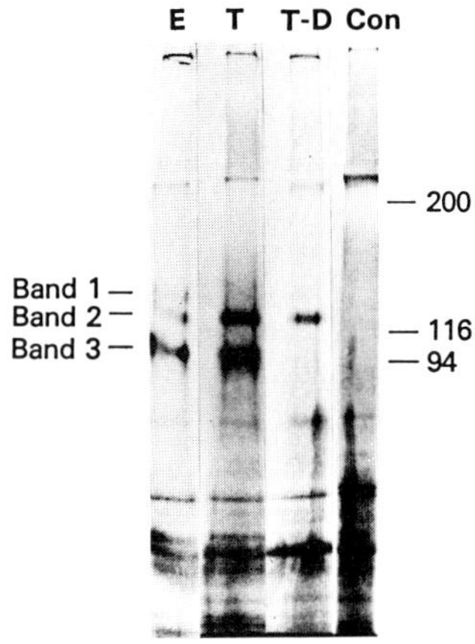

Fig. 9. a) SDS-PAGE profiles of NP40 extracts of ^{35}S methionine labeled chick tendon fibroblasts harvested with EDTA (E), trypsin (T) or trypsin-DTT (T-D) as described for Figure 8. Extracts were passed first over a nonrelevant Mab-Sepharose 4B column to identify nonspecifically-bound components (CON) and then over a CSAT-Mab Sepharose 4B column. Material eluted from the CSAT Mab column at pH 11.5 from the three preparations were analyzed by SDS-PAGE under nonreducing conditions. Protocols for conjugating antibodies to Sepharose and for isolating CSAT ag are as described in (32).

We have described a variety of approaches to understanding the role of the CSAT ag complex in cell-matrix adhesion. Adhesion and localization studies, along with more recent biochemical data suggest that this complex may serve as a ECM ligand receptor in several cell types. These studies have also pointed out that the adhesion mechanism in which the CSAT ag participates, although found in a wide variety of cell types, is but one of several possible adhesion mechanisms. Cell types differ in their ability to express or utilize these various mechanisms, thereby giving rise to a hierarchy of adhesiveness among the multiple cell types within a particular environment. The importance of this variety, amid the complex and changing environments found during development and during metastatic progression, may lie in its permitting a selective response by one cell type to an environmental signal to which all the cell types present are exposed.

In summary, using the same approach, we have identified two different antigens which perturb selectively either cell-cell or cell-matrix adhesion. Both antigens are expressed very early in development and maintain a wide distribution in the adult. They are both present on many kinds of transformed and tumor cells as well as normal tissues. By looking at early embryonic as well as more differentiated culture systems and by looking at defined as well as complex adhesion environments, we have been able to dissect multicomponent adhesive phenomena. As a result, we have gained some insight into the contributions of these two antigens to the overall process of cell adhesion. One area in which information remains particularly deficient, however, concerns their fate during metastatic progression.

ACKNOWLEDGEMENTS

The authors appreciate gifts of antibody against talin from Dr. Keith Burridge (39) and against fibronectin from Dr. Douglas Fambrough (46). Work carried out in the authors' laboratories was supported by: CA19144; CA32311; CA27909; HD15663; GM23244; CA07572. The technical assistance of Sena Smith, Eileen Crowley, Liesel Hud-Broderick and Cindi Decker, and manuscript preparation assistance by Marie Lennon are gratefully acknowledged.

REFERENCES

1. Couchman, J.R., Hook, M., Rees, D.A. and Timpl,R. J. Cell Biol. 96: 177-183, 1983.
2. Decker, C., Greggs, R., Duggan, K., Stubbs, J. and Horwitz, A. F. J. Cell Biol. 99: 1398-1404, 1984.
3. Damsky, C.H., Knudsen, K.A. and Buck, C.A. In: Biology of Glycoproteins (R. Ivatt, ed.) Plenum Press, New York pp 1-64.
4. Takeichi, M., Atsumi, J., Yoshida, K. and Okada, T. Dev. Biol. 87: 346-350, 1981.
5. Brackenbury R., Rutishauser, U. and Edelman, G. Proc. Natl. Acad. Sci. 78: 387-391, 1981.
6. Thomas, W.A. and Steinberg, M.S. Dev. Biol. 81: 106-114, 1982.
7. Cook, J.H., Pratt, R.S. and Lilien, J. Biochemistry 23: 899-904, 1984.
8. Thiery, J-P., Duband, J-L, Rutishauser, U. and Edelman, G.M. Proc. Natl. Acad. Sci. USA 79: 6737-6741, 1982.
9. Edelman, G. Science 219: 450-457, 1983.
10. Liotta et al. (this volume)
11. Brackenbury, R., Greenberg, M.E. and Edelman, G.M. J. Cell Biol. 99: 1944-1955, 1984.

12. Gerisch, G. Curr. Top. Deve. Biol. 14: 243-270, 1977.
13. Damsky, C.H., Knudsen, K.A., Dorio, R.J. and Buck, C.A. J. Cell Biol. 89: 173-184, 1981.
14. Knudsen, K.A., Rao, P., Damsky, C.H. and Buck, C.A. Proc. Natl. Acad. Sci. 78: 6071-6075, 1981.
15. Damsky, C. H., Richa, J., Solter, D., Knudsen, K. and Buck, C. Cell 34: 455-466, 1983.
16. Damsky, C.H., Wheelock, M.J., Richa, J., Damjanov, I. and Buck, C.A. 1985. In: Cells in Contact (G. Edelman, ed.) John Wylie and Sons, New York, 1985, in press.
17. Gallin, W.J., Edelman, G. and Cunningham, B. Proc. Natl. Acad. Sci. 80: 1038-1042, 1983.
18. Yoshida, C. and Takeichi, M. Cell 28: 217-224, 1982.
19. Hyafil, F., Babinet, C. and Jacob, F. Cell 26: 447-454, 1981.
20. Vestweber, D. and Kemler, R. Exp. Cell Res. 152: 169-178, 1984.
21. Boller, K., Vestweber, D. and Kemler, R. J. Cell Biol. 100: 327-332 1985.
22. Richa, J., Damsky, C.H., Buck, C.A., Knowles, B.K. and Solter, D. Dev. Biol. In press 1985.
23. Cohen, S., Gorbsky, G. and Steinberg, M.S. J. Biol. Chem. 258: 2621-2627, 1983.
24. Malinoff, H.L., and Wicha, M.S. J. Cell Biol. 96: 1475-1479. 1983.
25. Terranova, V.P., Rao, C., Kalebic, T., Margulies, I. and Liotta, L. Proc. Natl. Acad. Sci. 80: 444-449, 1983.
26. Mollenhaur, J. and Van der Mark, K. EMBO J. 2: 45-50, 1983.
27. Kurkinen, M., Taylor, A., Garrels, J. and Hogan, B.L.M. J. Biol. Chem. 259: 5915-5922, 1984.
28. Kjellen, K., Pettersson, I. and Hook, M. Proc. Natl. Acad. Sci. 78: 5371-5375, 1981.
29. Rapraeger, A. and Bernfield, M. J. Biol. Chem. 258: 3632-3636, 1983.
30. Wylie, D., Damsky, C.H. and Buck, C.A. J. Cell Biol. 80: 385-402, 1979.
31. Neff, N., Lowry, C., Decker, C., Tovar, A., Damsky, C., Buck, C. and Horwitz, A.F. J. Cell Biol. 95: 654-666, 1982.
32. Knudsen, K.A., Horwitz, A.F. and Buck, C.A. Exp. Cell Res. 157: 218-226, 1985
33. Damsky, C.H., Knudsen, K.A. and Buck, C.A. J. Cell. Biochem. 18: 1-13, 1982.
34. Knudsen, K.A., Damsky, C.H. and Buck C.A. J. Cell. Biochem 18: 157-167, 1982.
35. Herlyn, M., Balaban, G., Bennicelli, J., Guerry, D., Halaban, R., Herlyn, D., Elder, D.E., Maul, G., Steplewski, Z., Nowell, P.L., Clark, W.H. and Hilary Koprowski. J. Natl. Can. Inst., 1985, in press.
36. Damsky, C.H., Knudsen, K.A., Bradley, D., Buck, C. and Horwitz, A.F. J. Cell Biol. 100: in press, 1985.
37. Geiger, B., Tokuyasu, K.T., Dutton, A.H. and Singer, S.J. Proc. Natl. Acad. Sci. 77: 4127-4131, 1980.
38. Burridge, K. and Feramisco, F. Cell: 19: 587-596, 1980.
39. Burridge, K. and Connell, L. J. Cell Biol. 94: 359-367, 1983.
40. Chen, W-T, Haegawa, E., Hasegawa, T., Weinstock, C. and Yamada, K.M. J. Cell Biol. 100: in press, 1985.

41. Greve, J.M. and Gottlieb, D. J. Cell Biochem. 18: 221-229, 1982.
42. Buck, C.A., Knudsen, K.A., Damsky, C.H., Decker, C., Greggs, R.R., Duggan, K.E., Bozyscko, D., and Horwitz, A.F. In: Cells in Contact (G. Edelman, ed.) John Wylie and Sons: New York, 1985, in press.
43. Pytela, R., Pierschbacher, M.D., and Ruoslahti, E. Cell 40: 191-198, 1985.
44. Tarone, G., Galetto, G., Prat, M. and Comoglio, P. J. Cell Biol. 94: 179-186, 1982.
45. Giancotti, F., Tarone, G., Knudsen, K., Damsky, C. and Comoglio, P. Exp. Cell Res. 156: 182-190, 1985.
46. Chiquet, M. and Fambrough, D. J. Cell Biol. 98: 1926-1936, 1984.

4

PEPTIDE FRAGMENTS OF FIBRONECTIN AND LAMININ: ROLE IN CELL ADHESION AND INHIBITION OF EXPERIMENTAL TUMOR METASTASIS

L. Furcht, M. Basara, A. Norden-Skubitz, S. Palm, J. McCarthy, D. Sas

Department of Laboratory Medicine and Pathology, University of Minnesota, Minneapolis, Minnesota 55455

Metastasis is the major cause of morbidity and mortality in cancer patients. This process is extremely complicated and involves numerous cell biological and biochemical reactions. For convenience, the spread of tumor cells via a hematogenous route can be discussed in terms of a number of discrete, yet related steps. Metastasizing tumor cells must interact with and spread through extracellular matrices consisting of a number of molecular constituents. The cells must enter the capillaries or vessels at the primary site, get into the blood stream, travel to a distant site, adhere to endothelial or subendothelial structures, migrate through the basement membrane, and move and grow at this secondary locus (see reviews 1-4).

Adhesive interactions of cells are clearly important in a variety of aspects of the metastatic cell biology. For example, cell-cell adhesiveness may play a role in the release of cells from the primary site. Additionally, the adhesion of tumor cells to blood components such as platelets or proteins such as fibrinogen, may play a role in successful tumor spread (by providing growth factors such as PDGF from platelets) or unsuccessful spread, by promoting interactions with tumoricidal macrophages. The adhesion of tumor cells to subendothelial structures will also play a critical role in the overall biological process of metastasis. Importantly, these adhesive interactions need be reversable as the tumor cell in certain points of the metastatic cascade must release itself from one site and yet be able to form stable adhesions and set up growth at the secondary site. Therefore the migration of cells away from the primary site, the adhesion of tumor cells to endothelial or subendothelial structures at the distant site, and the ensuing invasion of basement membrane and migration into the tissue at the secondary site, are all intimately involved in metastatic spread (5). The subject of this report concerns the role of cell adhesion molecules and domains of these adhesion molecules in promoting the adhesion and migration of metastatic cells.

The cell matrix and basement membrane are composed of a variety of molecules including: genetically distinct collagens (see reviews 6,7) proteoglycans (see review 8), and noncollagen glycoproteins (see reviews 9-14) all of which potentially can serve as cellular adhesion molecules. The work presented herein will deal with the role of laminin and fibronectin, and domains within these molecules, that effect cell adhesion, cell migration, and tumor metastasis. Fibronectin is a high molecular weight glycoprotein, composed of nearly identical 220 Kd dimers, that is present in plasma, extracellular matrices, and basement membranes (see reviews 9-14). It can be produced by epithelial and mesenchymal cells. Fibronectin is a multi-domain molecule which can effect many cell functions including adhesion and migration of diverse cell types. In contrast, laminin is found in basement membranes throughout the body and appears to be produced by epithelial cells (15,16). Laminin is an >800 kd complex consisting of heavy and light chains, 400 Kd and 200 Kd, that assemble to form a cross-shape structure as seen by rotary shadowing. Laminin appears to play a role in the adhesion of epithelial cells to basement membranes and as an intermolecular bridge between epithelial cells and their underlying connective tissue matrices. Both fibronectin and laminin share certain characteristics relative to the ability to not only bind cells but also to bind to other constituents in the extracellular or basement membrane matrix. For example, fibronectin will bind well to a number of types of collagen, most notably denatured type I collagen. Laminin, on the other hand, will bind preferentially to type IV collagen (17). Both molecules will bind heparin and also heparan sulfate proteoglycan (18).

Cell migration is clearly important in the metastatic process and we have performed a variety of in vitro studies to evaluate the ability of cell adhesion molecules to effect the migration behavior of tumor cells. In these studies we examined the migration of high and low metastatic clonal variants of the 1735 murine melanoma cells (19,20) and the UV 2237 murine fibrosarcoma cells in microchemotaxis chambers. Studies have shown that tumor cells migrate largely in response to haptotactic stimulation of cells (19,20). That is, cells move quite effectively over a surface if it is coated with a particular attractant in the absence of fluid phase attractant, in contrast to chemotactic migration, which is the movement of cells in response to soluble attractants. Figure 1 shows the migration of B16F10 melanoma cells on fibronectin and laminin when the distal or lower surface is coated, or when both surfaces are coated. This data shows that the cells migrate best when the distal surface alone is coated with laminin, thus implying that a gradient of substratum-bound material provides the greatest stimulus for directed migration. In further support of this, studies have shown that during the course of a chemotaxis experiment, if one uses tritiated laminin in the lower well, there is a deposition of the laminin on the filter surface and that the cells are likely able to respond to this material deposited on the filter surface rather than the fluid phase material (Figure 2). Further confirmation of the haptotactic nature of tumor cell migration was achieved using fibrosarcoma cells cultered on rodamine isothiocyonate (RITC) conjugated fibronectin (Sas, McCarthy, & Furcht, submitted). These studies showed that lowly metastatic cells migrated on the fibronectin surface, removed the material, and degraded it from the surface over a period of 24 hours (Figure 3A). Interestingly, it was observed that the lowly metastatic cells formed dense fibrillar networks along the substratum from material that had coalesced (Figure 3B). In contrast to what was observed with the lowly metastatic cells, the highly metastatic clonal variants could apparently move around the surface well but failed to remove RITC-fibronectin from the substratum (Figure 3C).

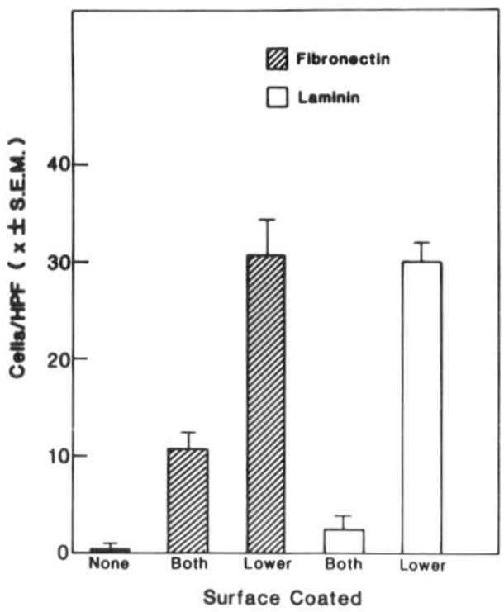

Fig. 1 - Haptotaxis of B16 mouse melanoma cell the lower, both or neither surface was coated with 100 µg/ml laminin or fibronectin. Migration assays were run for 4 hours.

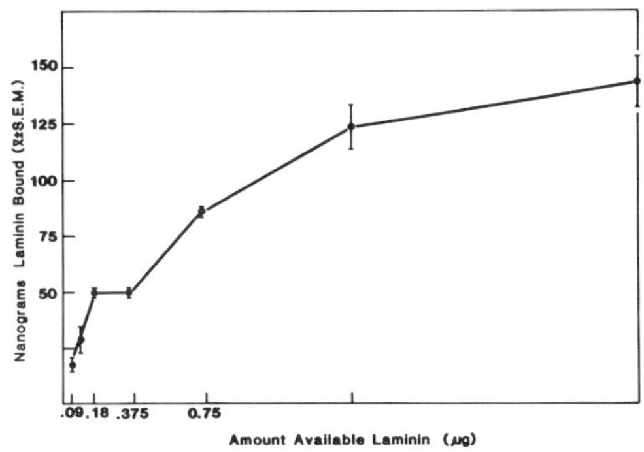

Fig. 2 - Laminin was labelled by reductive methylation and added to the bottom well of a microwell Boyden chamber. The binding to polycarbonate filters was counted after 90 min. Graph shows binding of soluble laminin to the filter surface

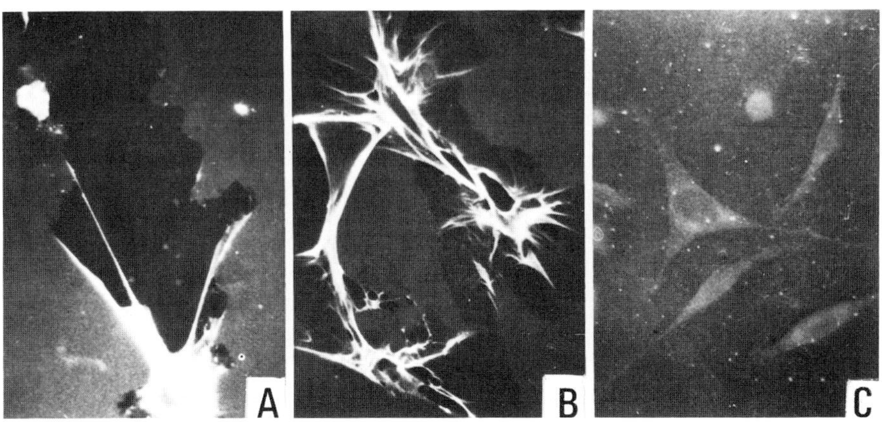

Fig. 3 - Glass coverslips were coated with 20 µg/ml fibronectin and conjugated with rhodamine. Various clones of mouse fibrosarcoma 2237 were plated and examined after 18-24 hr; A) lowly metastatic, B) lowly metastatic, C) highly metastatic.

The multidomain nature of fibronectin relative to its diverse biological activities was next evaluated. Fibronectin is composed of similar but not identical 220 Kd chains linked by disulfide bonds at the carboxy terminus. We have described a series of these domains in a recent review based on the work of many laboratories including our own (9-14). These domains include (see Figure 4): domain I, which is a site that shows weak heparin binding, collagen and fibrin cross-linking, and binding to Staphylococcus aureus; domain II, containing the gelatin binding region; domain III, which consists of a region within the central portion of the molecule of 75 Kd that possess a free sulfhydryl, reacts to a monoclonal antibody termed 180-8, and has a potential adhesion and motility site for tumor cells (see below); domain IV, which is a region that will promote cell attachment and cell spreading for CHO cells, reacts with a hybridoma 3E3 (21) and contains a tetrapeptide sequence RGDS (arginine, glycine, aspartate, serine), which will promote cell attachment; domain V, a carboxy terminal heparin binding region which from the A chain of the molecule contains cell attachment and spreading activity (the B chain in this region does not); and domain VI, which contains a free sulfhydryl and intrachain disulfide bond, and demonstrates reactivity with monoclonal 2-8 (12).

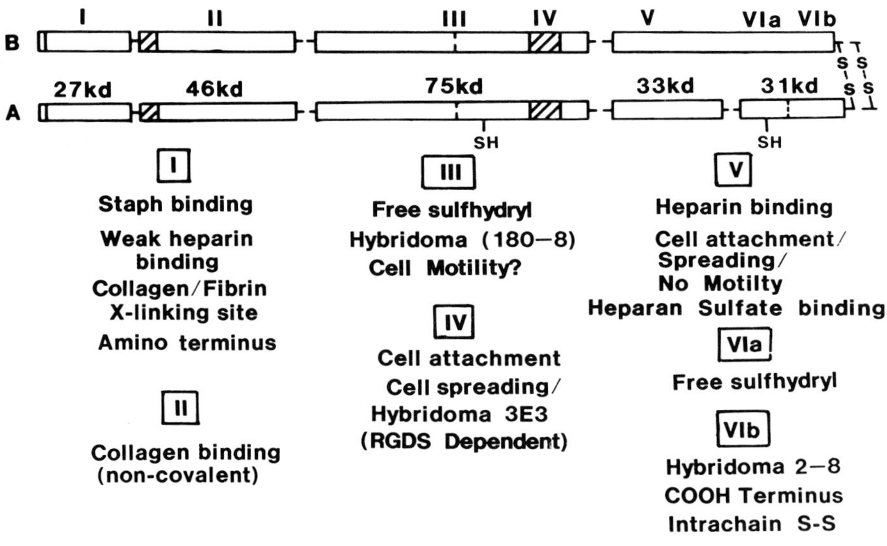

Fig. 4 - Schematic model of human plasma fibronectin consisting of two chains - A and B domains indicated in text above.

Studies were performed to isolate a central region of the molecule from tryptic digests that reacted with monoclonal antibody 180-8 (Figure 4). These studies showed that following 2 hrs trypsin digestion the proteolytic digest that bound to hybridoma 180-8 was highly effective in promoting the attachment of B16F10 melanoma cells compared to intact fibronectin. Analysis by SDS-PAGE showed that the 180-8 bound fraction consists of a 75 Kd proteolytic fragment that contains domains III and IV of fibronectin. In contrast to this, the material in the tryptic digest which did not bind to hybridoma 180-8 failed to promote the adhesion of these melanoma cells. Studies in progress suggest that a unique region in the fibronectin molecule promotes the attachment and haptotactic migration of melanoma cells (McCarthy, Hagen and Furcht, submitted). This region is most likely within domain III of the fibronectin model (Figure 4) and appears to be independent from the RGDS sequence existing in domain IV. That is, even at very high levels (10^{-3}M) of RGDS, relatively good adhesion could be observed for melanoma cells to the 75 Kd haptotaxis promoting fragment.

Studies pursued the possible existence of an additional cell adhesion promoting domain within fibronectin. These studies focused on an analysis of the carboxy-terminal 60-70 Kd of the molecule. It was observed that two heparin binding activities exist in the C-terminal region of the fibronectin molecule, one in the A chain and one in the B chain. It was further observed that if one segregated these two heparin binding activities by monoclonal antibody affinity chromatography, the A chain derived 33 Kd fragment would promote cell adhesion while the B chain derived 66 K fragment would not. The cell adhesion promoting capacity of this C-terminal heparin binding domain within the 33 K fragment is insensitive to the presence of high levels of the tetrapeptide RGDS (McCarthy, Hagen and Furcht, submitted). These studies corroborated others from our laboratory which have shown that there are two independent regions of the fibronectin molecule which will support neuronal adhesion and neurite outgrowth: one region is within the 75 Kd fragment, and a second region is within the C-terminal heparin binding domain V (22). This would suggest that there are unique sites within fibronectin that are distinct from the RGDS sequence and position, which can mediate attachment and effect phenotypic expression of different cell types.

A similar approach was taken in the analysis of the possible activity of laminin and laminin proteolytic fragments in modulating experimental metastatis. Laminin was fragmented with thrombin to produce a 600 Kd complex, with chymotrypsin to produce a 440 Kd complex, or thermolysin to produce 330 and 110 Kd bands (23-26). A schematic representation of the origin of these fragments is seen in Figure 5.

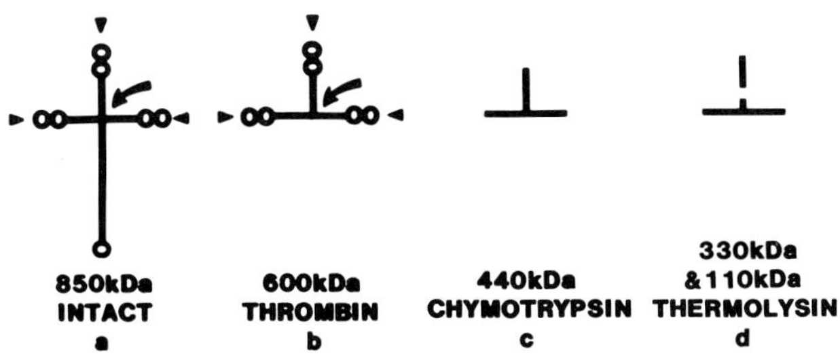

Fig. 5 - Schematic representation of intact laminin, the 600 Kd thrombin fragment 440 Kd chymotrypsin fragment and the 330/110 Kd thermolysin fragment.

In these studies either the highly metastatic clone M2 or M4 of the 1735 murine melanomas was used (27), or a highly metastatic clone of the UV2237 murine fibrosarcoma (28). In control studies, animals injected with tumor cells via the tail vein exhibited greater than 250 tumors per lung three weeks post injection. If the tumor cells were treated in vitro with intact laminin or the thrombin fragment of laminin prior to injection, they showed similar, extremely high levels of metastasis formation with any of the cell lines (the data for the UV 2237 fibrosarcoma is shown Figure 6). In contrast, if tumor cells were pretreated with the chymotrypsin generated 440 Kd fragment or with the thermolysin generated fragments (330 and 110 Kd) of laminin, a profound decrease in the number of pulmonary tumor nodules was observed; the number of tumors fell as low as 0-5 tumors for the 1735 melanomas to essentially none for the fibrosarcoma(s) (Figure 6). Essentially, there was a consistent, profound inhibition of experimental metastasis of both the murine melanoma and fibrosarcoma cells following pretreatment in vitro with laminin fragments generated by either chymotrypsin or thermolysin digestion.

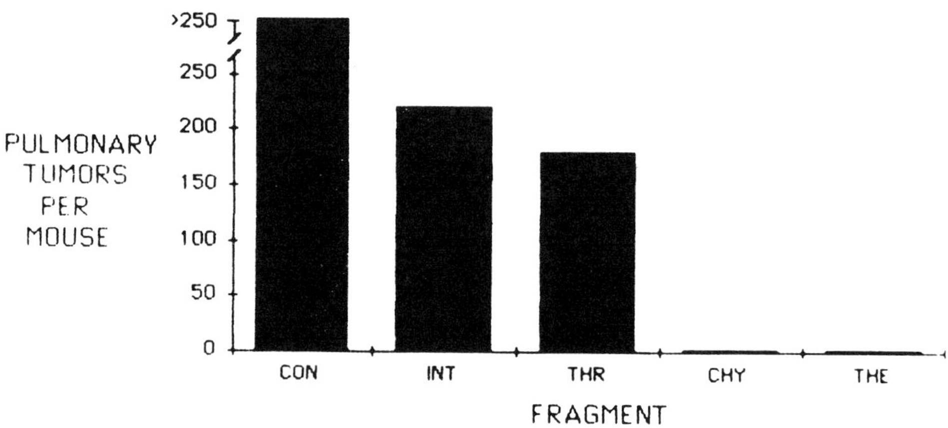

Fig. 6 - Experimental metastasis assay showing the number of pulmonary tumors formed per mouse at 3 weeks after tail vein injection of mouse fibrosarcoma cells. Cells were preincubated with medium - con; intact laminin - int, the thrombin fragment - Thr, the chymotrypsin fragment - Chy or the thermolysin fragment - the. Cells were nucleated with 100 µg/ml laminin or molar equivalents of fragments at 37°C and washed and then injected via tail vein.

Numerous studies were performed to evaluate the mechanism of what to us was an exciting observation. First, the cells were still viable and would grow in vitro. Second, the cells still remained tumorigenic when injected either into the muscle or subcutaneously after treatment with the various fragments. Preliminary studies were also done to determine whether there might be some differential sequestration of cells in the lung as the potential explanation for the profoundly lower metastasis seen in laminin inhibitory fragment pretreated cells. In these studies cells were labelled with ^{125}IudR in vitro, then injected via tail vein. Subsequent to this, lungs were harvested at 10 min, 4 hr, and 24 hr and radiolabelled cells remaining in the lung were quantitated with a gamma counter. These studies showed that there was no difference in the pulmonary retention of tumor cells when comparing control, intact laminin, or chymotrypsin fragment treated cells (Figure 7).

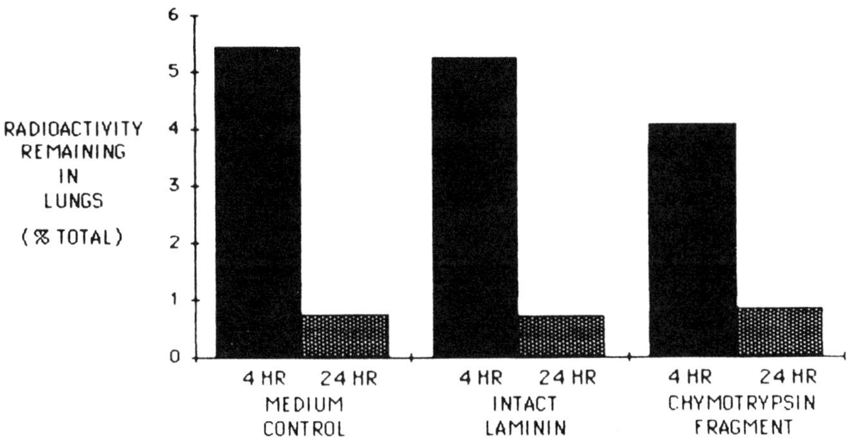

Fig. 7 - Pulmonary retention of ^{125}IudR labeled fibrosarcoma cells following tail vein injection; expressed as radioactivity remaining in lungs as a percentage of the total injected. Cells were pretreated with nothing (medium control); intact laminin or the chymotrypsin fragment of laminin (100 µg/ml); cells were washed and then injected and lungs harvested at 4 and 24 hr.

The next studies searched for a potential mechanism by which laminin fragments are able to inhibit experimental metastasis formation. In contrast to studies performed by other laboratories, it appeared that highly metastatic cells would adhere well to type IV collagen. Although laminin would in fact increase the adhesion to type IV collagen, the chymotrypsin or metastasis inhibitory fragments of laminin would not block adhesion of the cells to type IV collagen in vitro. This is not surprising since the cells did not exhibit an obligatory requirement for laminin in their adhesion to type IV collagen (Table 1). Also, studies were performed to look at the haptotactic migration of tumor cells on basement membrane proteins as a possible model for invasion. These studies showed that the metastatic cells migrated well in vitro on laminin and that the laminin fragments inhibitory for metastasis did not modify the migration of tumor cells in vitro to substratum attached laminin.

TABLE 1

	Cell Adhesion to Type IV Collagen (% of Total Applied)		
	Control	Intact Laminin	Chymotrypsin Laminin Fragment
Melanoma 1735-M2	40%	55%	38%
Melanoma 1735-M4	12%	30%	13%
Fibrosarcoma 2237-MM	12%	35%	14%

Legend: Cell adhesion was performed on microtiter wells coated with type IV collagen (100 μg/ml). Cells were pretreated in suspension with medium-control, intact laminin (100 μg/ml) or the chymotrypsin fragment of laminin (44 μg/ml). Cells were washed and then 20,000 cells were allowed to attach for 60 min and the percentage adhering was calculated (standard error of the mean was <10%).

Though the radiolabelled tumor cell retention showed no difference, the next studies further examined this possibility. Tumor cells were grown on Cytodex beads, then incubated with: a) medium alone, b) intact laminin, or c) the chymotrypsin fragments of laminin. The Cytodex beads with attached cells were injected via the tail vein and since the beads are relatively large, being 100-300 μ in diameter, they were physically trapped in the lung. Control, untreated beads showed diffuse growth of tumor cells from the beads forming large tumor masses. Cells on beads treated with laminin inhibitory fragments grew as concentric, tight swirls of cells around the bead. Figure 8 shows surprisingly that even

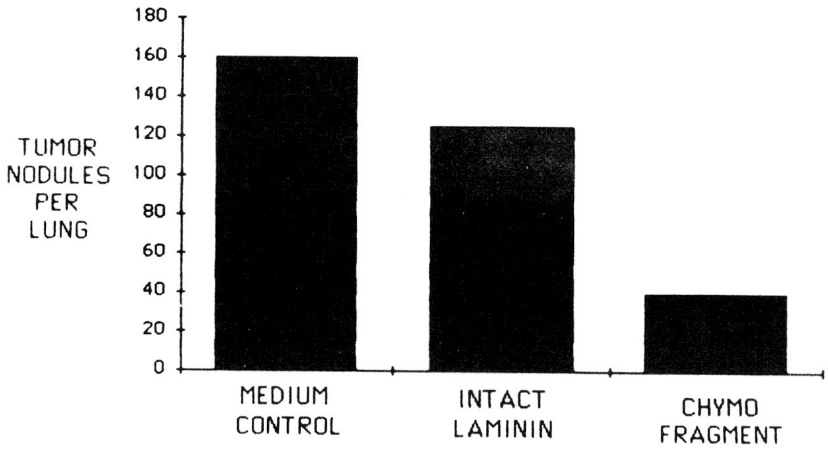

Fig. 8 - Experimental metastasis assay following injection of Cytodex beads which had 1735-M2 highly metastatic melanoma cells grown on them prior to injection. Cells on Cytodex beads were incubated with media, intact laminin or the chymotrypsin fragment of laminin; washed and injected. Pulmonary tumor formation at 3 weeks was determined visually.

when there was equivalent physical trapping of tumor cells in the lung by virtue of being grown on these Cytodex beads, there was a substantial decrease in the rate of tumor formation due to pretreatment of cells on beads with chymotrypsin fragments of laminin. As before, if the cells were grown on beads, trypsinized and released, the cells would still grow in vitro, were still tumorigenic, and had not been killed by being treated with laminin fragments.

There are a number of potential explanations for the dramatic inhibition of metastasis that we have demonstrated following treatment of cells with inhibitory fragments of laminin. The most simplistic explanation, that the cells were killed or damaged by the laminin fragments, does not seem to be the case. Additionally, there does not appear to be an induction of natural killer cell activity to account for the loss of metastatic ability. The other somewhat simplistic explanation, being a differential retention of tumor cells in the lung, also did not hold up. There was no apparent difference in the retention of radiolabelled cells in lungs, and when tumor cells were physically embedded in the lung on Cytodex beads there was still decreased tumor formation by the laminin chymotrypsin fragment treated cells. Also, no increase in extrapulmonary tumors was observed in either of these cases where cells had been treated with laminin fragments.

The hypothesis we are testing at the moment is that there is a reversible phenotypic change which occurs in these cells following treatment with certain laminin fragments, such that they become incapable of manifesting the fully developed phenotype of the highy metastatic cell. It would not be surprising that this phenomenon is reversible nor that it might require continued exposure of cells to laminin inhibitory peptides to manifest the full long term inhibition of metastasis in these systems. Future studies will assess this hypothesis and determine whether a rational approach for potential in vivo treatment of animals with proteolytic fragments or peptides derived from laminin in conjunction with other therapies could prove useful in inhibiting experimental and spontaneous metastasis.

Grant Support

This work was supported by NIH grant CA 29995 and CA 21463 from the National Cancer Institute. L.T.F. is a recipient of career development award CA 00651 from the NCI, and is a Stone Professor of Pathology.

References

1. Nicolson, G.L. Biochim. Biophys. Acta 695:113-170, 1982.

2. Weiss, L. In: K.V. Honn and B.F. Sloane (eds.) Homeostatic Mechanisms and Metastasis, pp. 15-38. Boston, Martinus Nijhof, 1984.

3. Nicolson, G.L., Poste, G. Int. Rev. Exp. Path. 25:78-181, 1984.

4. Hart, I.R., Fidler, I.J. Biochim. Biophys. Acta 651:37-50, 1981.

5. Weiss, L., Ward, P.M. Cancer Met. Rev. 2:111-123, 1983.

6. Kuhn, K. In: Furthmayr, H. (ed.) Immunochemistry of the Extracellular Matrix, Vol. I, pp. 1-30, Boca Raton, FL, CRC Press, 1982.

7. Bornstein, P., Sage, H. Ann. Rev. Biochem. 49:957-1003, 1980.

8. Hook, M., Kjellen, L., Johansson, S., Robinson, J. Ann. Rev. Biochem. 53:847-869, 1984.

9. Yamada, K.M. Ann. Rev. Biochem. 52:761-799, 1983.

10. Hay, E.D. Cell Biology of Extracellular Matrix. New York, Plenum Press, 1982.

11. Kleinman, H.K., Klebe, R.J., Martin, G.R. J. Cell Biol. 88:473-486, 1981.

12. Furcht, L.T. Mod. Cell Biol. 1:53-117, 1983.

13. Hynes, R.O., Yamada, K.M. J. Cell Biol. 95:369-377, 1982.

14. Mosher, D.F. Ann. Rev. Med. 35:561-575, 1984.

15. Foidart, J.M., Bere, E.W., Yaar, M., Rennard, S.I., Gullino, M., Martin, G.R., Katz, S.I. Lab. Invest. 42:336-342, 1980.

16. Timpl, R., Engel, J., Martin, G.R. Trends in Biochem. Sci. 8:207-209, 1983.

17. Woodley, D.T., Rao, C.N., Hassell, J.R., Liotta, L.A., Matin, G.R., Kleinman, H.K. Biochim. Biophys. Acta 761:278-283, 1983.

18. Sakashita, S., Engvall, E., Ruoslahti, E. FEBS Lett 116:243-246, 1980.

19. McCarthy, J.B., Palm, S.L., Furcht, L.T. J. Cell Biol. 97:772-777, 1983.

20. McCarthy, J.B., Furcht, L.T. J. Cell Biol. 98:1474-1480, 1984.

21. Pierschbacher, M.D., Ruoslahti, E. Nature 309:30-33, 1984.

22. Rogers, S.L., McCarthy, J.B., Palm, S.L., Furcht, L.T., Letourneau, P.C. J. Neuroscience 5:369-378, 1985.

23. Rao, C.N., Margulies, I.M., Tralka, T.S., Terranova, V.P., Madri, J.A., Liotta, L.A. J. Biol. Chem. 257:9740-9744. 1982.

24. Palm, S.L., McCarthy, J.B., Furcht, L.T. Biochemistry (in press), 1985.

25. Rao, C.N., Margulies, I.M.K., Goldfarb, R.H., Madri, J.A., Woodley, D.T., Liotta, L.A. Arch. Biochem. Biophys. 219:65-70, 1982.

26. Ott, U., Odermatt, E., Engvall, J., Furthmayr, H., Timpl, R. Eur. J. Biochem. 123:63-72, 1982.

27. Fidler, I.J., Gruys, E., Cifone, M.A., Barnes, Z., Bucana, C. J. Natl. Can. Inst. 67:947-956, 1981.

28. Fidler, I.J., Cifone, M.A. Am. J. Path. 97:633-648, 1979.

5

ROLE OF *ras* ONCOGENES IN EXPERIMENTAL MODELS OF METASTASIS

U.P. THORGEIRSSON, T. TURPEENNIEMI-HUJANEN, M.E. SOBEL,
J.E. TALMADGE*, AND L.A. LIOTTA

Laboratory of Pathology, National Cancer Institute, National Institutes of Health *Preclinical Screening Laboratory, NCI-Frederick Cancer Research Facility

INTRODUCTION

The most important characteristic of malignant tumors is their ability to metastasize. Although some progress has been made in understanding how cancer cells disseminate, the underlying mechanism is still unknown. Primary tumors contain heterogenous cell populations that differ in their metastatic capacity (1, 2). Metastatic and nonmetastatic tumor cell clones have been established from the same primary tumor (3), but so far no genetic elements specific for the metastatic phenotype have been identified. Somatic cell hybridization *in vitro* has been used extensively to study expression and regulation of the malignant phenotype (4). It has also been used to analyze some of the factors involved in the metastatic process (5, 6). Some reports have appeared on *in vivo* tumor-tumor cell or tumor-host cell hybridization, which in some instances can lead to an increase in the metastatic potential (4). However, most studies on *in vitro* hybridization between tumor cells and normal cells have demonstrated that the transformed and the metastatic phenotypes are recessive in nature (7, 8). Cytogenetic analyses have associated higher ploidy and more chromosomal abnormalities with advanced stages of malignancy (9). The acquisition of high chromosome numbers may lead to tumor progression through increased dosage of genes that may be involved in metastasis. Chromosomal changes consistent with gene amplification seen with increased malignancy include: double minutes (10), abnormally banded regions (11), and homogeneously stained regions (12).

Recent advances in molecular biology have provided new, powerful tools to study the genetic mechanisms involved in metastasis. The DNA transfection technique which allows a gene or genes to be incorporated into the genome of a recipient cell has led to the discovery of transforming cellular genes (oncogenes) in various human and tumor cell lines. To induce a transformation, the oncogene must either undergo a point mutation or be expressed at a higher level than the normal homologue (proto-oncogene) (13). Most of the mutated cellular oncogenes discovered in human malignancies so far belong to the ras family. This family includes Harvey (H), Kirsten (K), and N-ras oncogenes (14, 15, 16, 17). All three members of the ras family encode a 21-kilodalton protein, known as p21, which is located at the inner surface of the plasma membrane (18). The point mutation of the ras gene is usually found in that portion which encodes either the twelfth or the sixty-first amino acid residue of the p21 protein (19). As a consequence of the point mutation, the p21 protein shows slightly altered electrophoretic mobility.

Activation of the ras genes has been demonstrated in chemically transformed tumors (20). In N-nitrosomethylurea (NMU)-induced rat mammary carcinomas, a point mutation is present in the twelfth codon of the H-ras gene, resulting in glutamic acid instead of glycine in the twelfth amino acid residue of the p21 protein (21). NMU is a potent alkylating agent that covalently binds to DNA, and is carcinogenic without metabolic activation. It produces mammary carcinoma in sexually mature rats at a high frequency, with palpable tumors appearing 2 to 3 mo after a single intravenous injection of the chemical (22).

The purpose of our study was to find out a) whether the metastatic phenotype can be introduced into benign cells through transfection of DNA from malignant cells or isolated ras oncogenes, and b) whether amplification or increased expression of the ras gene plays a role in induction of the metastatic phenotype. For this purpose we used NIH/3T3 cells transfected with either tumor DNA or cloned activated c-H-ras oncogene, as well as autochthonous metastatic NMU-induced tumors containing an activated c-H-ras oncogene.

INDUCTION OF THE METASTATIC PHENOTYPE IN NIH/3T3 CELLS BY TRANSFECTED TUMOR DNA AND ras ONCOGENES

The nonneoplastic NIH/3T3 mouse cell line has been used extensively as recipient in transfection assays in which the donor DNA is added in the form of a calcium phosphate precipitate (23). Two to three rounds of transfection are usually performed, using the primary transfectant's DNA as a donor for the second round. It has been shown that the same transforming sequences are passaged in primary and secondary transfection assays (24), and that NIH/3T3 cells transfected with human tumor DNA contain human repetitive (Alu) sequences (25).

The oncogenic potential of NIH/3T3 transfectants has been mostly demonstrated by their tumorigenicity in nude mice, but little attention has been paid to the degree of malignancy. In order to identify genetic determinants of metastasis we have used the DNA transfection system. We first tested whether various transfected DNAs could confer the metastatic phenotype upon the NIH/3T3 cells. We focused on the ras oncogenes, since they have been most commonly implied in human malignancies. NIH/3T3 cells transfected with DNA from malignant cells which contained activated ras genes or NIH/3T3 cells transfected with cloned ras genes were tested. DNA was extracted from freshly isolated leukemic cells (acute lymphocytic (ALL) or acute myelogenous (AML) leukemia), and transfected into NIH/3T3 cells (a gift from Dr. E. Westin, NIH). Activated N-ras oncogene and human Alu sequences were demonstrated in the secondary ALL and AML transfectants (26). The second set of NIH/3T3 cells tested had been transfected with either genomic DNA from the T24 human bladder carcinoma (which contains an activated c-H-ras oncogene), or with the cloned T24 c-H-ras gene (a gift from Dr. S. Aaronson, NIH). For comparison, normal, spontaneously transformed mouse embryonic DNA as well as v-H-ras transfected NIH/3T3 cells were used. All of the transfectants containing either tumor DNA or ras genes alone, as well as the spontaneously transformed NIH/3T3 cells, were tumorigenic in NIH/3T3 nude mice, but the control and the mouse embryo transfected NIH/3T3 cells did not form tumors. With the exception of the spontaneous transformant, the different groups of the tumorigenic NIH/3T3 cells produced experimental metastases when injected into the tail vein of NIH nude mice (Table 1). However, after intramuscular injection, only the two transfectants containing the activated T24 c-H-ras produced spontaneous metastases. No difference in metastatic capacity was detected between the T24 tumor DNA and the cloned T24 c-H-ras transfectants. All of the transfectants produced rapidly growing fibrosarcomas with high cellularity and moderate degree of atypia. Sections of the metastatic lung nodules revealed the same histological picture. From these *in vivo* studies, it was apparent that the NIH/3T3 cells had acquired the metastatic potential through transfection of human tumor DNA containing activated ras genes, or the ras genes alone. This change in phenotype was not merely due to chromosomal integration of any type of foreign DNA, since transfection of mouse embryonic DNA into the NIH/3T3 cells had no effect. Furthermore, the spontaneous transfectant that did not contain any extraneous DNA was not metastatic.

Table 1. Experimental and spontaneous metastases in NIH nude mice injected with NIH/3T3 transfectants.

Cell type	Source of transfected DNA (oncogene)	Tumorigenicity (+/group)	Experimental metastasis (+/group)	Spontaneous metastasis (+/group)
NIH/3T3	-	0	0/20	
NIH/3T3	-	0		0/15
NIH/3T3	-		0/10	
(spont.)	-	10/10		0/10
NIH/3T3 (emb.)	mouse embryo	0	0/10	
NIH/3T3	ALL		16/20	
(ALL)	(N-ras)	15/15		0/15
NIH/3T3	AML		10/20	
(AML)	(N-ras)	15/15		0/15
NIH/3T3	v-H-ras		8/20	
(v-H-ras)		15/15		0/15
NIH/3T3	T24 human bladder		6/10	
(T24)	ca. (c-H-ras)	10/10		3/10
NIH/3T3	T24 c-H-ras		5/10	
(c-H-ras)		10/10		3/10

The cells (10^5/animal) were injected in a single-cell suspension either into the lateral tail vein of NIH nude mice or intramuscularly into the right thigh. The mice that received the intravenous injections were sacrificed after 4 weeks and and the others after 6 weeks. The mice were autopsied, the lungs were inflated in Bouin's fixative, and the metastatic nodules counted under a dissecting microscope. Microscopic examination was performed on all major organs.

In order to find out if an immunological mechanism was responsible for the metastatic capacity of the ras transfected NIH/3T3 cells in nude mice, they were tested in both NK cell and macrophage cytotoxity assays (27, 28). In both types of assays the transfectants were more sensitive to the immune cell killing than the control NIH/3T3 cells. This was supported by *in vivo* findings showing that the number of metastases produced by the ALL transfectants was decreased in immunostimulated nude mice. Thus, it was ruled out that the acquired metastatic capacity of the transfectants was merely due to escape of the nude mouse immune system.

Tumor cell invasion and dissemination is facilitated by proteolytic enzymes that break down connective tissue barriers in the migration path of the invasive cells (29). One of these enzymes, type IV collagen degrading metalloproteinase (type IV collagenase), specifically cleaves type IV collagen, a major structural component of basement membranes (BM) (30, 31). Migration of malignant cells through BM is a critical step in the metastatic process, i.e. during transition from in situ to invasive carcinoma and dissemination through vascular channels. A correlation has been demonstrated between secretion of type IV collagenase in vitro and metastatic capacity of different tumor cells lines (32). The NIH/3T3 transfectants were tested for type IV collagenase secretion in vitro. When compared with the untransfected control NIH/3T3 cells, we detected an increase in the secreted enzyme activity in the transfected cells that corresponded well with their metastatic behaviour in nude mice (Table 1). The control NIH/3T3 cells demonstrated minimal type IV collagenase activity (11 ng/10^6), whereas the increase in enzyme activity of the transfectants ranged from 200% (v-H-ras) to 700% (c-H-ras). The tumorigenic spontaneously transformed NIH/3T3 cells that were nonmetastatic in nude mice had minimal enzyme activity (8 ng/10^6 cells) (Table 2).

Table 2. Type IV collagenolytic activity in supernatants from NIH/3T3 transfectants.

Cell type	Type IV collagenase activity (ng/10^6 cells)
NIH/3T3	11
NIH/3T3 (spontaneous)	8
NIH/3T3 (ALL)	62
NIH/3T3 (AML)	58
NIH/3T3 (v-H-ras)	25
NIH/3T3 (T24)	23
NIH/3T3 (c-H-ras)	78

Type IV collagenolytic activity was measured in ammonium sulfate-concentrated culture supernatants as previously described (6). The data shown are means of three to five separate experiments with duplicate samples for each assay. The increase in the collagenolytic activity of the transfectants was significant, $P < 0.01$ (paired Student's t test).

NMU-INDUCED RAT MAMMARY CARCINOMA AS A MODEL TO STUDY THE EXPRESSION OF ACTIVATED *ras* ONCOGENES IN METASTASIS

The data from the NIH/3T3 transfectants demonstrate that the *ras* gene can induce the metastatic phenotype by a nonimmunological mechanism. We next addressed the question whether amplification or increased expression of the *ras* gene was responsible for the progression of the malignant phenotype. For this purpose an autochthonous rat tumor model was used, i.e. NMU-induced metastatic rat mammary carcinoma. Sukumar and coworkers demonstrated a point mutation in the H-*ras* gene of the NMU-induced rat carcinomas, encoding glutamic acid instead of glycine in the twelfth amino acid residue of the p21 protein (21). NMU induces several primary mammary tumors in each rat which usually are nonmetastatic, well differentiated adenocarcinomas. However, in rare instances, metastatic tumors have been isolated. We have studied four metastatic NMU tumors, one autochthonous freshly isolated tumor from a Sprague Dawley rat, and three tumors that were passaged and had maintained their metastatic pattern during multiple passages in Buffalo rats (a gift from Dr. P. M. Gullino, NIH). A Sprague-Dawley rat, carrying a single autochthonous primary tumor, was sacrificed 9 months after a single intravenous dose of NMU. It was found to have numerous pulmonary metastases at the time of autopsy. Ten of the metastases were individually resected and the remaining metastatic tumor was pooled and frozen. Histological sections of the primary tumor demonstrated a well differentiated adenocarcinoma (infiltrating duct type) with prominent desmoplasia (Fig. 1a). The microscopic appearance of the individually resected metastases, however, revealed a less differentiated adenocarcinoma with sheets of highly atypical cells and minimal desmoplasia (Fig. 1b).

DNA was isolated from the primary NMU tumor and multiple pooled fragments of metastases. Southern blot analysis (33) of EcoRI digested DNA revealed a single 2.3 kb H-*ras* specific fragment in both the primary and the metastatic tumor, but not in the NIH/3T3 cells (Fig. 2). As seen in Fig. 2, the primary tumor DNA contains clearly higher levels of the *ras* DNA sequences than the metastatic fragments. When quantitated by slot blot analysis (34), the primary tumor had ten-fold higher H-*ras* DNA levels. We next examined whether this finding would be repeated when multiple metastases from a single primary were tested. Ten separate lung metastases from a single primary were transplanted subcutaneously and expanded in nude mice for DNA and RNA isolation. Slot blot analysis of the DNA and RNA from the nude mice transplants of the primary and the metastases revealed a variability

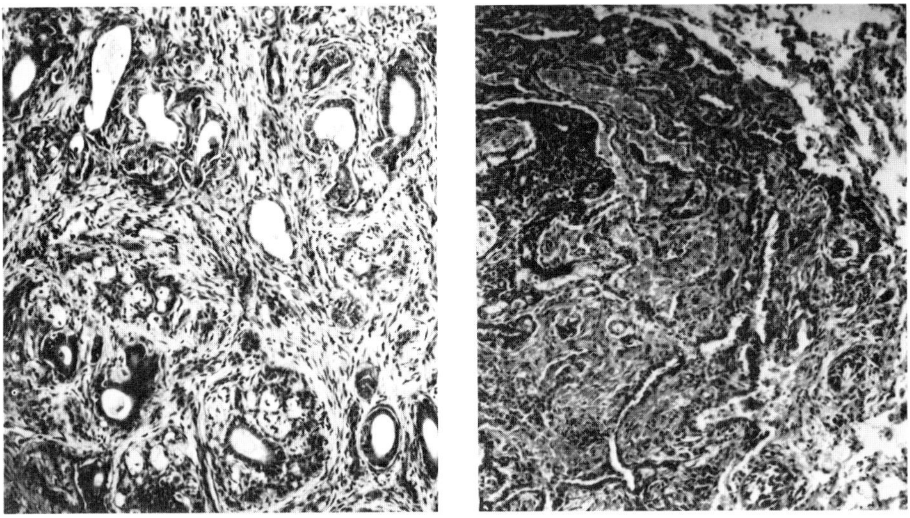

Fig. 1. NMU-induced rat mammary carcinoma: a) primary tumor, b) lung metastasis, H&E (x40).

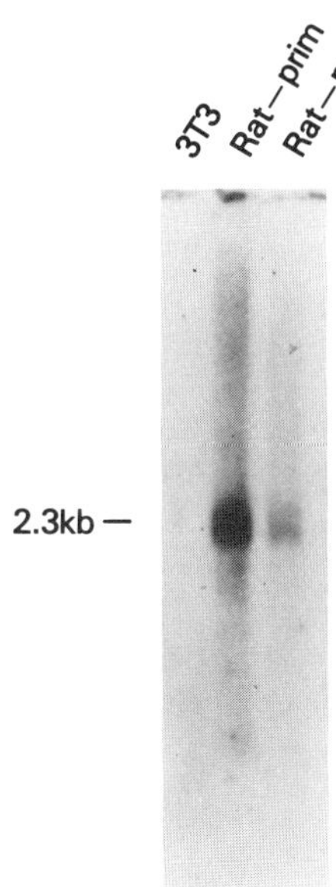

Fig. 2. Southern blot analysis on 20 μg of DNA isolated from a primary NMU tumor and lung metastases (pooled). A nick-translated, ^{32}P-labeled v-H-ras probe was hybridized to the nitrocellulose filter. DNA from NIH/3T3 cells served as a negative control.

in both the DNA levels and expression of the H-ras gene. Some were lower, others higher than the primary tumor (data not shown).

Multiple individual primary tumor isolates and metastases were isolated from a single passage of a NMU tumor that had been previously passaged numerous times in Buffalo rats. These were tested for H-ras sequences. Slot blot analysis of DNA from five primaries and six metastases is demonstrated in Fig. 3a, b. There is minimal variability within each group, as well as between the primary and the metastatic groups. Furthermore, when metastatic and nonmetastatic primary tumors were compared, less a than two-fold difference in the H-ras specific DNA levels were detected between the highest and the lowest of the two groups (Fig. 3c, d).

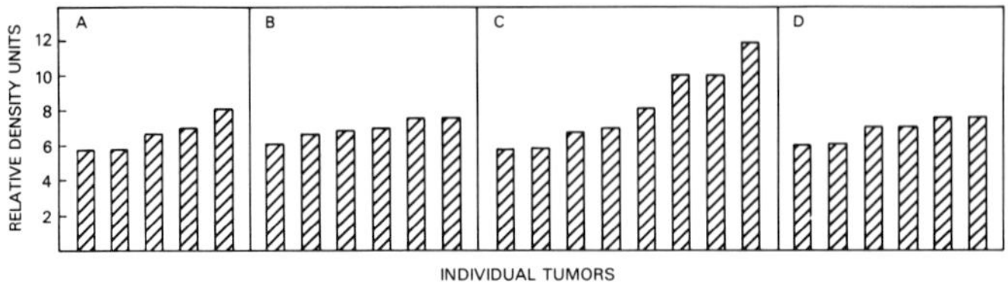

Fig. 3. Slot blot analysis of total cellular DNA from a) primary NMU tumors, b) NMU metastases, c) different metastatic primary NMU tumors, d) different nonmetastatic NMU tumors. The bars represent the densitometric readings of H-*ras* specific sequences in 5 μg of DNA.

CONCLUSIONS

The metastatic phenotype can be induced by a nonimmunological mechanism in NIH/3T3 cells through transfection with DNA from malignant cells containing activated *ras* oncogenes or with molecularly cloned *ras* genes alone. These *in vivo* findings are supported by an *in vitro* parameter, associated with the malignant phenotype, i.e. increased secretion of type IV collagenase by the transfected cells in culture. Data from NIH/3T3 transfectants and NMU-induced rat mammary carcinoma suggest that amplification or increased expression of the *ras* oncogenes does not play a role in the development of metastases. It is unlikely that the activated H-*ras* gene is the only gene responsible for carrying out the multistep metastatic process, which probably involves an interplay between multiple genes. In the NIH/3T3 transfection system the *ras* gene may either activate cellular genes or cooperate with already activated cellular genes that are necessary for the completion of the metastatic process. The DNA transfection system offers a powerful tool to search for genetic determinants of the metastatic phenotype.

References

1. Nowell, P.C. Science 194:23-28, 1976.
2. Poste, G., Doll, J. and Fidler, I.J. Proc. Natl. Acad. Sci. USA 78:6226-6230, 1981.
3. Fidler, I.J. and Kripke, M.L. Science 197:893-895, 1977.
4. Larizza, L. and Schirrmacher, F. Cancer Metastasis Rev. 3:193-222, 1984.
5. Ramshaw, I.A., Carlsen, S. Wang, H.C. and Badenoch-Jones, P. Int. J. Cancer 32:471-478, 1983.

6. Turpeenniemi-Hujanen, T., Thorgeirsson, U.P., Hart, I.R., Grant, S. and Liotta, L.A. J. Natl. Cancer Inst. 75:99-103, 1985.
7. Harris, H., Miller, O.J., Klein, G., Worst, P. and Tachibana, T. Nature 223:363-368, 1969.
8. Stanbridge, E.J., Der, C.J., Doersen, C.J.,Nishimi, R.Y., Peehl, D.M., Weissman, B.E. and Wilkinson, J.E. Science 215:252-259, 1982.
9. Wolman, S.R. Cancer Metastasis Rev. 2:257-293, 1983.
10. Cowell, J.K. Annu. Rev. Genet. 16:21-59, 1982.
11. Nowell, P., Finan, J., Dalla-Favera, R., Gallo, R.C., ar Rushdi, A., Romanczuk, H., Selden, J.R., Emanuel, B.S., Rovera, G. and Croce, C.M. Nature 306:494-497, 1983.
12. Biedler, J.L. and Spengler, B.A. Science 191:185-187, 1976.
13. Bishop, J.M. Annu. Rev. Biochem. 52:301-354, 1983.
14. Der, C.J., Krontiris, T.G. and Cooper, G.M. Proc. Natl. Acad. Sci. USA 79:3637-3640, 1982.
15. Reddy, E.P., Reynolds, R.K., Santos, E. and Barbacid, M. Nature 300:149-152, 1982.
16. Shimizu, K., Birnbaum, D., Ruley, M.A., Fasano, O., Suard, Y., Edlund, L., Taparowsky, E., Goldfarb, M. and Wigler, M. Nature 304:497-500, 1983.
17. Taparowsky, E., Shimizu, K., Goldfarb, M. and Wigler, M. Cell 34: 581-586, 1983.
18. Willingham, M.C., Pastan, I., Shih, T.Y. and Scolnick, E.M. Cell 19: 1005-1014, 1980.
19. Tabin, C.J., Bradley, S.M., Bargmann, C.I., Weinberg, R.A., Papageorge, A.G., Scolnick, E.M., Dhar, R., Lowy, D.R. and Chang, E.H. Nature 300: 143-149, 1982.
20. Balmain, A. and Pragnell, I.B. Nature 303:72-74, 1983.
21. Sukumar, S., Notario, V., Martin-Zanca, D. and Barbacid, M. Nature 306:658-661, 1983.
22. Gullino, P.M., Pettigrew, H.M. and Grantham, F.H. J. Natl. Cancer Inst. 54:401-414, 1975.
23. Perucho, M., Goldfarb, M., Shimizu, K., Lama, C., Fogh, J. and Wigler, M. Cell 27:467-476, 1981.
24. Lane, M.A., Sainten, A. and Cooper, G.M. Proc. Natl. Acad. Sci. USA 78:5185-5189, 1981.
25. Rubin, C.M., Houck, S.M., Deininger, P.L., Friedmann, T. and Schmid, C.W. Nature 284:372-374, 1980.
26. Thorgeirsson, U.P., Turpeenniemi-Hujanen, T., Williams, J.E., Westin, E.H., Heilman, C.A., Talmadge, J.E. and Liotta, L.A. Mol. Cell. Biol. 5:259-262, 1985.
27. Raz, A., Fogler, W.E. and Fidler, I.J. Cancer Immunol. Immunother. 7:157-163, 1979.
28. Talmadge, J.E., Meyers, K.M., Prieur, D.J. and Starkey, J.R. Nature 284:622-624, 1980.
29. O'Grady, R.L., Harrop, P.J. and Cameron, D.A. Pathology 14:135-138, 1982.
30. Salo, T., Liotta, L.A. and Tryggvason, K. J. Biol. Chem. 258: 3058-3063, 1983.
31. Fessler, L.I., Duncan, K.G. and Fessler, J.H. J. Biol. Chem. 259: 9783-9789, 1984.
32. Liotta, L.A., Tryggvason, K., Garbisa, S., Hart, I., Foltz, C.M., and Shafie, S. Nature 284:67-68, 1980.
33. Southern, E.M. J. Mol. Biol. 98:503-517, 1975.
34. Gasser, C.S., Simonsen, C.C., Schilling, J.W., and Schimke, R.T. Proc. Natl. Acad. Sci. USA 79:6522-6526, 1982.

6

EXPRESSION OF p21^{ras} GENE PRODUCTS IN FRESH PRIMARY AND METASTATIC HUMAN TUMOR TISSUE

G.E. GALLICK[1], R. KURZROCK[2], and J.U. GUTTERMAN[2]

[1]Departments of Tumor Biology and [2]Clinical Immunology, M.D. Anderson Hospital and Tumor Institute, Houston, Texas 77030

INTRODUCTION

Activation of cellular genes, termed "proto-oncogenes" has been implicated as significant in the development of many types of neoplasias. However, the role of "activated" oncogene expression as a driving force in tumor progression and the emergence of metastatic cells remains unclear. Aberrant expression (or aberrant gene products) may be required throughout the stages of tumor progression; alternatively, they may be required only at a particular stage. These possibilities can be investigated by examining oncogene expression at various stages of tumors. Perhaps the oncogene family most prevalently linked to tumorigenesis, and therefore the most intensely studied, are members of the c-ras gene family. Recent evidence suggests that normal expression of c-ras genes are involved in regulation of adenylate cyclase (1), and may be critical for initiation of the S-phase of the cell cycle in NIH-3T3 cells (2). In NIH-3T3-transfection assays, transformation results from overexpression of "normal" c-ras genes (3), or by expression of "altered" c-ras genes which encode proteins with only a single amino acid change (4,5). Both types of expression have been observed in human cancers. In human colorectal carcinomas, both point mutation in c-ras genes (6,7) and altered c-ras expression (8) have been observed. To examine the relevance of such expression during tumor progression, we have taken advantage of the availabilty of fresh human tumor tissue at M.D. Anderson Hospital to measure p21^{ras} proteins at various stages of primary colorectal tumors and in metastases to several different sites. Our results show that overexpression of p21^{ras} frequently occurs in earlier stages of the malignancy, but is uncommon at later stages, and is consistently depressed in metastatic lesions. However, such changes were not observed in metastases to the lung from several other types of primary tumors. These results suggest that changes in the expression of c-ras gene products are uncommon in most metastases.

MATERIALS AND METHODS

Preparation of human tissue for immunoblotting. Immediately after collecting, tissue was cut into pieces of approximately 1 gram in size, and quick-frozen in liquid nitrogen. Tissue pieces were then lyophilized for 12h or stored in liquid nitrogen. Tissues stored in this manner have consistently yielded reproducible results in protein analyses. On the basis of pathology reports, the stage of primary colorectal carcinoma was determined according to the Astler-Coller modification of the Dukes' system with designations as follows: B_1-lesions confined within the bowel wall; B_2-lesions which penetrate through all layers of the bowel; C-lesions classified as either B_1 or B_2 with the addition of lymph node involvement; D-tumors that have metastasized beyond regional lymph nodes. Histopathological examination was performed on all tumors, and biopsies of metastatic tissue were compared to stored samples from the primary site to confirm the origin of the primary tumor.

Immunoprecipitation of tissue extracts. Immunoprecipitation of $p21^{ras}$ products was accomplished with a monoclonal antibody, designated Y13-259, developed in the laboratory of Dr. E. Scolnick (9), and donated by Dr. Scolnick, and was performed as previously described (8). Briefly, 2.5 mg of lyophilized tissue was dissolved in 1 ml of detergent-containing buffer as described by Naso et al. (10). Lysates were cleared by centrifuging at 10,000rpm for 10 min. The resulting lysates are reacted 3h with 10μl of the monoclonal antibody (1.7 mg/ml) at 4°C. Precipitates were next reacted for 2h with 10μl of rabbit antibodies against rat IgG (Cappel Laboratories, Cochranville, Pa.), which had been reconstituted according to the manufacturer's instructions. Immunoprecipitates were then reacted with 60μl of formalin-inactivated Staphylococcus aureus (Cowan strain) as described by Kessler (11). The precipitates were washed three times and proteins were resolved by SDS-polyacrylamide gel electrophoresis on 10.5% acrylamide gels.

Immunoblotting. Immunoblotting of proteins was as previously described (8). Briefly, immunoprecipitated samples (as described above) were resuspended in SDS-PAGE buffer containing 1% SDS, 1% 2-mercaptoethanol, 0.001M EDTA, 0.005M bromophenol blue, 5% glycerol, and 0.01M Tris-HCl, pH 8.0. Samples were boiled for 5min, cooled to 20°C and centrifuged at 6,000rpm for 10min. Supernatant fluid was run by standard SDS-PAGE techniques, after which proteins were electroblotted from the gel onto nitrocellulose filter paper for 45 min at 100 volts. The filter was rinsed in TNE/NP40 buffer (0.15M NaCl, 0.002M EDTA, 0.1% NP40, 0.05M tris, pH 7.5), then remaining reactive sites were blocked for 3h in a plastic bag containing 2 ml per gel lane TNE/NP40 with 3% BSA. After blocking, fresh buffer containing the monoclonal antibody to $p21^{ras}$ (1:100 dilution) was added and the binding of antibody was allowed to proceed for 12h at 4°C. Filters were washed five times with TNE/NP40 buffer, placed in a plastic bag and incubated for 1h with 0.1 µCi/ml ^{125}I-goat anti-rat IgG, specific activity 40 mCi/mg (Amersham, Chicago, IL), per gel lane in TNE/NP40 buffer containing 3% BSA. After incubation, filters were washed 4 times with TNE/NP40 then once in TNE/NP40 containing 1.2% glycerol. Filters were dried and autoradiographed at -70°C.

RESULTS

Expression of $p21^{ras}$ in Primary and Metastatic Colon Tumors

Recently we have examined expression of c-ras gene products from fresh tumor tissue at various stages of colorectal carcinoma, comparing $p21^{ras}$ expression to that observed in adjacent normal colonic tissue (8). These results, summarized in Table 1, indicate that elevation of c-ras proteins is a common occurrence in earlier stages of colorectal carcinoma, but at later stages, when distant metastases are present, c-ras gene expression is similar in tumor tissue and adjacent normal tissue. These results suggest that tumor levels of p21^{ras} gene products. Such a possibility is strongly supported by levels of p21^{ras} in metastases. In each metastatic lesion examined, p21^{ras} expression was dramatically reduced with respect to that observed in primary tumors or normal colonic tissue, as summarized in Table 1. Reduction of p21^{ras} was independent of the site of metastasis, because metastases to the liver, lung and omentum gave similar results. Levels of tubulin, measured as an internal control, were unchanged between tumor and normal tissues, suggesting that degradation due to increased proteolytic activity, or to necrosis, was unlikely to account for the decrease in p21^{ras} levels. Thus, reduction of c-ras expression appears to be characteristic of colon metastases.

Table 1. p21^{ras} Expression in Primary and Metastatic Colon Tumors

Dukes' Classification	Number of Patients	p21 Expression[1] Elevated[2]	Unchanged	Depressed[2]
B	8	6	2	0
C	4	2	1	1
D	5	1	4	0
metastases[3]	9	0	0	9

[1] p21 expression was determined by densitometry
[2] greater (or less) than 2-fold difference between tumor and adjacent normal tissue
[3] sites of metastases were: 4 to liver, 5 to lung, and 1 to omentum

Expression of p21^{ras} in Lung Metastases

To determine if suppression of p21^{ras} was a common occurrence in metastases of different origins, a series of metastases to the lung from diverse primary tumors were examined. Of fifteen metastases to the lung from primary tumors including several types of carcinomas, sarcomas, melanomas, and neuroblastomas, only metastases from the colon were found to be suppressed in c-ras gene expression. The results are summarized in Table 2. An example of p21^{ras} expression in a primary and lung metastasis from a breast carcinoma from one patient compared to a primary and lung metastasis from a colorectal carcinoma

TABLE 2. $p21^{ras}$ Expression in Metastatic Lesions In The Lung

		p21 Expression		
Primary Tumor	Number of Patients	Elevated[1]	Unchanged	Depressed[1]
Colorectal Carcinoma	4	0	0	4
Breast Carcinoma	1	0	1	0
Osteogenic Carcinoma	1	0	1	0
Osteosarcoma	3	2	1	0
Leiomyosarcoma	1	1	0	0
Uterine Sarcoma	1	0	1	0
Melanoma	3	1	2	0
Neuroblastoma	1	0	1	0
TOTAL	**15**	**3**	**8**	**4**

[1]greater (or less) than 2-fold difference between primary tumors and metastatic tumors

from another patient is shown in Fig. 1. While $p21^{ras}$ levels are strikingly reduced in the colon metastases as compared to the primary tumor or normal tissues, $p21^{ras}$ levels are similar among primary and metastatic breast tumor tissue, and adjacent normal lung tissue. In some tumors, for example 2 of 3 osteosarcomas, $p21^{ras}$ expression was somewhat elevated as compared to that observed in primary tumor tissue. However, an insufficient number of tumors have been examined to determine if elevation of $p21^{ras}$ gene products is a consistent feature of these metastases. In general, most of the metastases other than from colorectal carcinomas (8 of 11) showed no apparent difference in $p21^{ras}$ levels. Thus, while altered regulation of c-ras genes may be a significant aspect in the development of metastases of some tumors, such alterations are apparently not consistent with the development of many types of metastases.

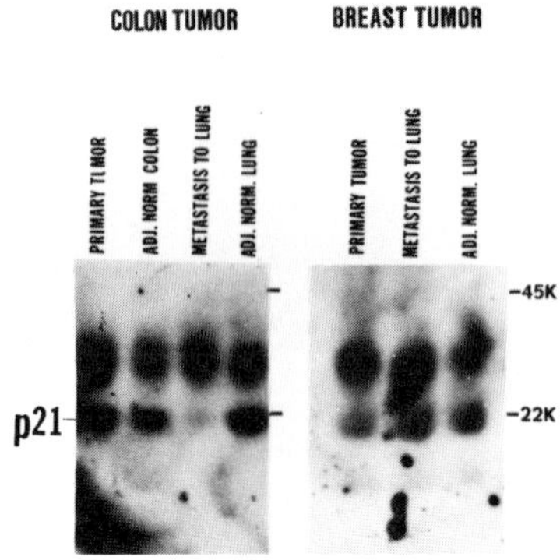

Fig. 1. Comparison of p21^{ras} Levels in a Primary and Metastastic Colon Tumor to a Primary and Metastastic Breast Tumor. Immunoblotting was performed as described in Materials and Methods. Tissue types are indicated in the Figure.

DISCUSSION

Aberrant expression of oncogenes has been associated with an ever increasing number of human malignancies. However, potential roles of these genes during tumor progression remain obscure. Single base changes in the c-ras gene family have been demonstrated to be "activating", i.e can induce the transformed phenotype in NIH-3T3 transfection assays. While such alterations have been found in a large number of diverse types of malignancies, they apparently occur at low frequency (12). Further, whether or not activated c-ras gene are sufficient to induce metastasis may depend on the model system employed. Evidence that altered c-ras genes expression may not be required for metastasis was derived by Albino et al. (13), who found no consistent presence of activated c-ras genes among cell lines derived from a primary human melanoma and from separate metastatic deposits. Such results suggest that altered c-ras genes which are detectable by transfection into NIH-3T3 cells may arise as a consequence of genetic and epigenetic factors which lead to tumor cell heterogeneity (14,15), and do not necessarily function in the generation of metastases. Alternatively, activation of c-ras genes may be important at early stages of tumorigenesis, setting in motion a cascade of molecular events which may proceed without further participation of the activated gene product (16).

To study expression of the c-ras gene family during stages of tumorigenesis, we have examined protein expression at various stages of human colorectal carcinoma, including distant metastases. In these tumors, no evidence of $p21^{ras}$ gene products with altered electropheretic mobility, characteristic of activated c-ras genes, have been detected, although such proteins are easily distinguished from "normal" proteins in cell lines harboring mutant c-ras genes. While mutated proteins might be missed by judging only the electropheretic mobility, these data agree with other investigators who find point mutations occur infrequently in tumor tissue and tumor cell lines (11). However, overexpression of c-ras gene products is common at earlier stages of the tumor. Tumor progression is toward decreasing levels of $p21^{ras}$ (8). In metastatic lesions, $p21^{ras}$ products are considerably reduced. These data suggest that while overexpression of c-ras proteins may be an important event (or correlate with) early stages of this cancer, activity of these gene products may not be important for metastasis. The apparent specific supression of these products suggest that another gene may be activated as part of the molecular changes leading to metastasis.

In contrast to colorectal metastases, other metastases, specifically to the lung, show little or no change in $p21^{ras}$ expression from that observed in primary tumors. These data lend further support to the hypothesis that activated c-ras genes may not play significant roles in the majority of metastases, and further suggest the disturbing possibility that the activation of different sets of genes may be important in tumor progression of different types of tumors. However, as evidence mounts that cellular proto-oncogenes are critically important in normal growth and regulation of cells, these genes remain likely candidates for triggering important stages of tumor progression and development of metastasis.

ACKNOWLEDGEMENTS

The authors gratefully acknowledge the excellent technical assistance of Ms. Nila Parikh and Ms. Marjorie Johnson. This work was aided by NIH Grant No. RR5511-21 awarded to G.E.G., and the research was conducted in part by the Clayton Foundation for Research and the James E. Lyon Foundation for Research. J.U.G. is a Senior Clayton Foundation Investigator.

REFERENCES:

1. Toda, T., Uno, I., Ishikawa,T., Powers, S., Kataoka, T., Broek, D.,Cameron, S., Broach, J., Matsumoto, K, and Wigler M. Cell 40:27-36, 1985.
2. Mulcahy, L.S., Smith, M.R. and Stacey D.W. Nature 313:41-243, 1985.
3. Chang, E.H., Furth, M.E., Scolnick, E.M. and Lowy, D.R. Nature 297:479-483, 1982.
4. Tabin, C.J., Bradley, S.M., Bargmann, C.I., Weinberg, R.A., Papageorge, A.T., Scolnick, E.M., Dhar, R., Lowy, D.R. and Chang, E.H. Nature 300:143-149, 1982.
5. Reddy, E.P., Reynolds, R.K., Santos, E. and Barbacid, M. Nature 300:149-152, 1982.
6. Der, C.J. and Cooper, G.M. Cell 32:201-208, 1983.
7. Capon, D.J., Seeburg, P.H., McGrath, J.P., Hayflick, J.S., Edman, U., Levinson, A.D. and Goeddel D.V. Nature 304:507-513, 1983.
8. Gallick, G.E., Kurzrock, R. and Gutterman, J.O. P . Natl. Acad. Sci. In Press, 1985.
9. Furth, M.E., Davis, L.J., Fluerdelys, B., and Scolnick, E.M. J. Virol. 43:294-304, 1982.
10. Naso, R.B., Arcement, L.J., and Arlinghaus, R.B. Cell 4:31-36, 1975.
11. Kessler, S.W. J. Immunol. 115:1617-1624, 1975.
12. Land, H., Parada, L.F. and Weinberg, R.A. Science 222:771-777, 1983.
13. Albino, A.P., Strange, R.L., Oliff, A.I., Furth, M.E. and Old, L.J. Nature 308:69-72, 1984.
14. Nowell, P.C. Science 194:23-28, 1983.
15. Heppner, G.H. and Miller, R.E. Canc. Met. Rev. 2:5-23, 1983.
16. Nicolson, G.L. Canc. Met. Rev. 3:25-42, 1984.

7

A ROLE FOR DIFFERENTIATION ARREST IN THE DEVELOPMENT OF NEURAL CREST TUMORS

Lee J. Helman, M.D.*, Carol J. Thiele, Ph.D.*, W. Marston Linehan, M.D.† and Mark A. Israel, M.D.*

*Pediatric Branch and †Surgery Branch, National Cancer Institute, National Institutes of Health, Bethesda, Maryland 20205

INTRODUCTION

All cancers are characterized by excessive growth and proliferation of tumor cells. Such malignant cells may arise from mature cells which have lost the ability to respond to normal growth control mechanisms, or alternatively, they may arise from immature cells which have never developed the ability to respond to these growth inhibitory signals. These origins should be reflected in the biologic characteristics of some tumors, and an understanding of the events regulating normal cellular differentiation should provide insight into the altered growth characteristics of these malignant tumors.

The involvement of differentiation arrest in oncogenesis has long been invoked for various tumors of hematopoietic and lymphoid origin. In myeloid differentiation, there is an easily recognized pathway of differentiation from myeloblast through promyelocyte and ultimately to mature granulocyte with each step in this pathway characterized by a morphologically distinct cell type. A tumor which may arise from alterations in this differentiation pathway is chronic myelogenous leukemia, a clonal expansion of immature myeloid forms which is believed to result from the proliferation of hematopoietic precursor cells arrested at an early stage in differentiation.

For solid tumors, however, the cell of origin has not generally been considered to be part of a self-renewing population derived from continually dividing stem cells. Therefore, the concept of differentiation arrest has not been widely applied to such tumors. Recent data, however, suggest that several solid tumors of tissues arising from the embryologic neural crest (pheochromocytoma, neuroblastoma, and neuroepithelioma) originate during the proliferation of cells occurring at various stages within the normal differentiation pathway. In the case of neuroblastoma, cellular proliferation is likely to be an important part of the normal physiologic behavior of neural crest cells at the stage of differentiation manifested by these tumor cells. The failure of these cells to progress to a more mature cell type may be an important aspect of the pathology underlying their behavior.

NEURAL CREST TUMORS

The neural crest is composed of cells that appear during gastrulation as a specialized longitudinal strip of neuroectoderm dorsal to the neural tube. Embryonal neural crest cells have several interesting biologic characteristics. These cells invade a variety of developing tissues to migrate over long distances during embryogenesis, an activity similar to the invasive properties of metastatic tumor cells (1). In neural crest cells, however, this capacity for migration appears to be precisely regulated during differentiation and is not found in mature neural crest cell derivatives. In addition, developing neural crest cells are pluripotent, ultimately forming the peripheral nervous system, certain neurosecretory cells, melanocytes, and a variety of connective tissues (2). The final phenotype of a neural crest cell is not solely genetically predetermined; cellular interactions during and after migration also contribute to the regulation of terminal differentiation. For example, using chick-quail chimeric transplants, Le Douarin has shown that quail cells from the neural crest cholinergic vagal region transplanted to the chick neural crest adrenal medulla region migrate into the chick adrenal medulla. These develop into catecholamine secreting cells, thus demonstrating the effects of the embryonic environment on phenotypic expression (3).

Although derived from the same embryonic tissue, different neural crest tumors demonstrate different biologic characteristics suggesting that they originate after differentiation has begun. These tumors vary in their clinical presentation, proliferative capacity, metastatic potential, and ability to undergo spontaneous differentiation along varying maturational pathways. Neuroblastoma, neuroepithelioma, and pheochromocytoma are three tumors derived from the neural crest which we believe may represent malignant proliferations of cells corresponding to different stages along the normal pathway of neuronal neural crest differentiation (Fig. 1). The study of these tumors may contribute important insights into the relationship between altered gene regulation, altered differentiation, and the neoplastic phenotype.

NEUROBLASTOMA

Neuroblastoma is the most common solid tumor of infants less than one year of age. This tumor, consisting of primitive neuronal cells, is found within the sympathetic nervous system, most commonly in the adrenal medulla. Several observations regarding the natural history of neuroblastoma suggest that differentiation arrest may play a significant role in its etiology. Spontaneous histologic differentiation of neuroblastoma into more mature-appearing ganglioneuroblastoma and even benign ganglioneuroma has been reported (4). In addition, rests of neuroblastoma cells in the adrenal glands of infants less than three months of age who were autopsied for reasons unrelated to malignancy have been seen at a frequency which is fifty times greater than the actual incidence of neuroblastoma (5), suggesting that a number of these "neuroblastomas in situ" undergo spontaneous regression. Spontaneous regression is also seen in Evans Stage IV-S neuroblastoma, a distinct clinical presentation with a small primary tumor and metastatic tumor involvement of the liver, skin, and/or bone marrow. These observations lend support to the hypothesis that neuroblastoma may result from a block in maturation of normal sympathetic ganglia precursor cells, and that under certain circumstances this block is overcome and the tumor regresses, perhaps by maturing into a more benign phenotype.

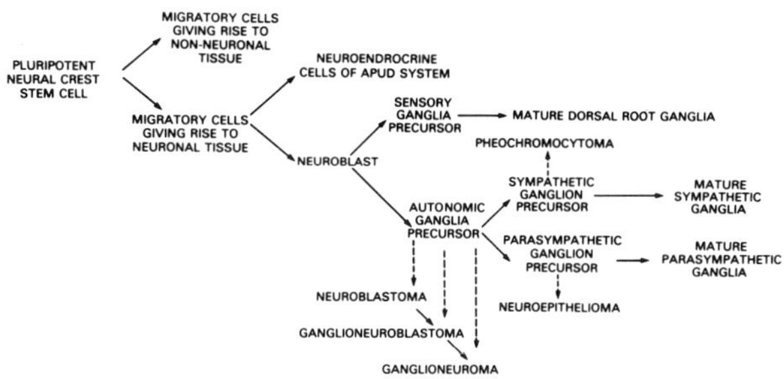

Fig. 1 - Postulated Differentiation Pathways of Neural Crest

An amplified DNA sequence with homology to the proto-oncogene c-myc was recently discovered in neuroblastoma cells and named N-myc (6,7). N-myc DNA has been found to be amplified and highly expressed in virtually all neuroblastoma cell lines (8, 9, 10). In addition, there is a striking correlation between N-myc amplification in neuroblastoma tumor specimens and the clinical stage of disease. Specifically, there is an approximately 50% incidence of N-myc amplification in tumor tissue obtained from patients with advanced Evans Stage III and IV disease compared to the rare occurrence of amplification of N-myc DNA in Evans Stage I and II disease (11). Since the two year survival for advanced stage neuroblastoma is less than 30% while the two year survival for early stage disease is between 75-90%, the presence of amplified N-myc in the genome of these tumors is likely to be correlated with a diminished chance for survival. Recently, N-myc expression was evaluated in paired cell lines obtained from two different patients with neuroblastoma. Each pair consisted of a cell line obtained at the patient's initial presentation and a cell line obtained from the same patient following progression of disease despite the initiation of treatment. In both cases, there was an approximately five-fold increase in N-myc expression in the cell line obtained after progression (12). These data provide additional support for the association of N-myc with tumor progression and poor prognosis.

Because of the clinical observations suggesting differentiation arrest as a possible etiologic factor in neuroblastoma and the association of increased expression of N-myc with tumor progression, our laboratory recently evaluated the effect of _in vitro_ induced differentiation of human neuroblastoma tumor cells on N-myc expression. In this experiment, several human neuroblastoma cell lines were induced to differentiate _in vitro_ by retinoic acid treatment, and the degree of morphologic differentiation was assessed by determining the percentage of cells exhibiting long, neuritic processes. N-myc RNA levels were determined in these cells at various intervals after the initiation of retinoic acid treatment by Northern blot analysis of total cytoplasmic RNA (13). As can be seen in Figure 2, a dramatic decrease in N-myc expression preceded morphologic differentiation in response to retinoic acid. We interpret these data as suggesting that high levels of N-myc RNA expression may inhibit terminal differentiation and thereby play a role in the pathogenesis of this tumor.

PHEOCHROMOCYTOMA

Pheochromocytomas, like neuroblastomas, are tumors of the sympathetic nervous system occurring most commonly in the adrenal medulla. Unlike neuroblastomas, however, these tumors rarely invade adjacent tissues or metastasize and are often manifested clinically by symptoms due to the excessive catecholamines which these tumors produce. High level catecholamine production and limited potential for tumor spread suggest a more differentiated phenotype than that of neuroblastoma (Fig. 1). Because these tumors both arise in adrenal medulla and consist of catecholamine-producing cells, however, we

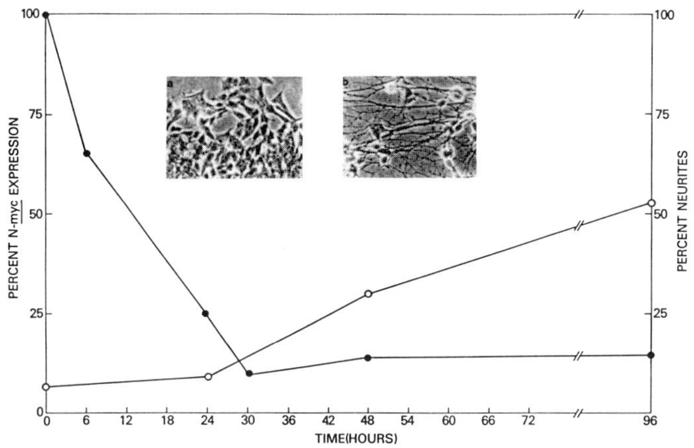

Fig. 2 - Comparision of N-myc Expression to Morphologic Differentiation of a Human Neuroblastoma Cell Line During Retinoic Acid Treatment. KCNR neuroblastoma cells were treated in vitro with 10 μM trans retinoic acid. Morphologic differentiation was assayed by determining the percentage of cells with neuritic processes (0——0). To assay the level of N-myc expression, cytoplasmic RNA was harvested from cells at various times after initiation of retinoic acid treatment. 20 μg total cytoplasmic RNA was examined by Northern blot analysis using a ^{32}P-labelled pNb-1 DNA probe specific for N-myc (Schwab et al.). N-myc RNA expression is plotted as the percent of N-myc RNA detectable in retinoic acid treated cells compared to untreated cells, as determined by densitometry of autoradiograms (●——●). Insert A, untreated KCNR cells; Insert B, KCNR cells after treatment with RA for 3 days.

examined several fresh pheochromocytoma tumor specimens for the amplification of N-myc DNA and expression of N-myc RNA (Fig.3) seen in neuroblastoma. In this experiment, high molecular weight DNA was extracted from pheochromocytoma tumor specimens and examined for N-myc amplification by Southern blot analysis. Cytoplasmic RNA was extracted from the same specimens and examined for N-myc expression by Northern blot analysis. The experiment in Panel A, Figure 3, indicates that approximately one copy of N-myc DNA per haploid genome was present in each of the tumors examined, and no evidence for amplification was found. The experiment in Panel B, Figure 3, demonstrates that we were unable to detect expression of this gene in total cytoplasmic RNA from any of the tumors examined. Our finding that N-myc is expressed in neuroblastoma but not in an apparently more differentiated tumor of the same tissue, pheochromocytoma, is consistent with the concept that N-myc expression is associated with an early stage in neural crest development.

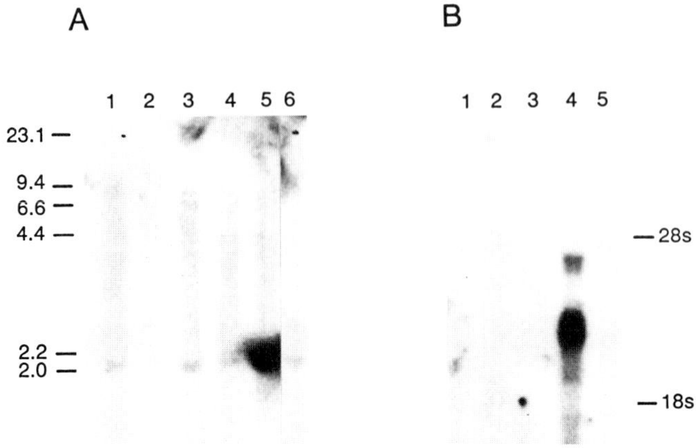

Fig. 3 - Analysis of N-myc DNA Structure and Expression in Pheochromocytoma Tumor Tissue.
Panel A: Southern Blot Analysis of N-myc DNA. High molecular weight DNA was extracted from four pheochromocytoma tumor specimens (lanes 1-4), a neuroblastoma cell line from patient SAN (lane 5), and an EBV transformed lymphocyte cell line from patient SAN (lane 6). 15 μg of DNA was digested with Eco RI and electrophoresed in a 1% agarose gel. DNA was transferred to nitrocellulose and hybridized with a ^{32}P-labelled pNb-1 DNA probe specific for N-myc under stringent conditions (42°C, 50% formamide). Molecular weight markers are from a Hind III digest of lambda DNA. Panel B: Northern Blot Analysis of N-myc RNA. Total cytoplasmic RNA was extracted from the same pheochromocytoma tumor specimens (lanes 1, 2, 3, & 5) and from the neuroblastoma cell line SAN (lane 4). 20 μg of total cytoplasmic RNA was electrophoresed in a 2.2 M formaldehyde/1% agarose gel. RNA was transferred to nitrocellulose and hybridized to the same N-myc specific probe as in Panel A under stringent conditions. The positions of the 28s and 18s ribosomal species are indicated.

Although the biologic activity of the protein encoded by N-myc is not known, it may play a role in enhancing the replication of neural crest cells. There are numerous human neuroblastoma cell lines, and virtually all have high levels of N-myc expression. Remarkably, these have all been derived from patients with advanced stage disease where N-myc amplification frequently occurs (11). To date, no human cell lines have been established from either pheochromocytoma or lower stage neuroblastoma where N-myc is not amplified despite numerous attempts by several laboratories to do so. Perhaps the overproduction of the N-myc gene product facilitates the establishment of cell lines from neuroblastoma cells by maintaining them in a proliferative state typical of the immature neural crest cells from which neuroblastoma arises.

NEUROEPITHELIOMA

The occurrence of N-myc amplification in neuroblastoma has led us to evaluate neuroepithelioma, a primitive round cell neuronal tumor which is morphologically indistinguishable from neuroblastoma. This tumor typically occurs on the trunk or extremities in tissues not known to be of sympathetic nervous system origin and generally presents in an older patient population than does neuroblastoma. For these reasons, neuroepithelioma has previously been referred to as peripheral or adult neuroblastoma; however, several recent biochemical, genetic, and molecular observations now clearly distinguish neuroepithelioma from neuroblastoma. These include a rcp(11;22) chromosomal translocation seen only in neuroepithelioma (14), high levels of c-myc expression in neuroepithelioma (15), and the presence of high levels of catecholamine synthetic enzymes only in neuroblastoma (R. Ross and M. Israel, unpublished observations). Cell lines derived from patients with neuroepithelioma do not have detectable levels of either dopamine-β-hydroxylase or tyrosine hydroxylase, enzymes important in the synthesis of catecholamine neurotransmitters. Instead, these cell lines contain high levels of choline acetyl transferase, an enzyme important in the synthesis of acetylcholine, the neurotransmitter of postganglionic synapses of the parasympathetic division of the autonomic nervous system (15). Based on the occurrence of neuroepithelioma at sites where cholinergic neuronal cell bodies might be expected, we have interpreted our results as suggesting that this tumor is likely to be a malignancy of postganglionic parasympathetic neurons (Fig. 1). Data in other experimental systems, however, have suggested that the cholinergic phenotype is expressed in neural crest cells at least transiently even in cells destined to become adrenergic (16, 17, 18). Therefore, it is possible that these tumors may also eventually be recognized as representing neural crest cells arrested at a early stage during differentiation.

DISCUSSION

N-myc is an oncogene which is amplified and highly expressed in neuroblastoma. One of the striking features of oncogenes is their marked sequence conservation over millions of years and across animal species. This conservation suggests that these genes encode proteins needed for critical cellular functions such as growth and differentiation. Although the function of N-myc is unknown at present, our data suggest that it may represent a regulatory gene involved in neural crest cell differentiation. The expression of this

Table 1. Tumors of the Autonomic Nervous System

VARIABLE FEATURES	NEUROEPITHELIOMA	NEUROBLASTOMA	PHEOCHROMOCYTOMA
Clinical Features at Presentation			
Age	adolescence	under 5 years	5 - 25 years
Complaint	extremity/chest wall mass	abdominal mass	hypertension
Useful Diagnostic Studies		urinary catecholamines; serum neuron specific enolase	urinary/serum catecholamines
Biologic and Pathologic Features			
Cell of Origin	unknown	neuroblast	chromaffin cell
Biochemical	cholinergic tissue	adrenergic tissue	adrenergic tissue
Microscopic Characteristics	small round cells, pseudorosettes	small round cells with rosettes and neurites	epithelioid cells, chromaffin positive
Ultrastructural Characteristics	neurosecretory granules, rare neurites with neurofilaments, neurotubules	abundant neurosecretory granules, frequent neurites with neurofilaments, neurotubules and occasional synapses with presynaptic vesicles	epithelioid cells with abundant epinephrine and norepinephrine containing granules
Genetic Features			
Cytogenetic Characteristics	rcp (11;22)(q24;q12)	$1p^-$, HSRs DMs	
Activated Oncogene	c-myc	N-ras, N-myc	

gene may be regulated during the physiologic differentiation of neural crest cells with high levels of expression found only in immature, proliferating cells. During normal embryologic development, downregulation of N-myc expression may allow terminal differentiation to occur. Perhaps it is the escape of neuroblastoma cells from the normal downregulation of N-myc expression that permits them to continue in a more primitive, proliferative state. In some instances, this escape may be due to the amplification of N-myc at the genomic level. Continued high level N-myc expression and the associated failure to terminally differentiate may be important genetic alterations leading to the malignant behavior of neuroblastoma.

Although other regulatory genes involved in the differentiation of the neural crest into mature cells have not yet been identified, we anticipate that these will be found and may play a prominent role in determining the biologic characteristics of pheochromocytoma and neuroepithelioma. Pheochromocytoma and neuroblastoma are tumors of cells which originate from a common neural crest precursor. The more mature phenotypic characteristics of pheochromocytoma suggest that this tumor arises in a cell which has progressed beyond the cell of origin of neuroblastoma in the differentiation pathway. If arrested differentiation is of biologic significance in the establishment of this tumor, it is likely that the pathologic lesion will be found to be a more distal block of terminal differentiation than the lesion in neuroblastoma.

Neuroblastoma, neuroepithelioma, and pheochromocytoma are compared and contrasted in Table 1. These tumors are comprised of cells which may correspond to different stages of differentiation along the pathway of neural crest cell development into mature autonomic ganglia (Fig. 1). In this regard they provide unique opportunities for examination of both oncogenesis and differentiation. We are particularly interested in pursuing the relationships between these processes, and towards this end we are seeking genes which may be important in the regulation of neural crest differentiation. The progression from immature proliferating cells to mature non-proliferating cells must be precisely regulated at the molecular level. During development, regulatory genes controlling early maturational processes may be turned off and other regulatory genes may become active as a cell progresses along a particular developmental pathway. Alterations in the expression of such genes at varying stages of differentiation may give rise to a pool of proliferating cells blocked at a specific point along a normal differentiation pathway and may, in turn, ultimately lead to tumor formation.

REFERENCES

1. Erickson, C.A., Tosney, K.W. and Weston, J.A. Develop Biol 77: 142-156, 1980.
2. Weston, J.A. In: Advances in Morphogenesis 8 (Eds. M. Abercrombie, J. Beachet and T.J. King), Academic Press, New York, 1970.
3. Le Douarin, N.M., Smith, J. and Le Lievre, C.S. Am. Rev. Physiol. 43: 653-671, 1981.
4. Evans, A.G. In: Advances in Neuroblastoma Research (Ed. B.A. Evans), Raven Press, New York, 1980.
5. Beckwith, M.B. and Perrin, E.V. Am. J. Pathol. 43: 1089-1104, 1963.
6. Schwab, M., Alitalo, K., Klempnauer, K.H. et al. Nature 305: 245-248, 1983.
7. Kohl, N.E., Kanda, N., Schreck, R.R. et al. Cell 35: 359-367, 1983.
8. Kanda, N., Schreck, R., Alt, F. et al. Proc. Natl. Acad. Sci. USA 80: 4069-4073, 1983.
9. Montgomery, K.T., Biedler, J.L., Spengler, B.A. et al. Proc. Natl. Acad. Sci. USA 80: 5724-5728, 1983.
10. Kohl, N.E., Gee, C.E. and Alt, F.W. Science 226: 1335-1337, 1984.
11. Brodeur, G.M., Seeger, R.C. and Schwab, M. Science 224: 1121-1124, 1984.

12. Rosen, N., Reynolds, C.P., Thiele, C.J., et al. In preparation.
13. Thiele, C.J., Reynolds, C.P. and Israel, M.A. Nature 313: 404-406, 1985.
14. Whang-Peng, J., Triche, T.J., Knutsen, T. et al. New Engl. J. Med. 311: 504-585, 1984.
15. Thiele, C.J., Ross, R., and Israel, M.A. In Preparation.
16. Le Douarin, N.M. and Teillet, M.A. Devel Biol. 41: 362-378, 1980.
17. Le Douarin, N.M., Renaud, D., Teillet, M.A. et al. Proc. Natl. Acad. Sci. USA 72: 728-739, 1975.
18. Le Lievre, C.S., Schweizer, G.G., Ziller, C.M. et al. Devel Biol. 77: 362-378, 1980.

8

GENETIC AND EPIGENETIC REGULATION OF THE METASTATIC PHENOTYPE: A BASIS FOR RESOLVING THE CONTROVERSY REGARDING ITS SELECTIVE OR RANDOM NATURE AND VARIABLE PHENOTYPIC STABILITY

R.S. KERBEL, P. FROST AND R.G. LITEPLO

Department of Pathology, Queen's University, Kingston, Ontario, Canada, and the Department of Cell Biology, University of Texas System Cancer Center, M.D. Anderson Hospital and Tumor Institute, Houston, Texas, U.S.A.

SUMMARY

Considerable controversy exists as to whether the metastatic phenotype is selective or random in nature, and if selective, whether it is always inherited and expressed in a stable, or unstable, manner. Which of the alternatives is correct has important implications and ramifications for virtually every aspect of the experimental study of metastasis. Based on analogous experimental situations (e.g., drug resistance due to gene amplification) and results from two separate experimental systems obtained in our laboratories, we conclude that the metastatic phenotype is indeed selective in nature, i.e., there are genotypically distinct subpopulations of metastatic variants. However, the ways in which the genetic information relevant to the metastatic phenotype is regulated and expressed will determine the relative stability of the phenotype: both unstable and stable variants can co-exist. We hypothesize that changes in DNA methylation (i.e., of 5-methylcytosine) is one of several possible mechanisms which can account for an unstably expressed phenotype. We also argue that the selection pressure required to isolate the stable variety of variant may be much stronger than what many investigators have previously used and can account for some of the discrepancies in the experimental metastasis literature.

INTRODUCTION: ON THE CONTROVERSIAL NATURE OF THE METASTATIC PHENOTYPE

In 1973 Fidler published his pioneering study on the selective nature of the metastatic phenotype (1). More than any other single study, it helped open up the field of experimental metastasis. Fidler showed that if so-called 'artificial' lung metastases were recovered from mice previously given injections of B16 melanoma cells, the tumor cells adapted to culture and then reinjected into new mice, sublines could eventually be derived which displayed a significantly greater lung colonization capacity than the unselected cells. This result was subsequently reproduced by Fidler and his colleagues using several other mouse or human tumor cell lines (2-4), and by a number of other laboratories as well (e.g., 5-7).

The results were interpreted in favour of the notion that metastases generally arise form minor subpopulations of tumor cells (i.e., "mutants" or "variants") which possess and express the 'proper' set of genes necessary to complete all of the steps of the so-called metastatic cascade. This is a very important concept that has broad implications for the biology of metastatic disease. It implies that metastatic cells may express a spectrum of phenotypic traits (e.g., certain antigens, receptors, enzymes, etc.) that are unexpressed, or expressed in a very different fashion, in non-metastatic cells (8-10). Acceptance of the validity of this concept has induced a considerable number of laboratories to seek out these special differences and to isolate the relevant genes involved in their expression (11-13). This has obvious implications, as well, for the treatment of metastases. For example, if a particular cell surface structure could be found only on metastatic cells as defined by, say, monoclonal antibodies (14), both the structure itself and the antibodies could be exploited for either diagnostic or therapeutic purposes.

Several authorities, however, have questioned the concept of the selective nature of the metastatic phenotype (15,16), and some have arrived at a very different conclusion, namely that the process is semmingly random or that the metastatic phenotype is extremely ephemeral, being lost as quickly as it is acquired. This could occur as a result of highly unstable genetic mutations, thus leading to a dynamic or rapidly changing phenotype (17,18). These conclusions are based largely on the findings, now reported by a number of laboratories (e.g., 19-21) that the metastatic aggressiveness of cells or cell lines from metastases is no greater than cells obtained from the 'primary' tumor. In other words, the progressive selection results of Fidler (1) and others could not always be reproduced.

Our view is that it is reasonable to suppose that metastases do indeed arise from genotypically distinct subpopulations of mutant or variant tumor cells, regardless of the tumor system. However, the manner in which the relevant genetic information is regulated and expressed will determine whether the metastatic phenotype is expressed in a transient, i.e., unstable manner, or in a heritable, highly stable fashion in a given tumor cell population. Essentially we have been investigating two hypotheses which we feel are <u>not</u> mutually exclusive. These hypotheses are:

(i) one of several mechanisms which can account for the metastatic phenotype being expressed in an unstable fashion is changes in the pattern and levels of 5-methylcytosine.

(ii) stable metastatic mutants (also) exist, but on the assumption they are rare (e.g., less than one in 10^{-5} or more cells) one would have to exert an effectively stringent selection pressure in order to successfully isolate them. We will argue that many of the selection protocols that have been used were not sufficiently stringent. Before summarizing our results and theories we will draw upon what we feel is an appropriate experimental analogy, involving a very different phenotype - drug resistance.

AN EXPERIMENTAL ANALOGY: STABLE VERSUS UNSTABLE DRUG RESISTANT MUTANTS CAUSED BY GENE AMPLIFICATION

It is now well known that multi-step exposure of neoplastic cells to certain toxic drugs such as methotrexate or vincristine can lead to the emergence and overgrowth of drug-resistant cells. The stability of the drug-resistant phenotype shows tremendous variability, such that if the drug-resistant cells are grown in absence of the drug, the drug-resistant phenotype may be lost in a matter of days or weeks - or it may be maintained in large part for periods of at least one - two years (22,23). In the case of methotrexate-resistance it is now known that extensive amplification of the dihydrofolate reductase (DHFR) gene copy number is responsible for development of most cases of resistance to methotrexate (22,23). Moreover, the gene can be amplified in essentially one of two different ways: either in the form of tiny pairs of acentric chromosomes called 'double minutes' (DMs) or alternatively, in the form of tandomly replicated genes within a chromosome called homogeneously staining regions (HSRs). In the former case, because they lack a centromere, DMs tend to be rapidly lost by segregation during mitosis whereas HSRs tend to remain, though their overall length may decrease over long periods of time in the absence of the appropriate selection pressure. Consequently, tumor cells which, for example, are many hundred of thousand-fold times more resistant to methotrexate (or vincristine) due to formation of DM-related amplified genes, may approach wild-type levels of drug sensitivity within a period of 10 days to several weeks when grown in the absence of the selective agent (22,23). This is not true for HSR related amplified genes where a high degree of drug resistance may be maintained in drug-free medium for many months or years (22).

The point is that the methotrexate-resistant state in both cases is due to a large scale genetic type of mutation involving the same gene - the one which codes for dihydrofolate reductase (DHFR) - but the stability or maintenance of the mutation is remarkably different in the two situations. Clearly it would be erroneous to conclude that the highly unstable type of methotrexate resistance is 'randomly' acquired and not the consequence of a genetic (mutational) event. Likewise, as pointed out by Hill, Ling et al. (17,18), the cells populating a metastasis may not be more metastatic than the cells populating a primary tumor mass since the relevant mutation(s) in a cell leading to its ability to establish a micro-metastasis may revert quickly and therefore not be expressed in the progeny of that cell. Hence, the transient nature of the metastatic phenotype in that type of situation. However, drawing upon, once again, the methotrexate-resistance situation, there are also highly stable (HSR-type) mutants. Could not the same be true for a more complicated phenotype such as metastasis, i.e., that there are both highly stable and unstable mutants? We feel the answer is affirmative and we will briefly review the evidence from our labortories which we feel supports this conclusion. The results derive from two ostensibly very different systems. The first is described in the next section.

THE POSSIBLE CONTRIBUTION OF DNA METHYLATION TO TUMOR CELL HETEROGENEITY, PHENOTYPIC INSTABILITY, AND THE TRANSIENT METASTATIC PHENOTYPE: INITIAL STUDIES

During the course of metastasis studies involving the use of cloned drug or lectin resistant mutant tumor cell lines that we had isolated from mutagenized mouse tumor cell populations, we noted a high proportion (e.g., 20%) of the clones did not give rise to progressive tumors when inoculated into normal syngeneic mice (24). However these clones grew progressively in highly immunosuppressed animals (24). Further studies demonstrated this was probably a reflection of the 'Boon' phenomenon. Boon and his colleagues first demonstrated mutagen (MNNG) treatment and cloning of various mouse tumor lines in vitro - without any further selection - led to the emergence of very high proportions (e.g., 10-30%) of clones which did not grow, or which did so transiently and then regressed, when inoculated into syngeneic mice (reviewed in 25,26). Such clones, designated "tum$^-$" would grow progressively when injected into highly immunosuppressed X-irradiated recipients. Hence the tum$^-$ phenotype appears to have an underlying immunological basis (25,26).

We were able to reproduce Boon's findings in our tumor systems using EMS (27-29), and we were struck in particular by two observations: (i) the frequency of immunogenic tum$^-$ clones in some cases exceeded 90% (28) and was greater than 10% even when the clones were obtained from a spontaneous non-immunogenic carcinoma (30). (ii) most of our tum$^-$ clones eventually reverted to a tumorigenic phenotype (29,30); this occurred if the clones were maintained in culture for periods of time ranging from several weeks to 6 months. This prompted us to ask what genetic or epigenetic mechanism(s) could bring about such extraordinarily high frequency, heritable, but ultimately unstable phenotypic alterations. Our feeling was that whatever the mechanism, it could clearly have relevance to a plethora of other biological phenomena, in particular to "phenotypic instability" e.g., the transient metastatic phenotype when it behaves in this manner.

The nature of the findings and a review of the literature on gene regulation immediatley implicated DNA methylation alterations as a possible mechanism. The reasoning for this was based on the following major points:

(i) changes in the pattern and level of 5-methylcytosine in a gene or its promotor region may regulate its transcription (31-35).

(ii) induction of DNA hypomethylation can sometimes lead to the transcriptional activation of genes and this can occur at very high frequencies of the treated cells (e.g., >10%) when they are treated with potent hypomethylating agents such as the drug 5-azacytidine (31-35).

(iii) methylation patterns are somatically heritable so that the pattern of 5-methylcytosines in CG doublets which exists on the parental strand of DNA can be passed on to the replicating daughter strand - but not necessarily with perfect fidelity (36,37). The occurrence of some de novo methylation raises the possibility that a gene that is transcriptionally activated by demethylation may be switched off again some time later. The stability of the phenotypic change may therefore reflect the rate at which the appropriate cytosines in a gene are remethylated. We will expand on this point more fully below.

(iv) many mutagens, including EMS and MNNG, have been shown to be able to induce alterations in 5-methylcytosine levels in DNA (38-40) in addition to inducing structural genetic changes, thus raising the clear possibility that such agents may alter gene expression (and therefore phenotypic traits) by gene activation/inactivation mechanisms - in addition to, or instead of, structural gene changes (41).

We therefore embarked on a series of experiments to determine if the "mutagen - tum$^-$" results could be mimicked by treating tumor cell populations with poorly mutagenic hypomethylating agents such as the cytidine analogue 5-azacytidine (28,29). Even low levels of incorporation of 5-azacytidine into replicating DNA can cause profound inhibitions in the levels of DNA 5-methylcytosine content, and this in turn is often accompanied by extraordinarily high frequency phenotype alterations in the treated cells, e.g., induction of thymidine kinase (TK) activity in 10% or more of TK deficient mutants (42,43), induction of interleukin-2 synthesis in about 50% of EL4 T lymphoma cells (44), induction of prolactin in 50% of cells from prolactin-deficient rat pituitary tumor cell mutants (41), induction of steroid hormone sensitivity in 10% of hormone resistant mouse lymphoma cells (45), and induction retroviral expression in 90% of mouse tumor lines (46), to cite but a few examples. Moreover, whereas some of these 5-azacytidine-induced changes are stable (42,43) many are not, such that the 'wild type' phenotype returns over a period of several weeks or months (44,45).

Our approach essentially consisted of treating several mouse tumor cell lines with minimally toxic, low concentrations (1-4 μm) of 5-azacytidine, cloning the cells, and then assessing their resultant tumorigenic and malignant behaviour in syngeneic mice (28-30). We also assessed the relative mutagenic effects of such treatment, the impact on levels of genomic 5-methylcytosine, and the stability of any induced changes. The experimental results quickly revealed that we could indeed mimic our prior EMS or MNNG results in any given tumor cell line by substituting 5-azacytidine. High frequencies of immunogenic tum$^-$ clones were again generated. These clones grew in athymic nude mice, could induce strong T cell immune responses and were phenotypically unstable (28-30). This, despite evidence that 5-azacytidine was essentially devoid of mutagenic potential as assessed by its ability to induce point somatic mutations at three independent gene loci in two different mouse tumor lines - in constrast to EMS which, as expected, was found to be strongly mutagenic (28).

We subsequently found in one tumor system (the TA3 mammary carcinoma) that out of 124 clones examined, obtained after 5-azacytidine treatment, 108 displayed a tum^{-} phenotype (29). Of the remaining 16 tumorigenic clones, three gave rise to spontaneous metastases in different organ sites, in contrast to the untreated TA3 cells or clones which showed absolutely no capacity to metastasize under identical conditions. However, once again, the metastatic phenotype of these three clones eventually was lost (29), i.e., it was phenotypically unstable.

ON THE STABILITY OR INSTABILITY OF HIGH FREQUENCY PHENOTYPIC ALTERATIONS INDUCED BY 5-AZACYTIDINE/DNA HYPOMETHYLATION: AN HYPOTHESIS

We subsequently attempted to repeat the metastasis results using several other non-metastatic mouse tumors or cell lines, including a spontaneous CBA strain mammary adenocarcinoma (called Sp1) and NIH 3T3 cells transfected with the human H-ras oncogene. We were still unable to detect metastases, or at least macroscopic metastases, by pretreating the cells with 5-azacytidine and then injecting them subcutaneously (unpublished observations). In retrospect this may not be surprising because of the instability in inheritance of induced hypomethylation levels and patterns (36,37). Thus, if a silent gene required for spontaneous metastasis is switched on by the drug treatment, it may be switched back off before it is required during the multi-step process of in vivo metastasis. And when the tumor cells are inoculated subcutaneously (s.c.), as opposed to intravenously (i.v.), the outcome of the metastatic process takes a great deal longer. Thus one might see an enhancement of lung metastases after 5-azacytidine treated cells are given i.v. but not when the same cells are given s.c. Indeed, this was found to be the case by Trainer, Poste, and their colleagues using the B16 melanoma (unpublished observations). Hart's laboratory has also reported enhancement of lung metastases after intravenous injection of human melanoma cells in nude mice (unpublished observations). This leads us to a consideration of the relative stability of demethylation alterations at the genome and single gene levels.

Christman's laboratory demonstrated recently that induction of hypomethylation in vitro in Friend virus leukemia cells induced their differentiation and this was accompanied by a switching on of the β -globin gene (47). However, normal levels of total 5-methylcytosine in genomic DNA returned within a day of the drug treatment. We have also noted a similar rapid reduciton and return to normal levels of DNA methylation in a variety of 5-azacytidine treated mouse tumor lines (Liteplo and Kerbel, unpublished observations).

When one surveys the literature there are numerous reports to be found which indicate that a particular high frequency phenotypic change induced in a particular cell line by 5-azacytidine treatment may be stably inherited, or unstably inherited. For example, induction of thymidine kinase (TK) activity from TK^- mutants (42,43) appears to be a highly stable change: it can persist for many months or longer (42). So too is the induction, by 5-azacytidine, of ornithine carbamoyl transferase (OCT) from OCT deficient hepatoma cells (48), of asparagine synthetase activity in rat sarcoma cells (49) to cite three examples. On the other hand, a number of other studies have indicated quite different results. These include induction of interleukin-2 synthesis in T lymphoma cells (44), corticosteroid sensitivity in steroid resistant lymphoma cells (45), the EBV genome in human B cell lymphomas (50) class I MHC antigens in mammary carcinoma cells (30) and increased cloning efficiency in human tumor cell populations (51). In all these cases the induced phenotype occurred in 10-90% of the 5-azacytidine treated clones, but was gradually lost in the clones within a few months. Moreover, in some cases (eg., see 45) the reversion was accompanied by a return to near normal levels of genomic DNA 5-methylcytosine.

What are the implications of these discrepant results and can they be reconciled? We feel the most plausible answer lies in the nature of the genes affected. An hypothesis we are currently testing is that the stability of expression of a gene switched on by hypomethylation in a tumor cell will depend on its normal mode of expression and regulation in the normal cell type from which it was derived. Thus, the TK gene is an example of a so-called 'housekeeping' gene which is always transcriptionally active in all somatic cells. In contrast the interleukin-2 gene, a T lymphocyte-associated gene product, is expressed transiently in certain types of dividing T cells (53): it is an example of a gene which is regulated in an "on-and-off" fashion. Hence when IL-2 deficient EL4 T lymphoma cells are treated with 5-azacytidine, 50% of the clones may produce IL-2 but this phenotype is lost within all the clones within two months (44). In contrast, the TK gene remains active in 5-azacytidine TK-deficient CHO cells (42) or mouse leukemia-like cells (43; and Liteplo, Frost et al, unpublished observations). In these two cases the mode of expression of the respective 5-azacytidine 'induced' genes in the tumor cell populations recaptiulates their usual modes of expression.

If this hypothesis is correct it would mean that - like gene amplification - DNA methylation has the potential to alter gene expression (and therefore induce new or altered phenotypes) in a way that they could be either stably or unstably inherited. This could explain why 5-azacytidine treatment of some non-metastatic tumor cell lines may result in the emergence of metastatic cells which behave in a phenotypically stable manner as shown by Olsson's laboratory with the Lewis lung carcinoma (14), or in a highly unstable fashion, as shown in our studies with the TA3 mammary carcinoma (29). It would simply depend on the nature of the gene that was hypomethylated and transcriptionally activated which endowed the respective tumor cell lines with metastatic capability.

It will clearly be difficult to rigorously evaluate this hypothesis in relationship to metastasis until we know what genes are actually involved in metastasis and the cDNA probes for them are available. Nevertheless we feel it is important to pursue this type of work and hypothesis further because of their implications for the therapy of cancer. In short, is it not possible that chemotherapeutic drugs may themselves alter gene expression of surviving tumor cells and/or their progeny by bringing about changes in DNA methylation? Indeed we have recently shown (Liteplo et al., submitted for publication) that exposure of TK^- mouse tumor cells *in vivo* to 5-azacytidine can induce high frequency activation of TK activity. Given the fact that many chemotherapeutic drugs can cause point mutations or induce gene amplification *in vivo* (23) we might also ask whether the expression of such mutated or amplified genes can be affected by the *same* drug's ability to alter normal DNA methylation patterns and levels? The potential of a drug to induce both alterations in structure and expression of genes, when combined with host selection, could clearly represent a powerful combination to promote the course of tumor progression.

ON THE ISOLATION OF PHENOTYICALLY STABLE METASTATIC TUMOR CELL VARIANTS: THE ROLE OF APPROPRIATE SELECTION PRESSURES

The two most important factors involved in the isolation of mutants are: (i) the frequency of the desired mutant, and (ii) the strength of the selection pressure applied to select the mutant. If a mutant exists in a high frequency (say 10^{-2}) the severity of the selection pressure required will be orders of magnitude less than if the same type of mutant exists in a far lower frequency (say 10^{-6}). On the assumption that phenotypically stable and unstable metastatic mutants co-exist and the frequency of the stable mutants is low (e.g., 10^{-5} to 10^{-6}) an appropriately stringent selection pressure will be necessary to successfully isolate them. Thus some of the lung metastases isolated after s.c. inoculation of a lung-metastasizing tumor cell population in an animal would more likely be stable than those obtained after i.v. inoculation. This is because more is required of the cells to reach, survive, and proliferate in the lungs after s.c. inoculation than i.v. inoculation.

On the basis of this kind of reasoning, we might predict that there would be a greater chance of spontaneous metastases recovered from the organs of adult, normal nude mice inoculated with human tumor cells displaying evidence of selectivity and stability of metastatic phenotype, as they would be the progeny of very 'rare events'. This is because human tumor metastases are rarely observed in adult nude mice after s.c. inoculation (54). There may be many reasons for this, one being the presence of endogenously high levels of natural killer (NK) cells in such animals (55). Thus the incidence of metastases can be enhanced by using very young (2-3 wk. old), i.e., NK-cell deficient, nude mice (55). On this basis we would argue that human tumor cells should not be administered i.v., nor should very young nude mice be used if one wanted to select stable metastatic human tumor cell variants in nude mice: both would serve to decrease the stringency of selection pressure. The idea is to maximize the selection pressure.

We adopted this approach using a human melanoma line (called MeWo) and found very persuasive evidence for the existence of highly selective and phenotypically stable subpopulations of metastatic cells (56,57). Moreover they could be selected in a single-step method (57). The cells were more aggressively metastatic on the basis of such parameters as: (i) extent of lung metastases, (ii) presence of extra-pulmonary metastases, and, (iii) their time of initial appearance (56,57). Karyotype analysis (57) revealed unequivocal evidence that spontaneous lung metastases (i.e., those obtained after s.c. injection) belonged to a distinct subpopulation of cells whereas this was not found to be the case for 'artificial' metastases (i.e., those obtained after i.v. injection).

We would suggest that the failure of other laboratories to detect and isolate such variants (e.g., 19-21) may, in many cases, be due to an inadequate in vivo selection pressure: the 'rare events' would be missed. Other factors are possible and these have been discussed recently by Talmadge and Fidler (58).

CONCLUSIONS

The nature of the multi-step process of metastasis combined with a knowledge of new aspects of gene structure, regulation and expression argue strongly for metastasis being a selective process that can be expressed either as a stable or unstable trait. An awareness of this (and the underlying genetic/epigenetic reasons why it is so) should help resolve many of the discrepancies in the metastasis literature as well as the controversy surrounding the 'random-versus-selective' nature of metastasis. Certainly it should be taken into account in the design of any experiment meant to explore the biochemical and genetic nature of the metastatic phenotype.

ACKNOWLEDGMENTS

Work by the authors summarized in this review was supported by the National Cancer Institute of Canada (to RSK) and by Grant #CA 28060 awarded by the National Cancer Institute, United States Department of Health and Human Sciences (to PF). RSK is a Research Associate of the National Cancer Institute of Canada. We thank Mrs. Beverley Fluhrer and Mrs. Beverley Muller for their excellent secretarial assistance. RGL's current address is: Department of Experimental Oncology, Ottawa Regional Cancer Centre, 501 Smyth Road, Ottawa, Ontario, Canada K1H 8L6.

REFERENCES

1. Fidler, I.J. Nature (New Biol.) 242: 148-149, 1973.
2. Raz, A., Hanna, N. and Fidler, J.J. J. Natl. Cancer Inst. 66: 183-189, 1981.
3. Talmadge, J.E. and Fidler, I.J. J. Natl. Cancer Inst. 69: 975-980, 1982.
4. Kozlowski, J.M., Hart, I.R., Fidler, I.J. and Hanna, N. J. Natl. Cancer Inst. 72: 913-917, 1984.
5. Honma, Y., Kasukabe, T. and Hozumi, M. Gann 72: 989-905, 1981.
6. Brunson, K.W., Beattie, G. and Nicolson, G.L. Nature 272: 543-545, 1978.
7. Tao, T.W., Matter, A., Vogel, K., and Burger, M.M. Int. J. Cancer 23: 854-857, 1979.
8. Fidler, I.J. and Hart, I.R. Science 217: 998-1003, 1982.
9. Poste, G. and Greig, R. Invasion and Metastasis 2: 137-176, 1982.
10. Nicolson, G.L. Clin. Exp. Metastasis 2: 85-106, 1984.

11. Chambers, A.F. and Ling, V. Cancer Res. 44: 3970-3975, 1984.
12. Thorgeirsson, U.P., Turpeeniemi-Hujanen, T., Williams, J.E., Westin, E.H., Heilman,C.A., Talmadge, J.E. and Liotta, L. Mol. Cell. Biol. 5: 259-262, 1985.
13. Bernstein, S.C. and Weinberg, R.A. Proc. Natl. Acad. Sci. USA 82: 1726-1730, 1985.
14. Olsson, L. and Forchhammer, J. Proc. Natl. Acad. Sci. USA 159: 1491-1501, 1984.
15. Weiss, L. Invasion and Metastasis 3: 193-207, 1983.
16. Alexander, P. Cancer Topics 4: 116-117, 1984.
17. Harris, J.F., Chambers, A.F., Hill, R.P. and Ling, V. Proc. Natl. Acad. Sci. USA 79: 5547-5551, 1982.
18. Hill, R.P., Chambers, A.F., Ling, V. and Harris, J.F. Science 224: 998-1001, 1984.
19. Weiss, L., Holmes, J.C. and Ward, P.A. Br. J. Cancer 47: 81-89, 1983.
20. Milas, L., Peters, L.J. and Ho, H. Clin. Exp. Metastasis 4: 309-315, 1983.
21. Giavazzi, R., Alessandri, G., Spreafico, F., Garattini, S., and Mantovani, A. Br. J. Cancer 42: 462-472, 1980.
22. Schimke, R.T. Cancer Res. 44: 1735-1740, 1984.
23. Schimke, R.T., Hill, A. and Johnson, R.N. Br. J. Cancer 51: 459-465, 1985.
24. Kerbel, R.S. Am. J. path. 79: 609-622, 1979.
25. Boon, T. Adv. Cancer Res. 39: 121-152, 1983.
26. Kerbel, R.S. and Frost, P. Immunol. Today 3: 34-35, 1982.
27. Frost, P., Kerbel, R.S., Bauer, E., Tartamella-Biondo, R. and Cefalu, W. Cancer Res. 43: 125-132, 1983.
28. Frost, P., Liteplo, R.G., Donaghue, T. and Kerbel, R.S. J. Exp. Med. 159: 1491-1501, 1984.
29. Kerbel, R.S., Frost, P., Liteplo, R., Carlow, D.A. and Elliott, B.E. J. Cell. Physiol. (Suppl) 3: 87-97, 1984.
30. Carlow, D.A., Kerbel, R.S., Feltis, T. and Elliott, B.E. J. Natl. Cancer Inst. - in press.
31. Razin, A. and Riggs, A.D. Science 210: 604-610, 1980.
32. Riggs, A.D. and Jones, P.A. Cancer Res. 40: 1-30, 1983.
33. Ehrlich, M. and Wang, R.Y.H. Science 212: 1350-1357, 1981.
34. Doerfler, W. Ann. Rev. Bioch. 52: 93-124, 1983.
35. Jones, P.A. Cell 40: 485-486, 1985.
36. Wigler, M., Levy, D. and Perucho, M. Cell 24: 33-40, 1981.
37. Schmookler-Reis, R.J. and Goldstein, S. Proc. Natl. Acad. Sci. USA 79: 3349-3953, 1982.
38. Boehm, T.L.J. and Drahovsky, D. Carcinogenesis 2: 39-42, 1981.
39. Boehm, T.L.J. and Drahovsky, D. J. Natl. Cancer Inst. 71: 429-435, 1983.
40. Boehm, T.L.J., Grunberger, D. and Drahovsky, D. Cancer Res. 43: 6066-6071, 1983.
41. Ivarie, R.D. and Morris, J.A. Proc. Natl. Acad. Sci. USA 79: 2967-2970, 1982.
42. Harris, M. Cell 29: 483-492, 1982.
43. Liteplo, R.G., Frost, P. and Kerbel, R.S. Exp. Cell Res. 150: 499-506, 1984.
44. Ballus, Z.K. J. Immunol. 133: 7-10, 1984.
45. Gasson, J.C., Ryden, T. and Bourgeois, S. Nature 302: 621-623, 1983.

46. Tennant, R.W., Otten, J.A., Myer, F.E. and Rascati, R.J. Cancer Res. 42: 3050-3055, 1982.
47. Creusto, F., Acs, G. and Christman, J.K. J. Biol. Chem. 257: 2041-2048, 1982.
48. Delers, A., Szpirer, J., Szpirer, C. and Saggioro, D. Mol. Cell. Biol. 4: 809-812, 1984.
49. Sugiyama, R.H., Arfin, S.M. and Harris, M. Mol. Cell. Biol. 3: 1937-1942, 1983.
50. Ben Sasson, S.A. and Klein, G. Int. J. Cancer 28: 131-135, 1981.
51. Olsson, L., Due, C. and Dumant, M. J. Cell. Biol. 100: 508-513, 1985.
52. Goldman, M.A., Holmquist, G.P., Gray, M.C., Caston, L.A. and Nag, A. Science 224: 686-692, 1984.
53. Paetkau, V. Nature 294: 689-690, 1981.
54. Sharkey, F.E. and Fogh, J. Int. J. Cancer 24: 733-738, 1979.
55. Hanna, N. and Fidler, I.J. Cancer Res. 4: 438-444, 1981.
56. Kerbel, R.S., Man, M.S. and Dexter, D. J. Natl. Cancer Inst. 72: 93-108, 1984.
57. Kerbel, R.S. and Man, M.S. Invasion and Metastasis (Suppl 1) 4: 31-43, 1984.
58. Talmadge, J.E. and Fidler, I.J. Nature 297: 593-594, 1982.

9

CYTOCHEMICAL CELL TYPING OF METASTATIC TUMORS ACCORDING TO THEIR CYTOSKELETAL PROTEINS

R. MOLL and W. W. FRANKE

Institute of Pathology, University of Mainz Medical School, D-6500 Mainz, and Institute of Cell and Tumor Biology, German Cancer Research Center, D-6900 Heidelberg, Federal Republic of Germany

INTRODUCTION

Conventional staining methods alone are often insufficient to make accurate histodiagnosis of malignant tumors. However, additional information about the cell type and the state of differentiation of tumor cells can be provided by several molecular markers. To be relevant diagnostically, these markers must be conservative, i.e. expressed not only in well differentiated tumors but also in poorly differentiated tumors and in metastases. These criteria are fulfilled by certain components of the cytoskeleton, notably the proteins forming the intermediate-sized filaments (IFs) and the desmosomal plaque proteins.

IFs can be divided into five classes according to their constituent polypeptides. The expression of these classes is strictly correlated to certain pathways of cellular differentiation and is maintained after malignant transformation (for review see 1). The major IF classes are: (1) Vimentin IFs which typically occur in mesenchymally derived cells as well as in sarcomas, lymphomas, and malignant melanomas. They also occur in cells with certain "epithelioid" features of morphology such as "eye lens epithelium" and Sertoli cells of seminiferous tubules (for refs. see 1), and are co-expressed with cytokeratin IFs in some special epithelia and carcinomas as well as in many cultured cell lines (1-4). (2) Desmin IFs are specific for most types of muscle cells as well as for rhabdo- and leiomyosarcomas. (3) Glial filaments are typical of certain glial cells, such as astrocytes and Bergmann glia cells, and astrocytomas. (4) Neurofilaments are present in most neurons as well as in ganglioneuroblastomas, pheochromocytomas and some neuroblastomas. (5) Cytokeratin (CK) IFs are specific for epithelial cells and carcinomas. The availability of antibodies specific for these IF proteins has allowed typing of IFs to become a valuable tool in the histodiagnosis of poorly differentiated and metastatic tumors.

Biochemically, the CKs are a multigene family of distinct polypeptides which can be divided into two subfamilies, the acidic and small (type I)

CKs and the basic and large type II CKs (5-7). In human tissues, eight type II CKs (CKs 1-8) and eleven type I CKs (CKs 9-19) have been identified (Fig. 2a; 7,8). The CK subunits are always expressed as "pairs" of type I and type II molecules forming defined complexes of coiled-coil α-helices that are assembled into protofilaments and IFs (9-11). The CK-IF system of a given cell consists of at least one but sometimes several CK pairs. Certain combinations of CK polypeptides are characteristic of the specific type of differentiation of an epithelial cell, such as squamous or glandular differentiation.

The desmoplakins, which are major protein constituents of the desmosomal plaques, represent another marker of cellular differentiation (12). Using specific antibodies, desmoplakin-containing desmosomes have been identified in all true epithelial cells examined, in myocardiac cells and certain arachnoidal cells, and the expression of desmoplakins is also maintained in most carcinomas and in meningiomas (13,14).

In this study we show examples of immunohistochemical analyses of IF proteins and desmoplakins in metastatic tumors. We also present a summary of CK polypeptide patterns of diverse kinds of primary and metastatic carcinomas and discuss the relevance of the CK typing for tumor diagnosis.

MATERIALS AND METHODS

Immunofluorescence microscopy using various antibodies against IF proteins and desmoplakins was performed on frozen sections as described previously (3, 4,8,13-16). CK polypeptide patterns were analyzed by two-dimensional gel electrophoresis, immunoblot experiments and tryptic peptide mapping, using cytoskeletal material mostly prepared by microdissection of frozen sections as described (8,17,18).

RESULTS AND DISCUSSION

Immunohistochemical analysis of intermediate filament proteins and desmoplakins in metastatic tumors.

IF typing facilitates the classification of many epithelial and non-epithelial tumors, including metastases (1). For example, we present three typical cases of metastatic carcinomas (Fig. 1). Single tumor cells present in lymph node sinus were positive with CK antibodies (Fig. 1a) and thus unequivocally identified as carcinoma cells. In a poorly differentiated carcinoma metastasis, delicate CK fibrils were detected in most tumor cells, although they displayed little cytoplasm (Fig. 1b). Desmoplakin antibodies specifically decorated small dots along the cell boundaries of carcinoma cells (Fig. 1c), each dot representing a single desmosome. Desmoplakin can be used as a marker for carcinomas and meningiomas independently of the IF system (13,14).

Cytokeratin polypeptide patterns of primary and metastatic carcinomas.

Diversity in expression of CK polypeptides is found not only in normal epithelia but also in different kinds of carcinomas. The 19 human CK polypeptides (Fig.2a) may be divided into CKs that are characteristically expressed in, but not exclusive to, stratified squamous epithelia ("stratified epithelium-type CKs": CKs 1-6 of type II and 9-17 of type I) and CKs that are typical of simple, one-layered and glandular epithelia ("simple epithelium-type CKs": CKs 7 and 8 of type II and 18 and 19 of type I). In carcinomas, 15 CKs have so far been detected (Fig. 2a). The profound differences of the CK patterns of different kinds of carcinomas are illustrated for two metastases in Figure 2b and c. While a substantial amount of data of CK analyses from primary carcinomas is available there have been very few studies on metastases. In the following we discuss the CK patterns of primary carcinomas and, where possible, those of related metastatic tumors.

The following tables summarize CK polypeptide patterns determined in various cases of primary and metastatic carcinomas from different organs and allow the comparison with the patterns of the corresponding normal epithelia.

Simple and glandular epithelia of the gastrointestinal tract contain exclusively the simple epithelium-type CKs 7, 8, 18 and 19. However, the different epithelia within the tract possess cell-type specific combinations of

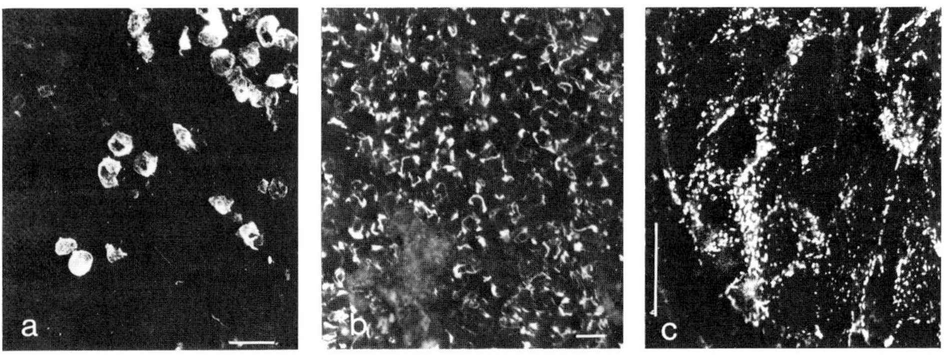

Fig. 1. (a) Immunofluorescence microscopy (IM) of a lymph node metastasis of gastric adenocarcinoma using CK antibody K_G 8.13 (19). (b) IM of a skin metastasis of oat cell carcinoma of lung using CK antibody K_S 18.18 (cf. 20). (c) IM of a lymph node metastasis of squamous cell carcinoma of the vulva using monoclonal antibodies to desmoplakins (P. Cowin et al., manuscript in preparation). Bars, 20 μm.

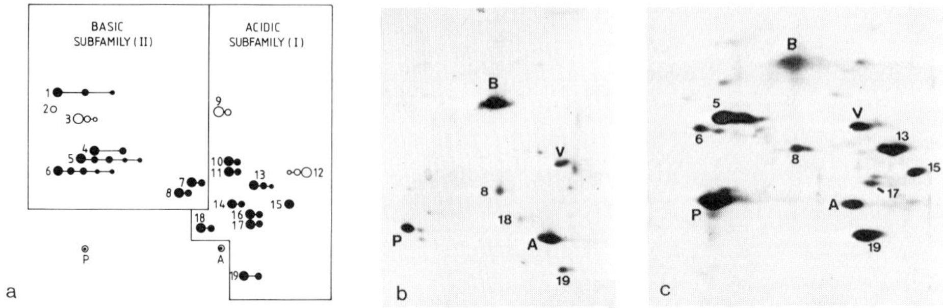

Fig. 2. (a) Positions of human CK polypeptides (nos. 1-19) in two-dimensional gels. Full circles denote CKs so far found in carcinomas. Isoelectric pH values and relative molecular weights (M_r) are given in Table 1. (b) Two-dimensional gel electrophoretic pattern of a metastasis of adenocarcinoma of colon (case no. 4 of Table 1). Note the simple CK composition consisting of only simple epithelium-type CKs. (c) Pattern of a lymph node metastasis of squamous cell carcinoma of larynx. Note the highly complex CK pattern. Markers used for co-electrophoresis: P, phosphoglycerokinase; B, bovine serum albumin; A, actin.

these CKs (Table 1; cf. 18). In a variety of gastrointestinal carcinomas essentially the same CK subset has been detected (Table 1). Conspicuously, the very simple pattern of the CK pair 8:18 expressed in hepatocytes is maintained in hepatocellular carcinomas, and the absence of CK 7 in intestinal epithelium is also seen in various cases of primary and metastatic adenocarcinomas of the colon. These data show the highly conservated expression of CKs in primary and metastatic carcinomas of the gastrointestinal tract.

In the simple epithelia of the upper female genital tract, again the simple epithelium-type CKs 7, 8, 18 and 19 are found, and their expression is maintained in most corresponding primary and metastatic carcinomas of the ovary and the endometrium (Table 2; cf. 3, 21). The finding of small amounts of the stratified epithelium-type CK 5 in an endometrial adenocarcinoma (no. 3) may be correlated to the presence of foci of squamous metaplasia present in this tumor.

Normal mammary gland ducts are characterized by CKs of both the simple epithelial and the stratified epithelial subset (Table 3; 18). This complexity probably reflects the cell type heterogeneity of these ducts since antibodies selective for the simple epithelium-type cytokeratin 18 stain the luminal but not the myoepithelial cells (22). In contrast to the cytokeratin pattern of

Table 1. Cytokeratin patterns of epithelia and carcinomas of the gastrointestinal tract.

SUBFAMILY	Type II (basic)						Type I (acidic)							
CYTOKERATIN NO.	1	4	5	6	7	8	10/11	13	14	15	16	17	18	19
MOLECULAR WEIGHT ($\times 10^{-3}$)	68	59	58	56	54	52.5	56	54	50	50	48	46	45	40
Hepatocytes						**+**							**+**	
Gall bladder epithelium					+	**+**							+	**+**
Pancreatic ducts					+	**+**							+	**+**
Small intestine epithelium						**+**							+	**+**
Colon epithelium						**+**							+	**+**
Hepatocellular carcinoma														
no. 1						**+**							**+**	
no. 2						**+**							**+**	
Cholangiocarcinoma of liver					**+**	**+**							**+**	**+**
Adenocarcinoma of gall bladder					**+**	**+**						+	**+**	**+**
Adenocarcinoma of stomach														
no. 1 (well differentiated),														
primary tumor					+	**+**							+	**+**
lymph node metastasis					+	**+**							+	**+**
no. 2 (poorly differentiated)						**+**							**+**	**+**
Adenocarcinoma of colon														
no. 1 (well differentiated)						**+**							**+**	**+**
no. 2 (well differentiated)						**+**							**+**	**+**
no. 3 (well differentiated)						**+**							**+**	**+**
no. 4 (well differentiated),														
primary tumor						**+**							**+**	**+**
lymph node metastasis						**+**							**+**	**+**
no. 5 (well differentiated),														
liver metastasis						**+**							**+**	**+**
no. 6 (moderately differentiated)						**+**							**+**	**+**
no. 7 (moderately differentiated),														
primary tumor						**+**						+	**+**	**+**
lymph node metastasis						**+**						+	**+**	**+**

Explanation of symbols: Large +, major amount; small +, minor amount; (+), trace amount of the respective CK polypeptide.

total ducts, CKs of the vast majority of breast carcinomas are restricted to the simple epithelial subset (Table 3; 18), similar to those of the adenocarcinomas of the gastrointestinal and the female genital tract. Again, no differences are observed between primary and metastatic carcinomas. If the assumption that the stratified epithelium-type CKs are restricted to myoepithelial cells of the normal breast is true the finding of only simple

Table 2. Cytokeratin patterns of epithelia and carcinomas of the ovary and the endometrium.

CYTOKERATIN NO.	1	4	5	6	7	8	10 11	13	14	15	16	17	18	19
Ovarian surface epithelium					(+)	**+**							**+**	**+**
Endometrium					+	**+**							**+**	**+**
Adenocarcinoma of ovary														
no. 1 (papillary serous)					+	**+**							**+**	**+**
no. 2 (mucinous)						**+**							**+**	**+**
no. 3 (mucinous and endometrioid)						**+**							+	**+**
no. 4 (endometrioid),														
primary tumor					+	**+**							**+**	**+**
peritoneal metastasis					+	**+**							**+**	**+**
no. 5 (poorly differentiated)					**+**	**+**							+	**+**
no. 6 (poorly differentiated)					+	**+**							**+**	**+**
Adenocarcinoma of endometrium														
no. 1 (well differentiated)					**+**	**+**							**+**	**+**
no. 2 (well differentiated)					+	**+**							**+**	**+**
no. 3 (well differentiated, with foci of adenoacanthoma)			+		+	**+**							**+**	**+**
no. 4 (moderately differentiated)					+	**+**							**+**	**+**
no. 5 (poorly differentiated)						**+**							**+**	+

epithelium-type CKs in most breast carcinomas might suggest that they are not derived from myoepithelial cells.

The urothelium displays a unique CK pattern which in addition to the complete simple epithelium-type subset contains conspicuous amounts of CK 13, besides some CK 4 and 5 (8,18,20). It is not yet clear if this complexity reflects the complexity of morphologically distinct cell layers (20). Major features of the normal urothelial pattern are conserved in transitional cell carcinomas, including one case of a lymph node metastasis (Table 4; cf. 20); however, some changes in the relative amounts of the CK polypeptides present are also evident. Whether less differentiated tumors are devoid of certain CKs, as suggested from case no. 5, is currently under study.

The stratified squamous epithelia of the skin express characteristic CK patterns exclusively consisting of polypeptides of the stratified epithelium-type; among them CKs 1, 10 and 11 appear to be restricted to the suprabasal layers of the epidermis (Table 5; 7, 17,23). All basal cell epitheliomas studied show a prominence of CKs 5, 14 and 17 (Table 5; 17). In contrast, in squamous cell carcinomas of the skin (24) the CK pair 6:16 usually is more pronounced, and small but significant amounts of the pair(s) 1:10,11 are also

Table 3. Cytokeratin patterns of epithelia and carcinomas of the female breast.

CYTOKERATIN NO.	1	4	5	6	7	8	10/11	13	14	15	16	17	18	19
Mammary gland ducts			+		+	+			+	+		+	+	+
Invasive ductal carcinoma														
no. 1, primary tumor					+	+							+	+
lymph node metastasis					+	+							+	+
no. 2					+	+							+	+
no. 3					+	+						+	+	+
no. 4					+	+							+	+
no. 5					(+)	+							+	+
no. 6						+							+	+
Invasive lobular carcinoma														
no. 1					+	+							+	+
no. 2					+	+							+	+
Medullar carcinoma						+							+	+

Table 4. Cytokeratin patterns of urothelium and urinary bladder carcinomas.

CYTOKERATIN NO.	1	4	5	6	7	8	10/11	13	14	15	16	17	18	19
Urothelium		(+)	+		+	+		+					+	+
Transitional cell carcinoma														
no. 1 (grade 1 - 2)			+		+	+		+				+	+	+
no. 2 (grade 2)			+	+		+		+					+	+
no. 3 (grade 2)			(+)		+	+		+				+	+	+
no. 4, lymph node metastasis			+		+	+		+				+	+	+
no. 5 (grade 3)					+	+						+	+	+

found in many tumors of this kind, indicative of some terminal epidermal differentiation.

The CK patterns of the stratified squamous epithelia of the upper digestive tract studied so far differ from those of the stratified epithelia of the skin by the absence of the CK pair(s) 1:10,11 and by the additional presence of the CK pair 4:13 and minor proportions of CK 19 (Table 6; 8,18). Like the corresponding normal epithelia, the CK patterns of the squamous cell carcinomas are dominated by CKs of the stratified epithelium-type. Interestingly, CK 13 which is prominent in the normal epithelia has not yet been found in the tumors. Preliminary data using antibodies selective for CK 13 suggest that the basal cells of the tongue and esophagus mucosa, from which these carcinomas

Table 5. Cytokeratin patterns of squamous epithelia and tumors of the skin.

CYTOKERATIN NO.	1	4	5	6	7	8	10/11	13	14	15	16	17	18	19
Epidermis	**+**		**+**				**+**		**+**	+				
Outer root sheath of hair follicle			**+**	**+**					**+**	+	**+**	**+**		
Basal cell epithelioma														
no. 1			**+**			+			**+**	(+)		**+**		
no. 2			**+**			+			**+**	(+)		**+**		
no. 3			**+**			+			**+**	(+)		**+**		
no. 4			**+**	(+)		(+)			**+**	(+)	(+)	**+**		
no. 5			**+**	+		+			**+**	(+)	+	**+**		
no. 6			**+**	+		+			**+**	(+)	+	**+**		
Squamous cell carcinoma of skin														
no. 1 (well differentiated)	+		**+**	**+**			+		**+**		**+**	+		
no. 2 (well differentiated)	+		**+**	**+**			+		**+**		**+**	**+**		
no. 3 (well differentiated)	+	**+**	**+**	**+**			+		**+**		**+**	**+**		
no. 4 (poorly differentiated)			**+**	**+**					**+**	+	+	**+**		

Table 6. Cytokeratin patterns of squamous epithelia and carcinomas of the upper digestive tract.

CYTOKERATIN NO.	1	4	5	6	7	8	10/11	13	14	15	16	17	18	19
Tongue epithelium		**+**	**+**	**+**				**+**	**+**	+	**+**	+		+
Esophageal epithelium		**+**	**+**	**+**				**+**	+	(+)	+	(+)		+
Squamous cell carcinoma (SCC) of tongue (two cases)			+	**+**					**+**		**+**	**+**		
SCC of esophagus														
no. 1			**+**	**+**		+			+	+	+	**+**	(+)	**+**
no. 2		**+**		**+**		+	+		**+**		+	**+**	(+)	**+**

probably arise, also lack CK 13. The esophageal carcinomas seem to display elevated levels of CK 19, and CKs 8 and 18 found in these tumors have not been detected, by gel electrophoresis (8) and monoclonal antibodies specific for CK 18 (22,25), in normal esophageal mucosa. Whether the latter finding reflects the expression of new CKs or the selection of a minor cell type of the normal epithelium during tumorigenesis remains to be established.

Table 7. Cytokeratin patterns of epithelia and carcinomas of the cervix uteri.

CYTOKERATIN NO.	1	4	5	6	7	8	10 11	13	14	15	16	17	18	19
Exocervix, major portion	**+**	**+**	**+**	**+**			+	**+**	**+**	+	+	(+)		+
Exocervix at endo-exocervical junction			**+**					+				**+**		**+**
Endocervix with squamous metaplasia			**+**		**+**	**+**						**+**	**+**	**+**
Endocervix					+	**+**							**+**	**+**
Squamous cell carcinoma of cervix														
no. 1 (moderately differentiated)			**+**	**+**				**+**	**+**		+	**+**		+
no. 2 (moderately differentiated), peritoneal metastasis			**+**	+				**+**	+	+	**+**	+		+
no. 3 (moderately differentiated)			**+**	**+**		**+**		**+**	**+**	**+**	+	**+**	+	**+**
no. 4 (poorly differentiated)			**+**	**+**		**+**				**+**	**+**	**+**	+	**+**
no. 5 (poorly differentiated)			**+**	**+**	**+**	+		**+**	**+**	+	**+**	**+**	+	**+**
no. 6 (poorly differentiated)			**+**	**+**	+	**+**		**+**		+	+	**+**	+	**+**
Adenocarcinoma of cervix														
no. 1					**+**	**+**						**+**	+	**+**
no. 2					**+**	**+**							**+**	**+**

The complex morphological features of the transition zone of the exocervix and the endocervix of the uterus are reflected in the corresponding CK patterns (Table 7; for details see 21). Squamous cell carcinomas arising at this site often display unusually complex CK patterns (Table 7). Characteristically, CK 13 is found in most of the cervical carcinomas, including a metastatic tumor. The expression of simple epithelium-type CKs varies among these tumors and may be correlated with the degree of differentiation; they appear more prominent in the less differentiated tumors. Antibodies selective for CK 18 stain these tumors heterogeneously (26). Cervical adenocarcinomas produce simple epithelium-type CKs, and some of them also express CK 17.

The CK compositions of the epithelia and carcinomas of the respiratory tract (Table 8; 15,16) display certain analogies to the female genital tract; in both body regions, metaplastic changes play important roles in carcinogenesis. The complex CK pattern of the respiratory epithelium probably reflects differences of CK expression between bronchial basal cells and ciliated and goblet cells (the latter synthesize simple epithelium-type CKs; see also 15). Squamous cell carcinomas of the lung reveal very complex CK patterns which, however, are dominated by stratified epithelium-type CKs, notably CKs 5 and 6. Most tumors studied also contain substantial amounts of simple epithelium-type CKs and minor amounts of CK 13. This pattern is quite similar to that of the histologically similar cervical tumors (Table 7). Remarkably, this CK

Table 8. Cytokeratin patterns of epithelia and carcinomas of the lung.

CYTOKERATIN NO.	1	4	5	6	7	8	10/11	13	14	15	16	17	18	19
Trachea			**+**	+	(+)	+		+	(+)	+		+	(+)	**+**
Bronchus			**+**	+	+	**+**				+		+	+	**+**
Lung alveoli					**+**	**+**							**+**	**+**
Squamous cell carcinoma of lung														
no. 1 (moderately differentiated)			**+**	**+**	+	**+**			+			**+**	+	**+**
no. 2 (poorly differentiated)			**+**	**+**		**+**		+		+		**+**	+	**+**
no. 3 (poorly differentiated)			**+**	**+**		**+**		+	+	+		(+)	+	**+**
no. 4 (poorly differentiated)		**+**	**+**	**+**		**+**		+	+	+		**+**	+	**+**
no. 5 (spindle cell type)			**+**	**+**	+	**+**		(+)		**+**		**+**	+	**+**
Adenocarcinoma of lung														
no. 1					**+**	**+**							**+**	+
no. 2					**+**	**+**							+	**+**
no. 3					**+**	**+**							+	**+**
no. 4 (bronchiolo-alveolar carcinoma)					+	**+**							**+**	**+**
Large cell carcinoma of lung														
no. 1				**+**	**+**	**+**						**+**	**+**	+
no. 2					+	**+**							**+**	**+**
no. 3, adrenal metastasis						**+**							**+**	**+**
Carcinoid tumor														
no. 1						**+**							**+**	+
no. 2*						**+**							**+**	
no. 3*						**+**							**+**	
Small cell carcinoma of lung, intermediate cell type														
no. 1					+	**+**							**+**	**+**
no. 2					+	**+**							**+**	+
Small cell carcinoma of lung, oat cell type**														
no. 1, hilar lymph node metastasis						**+**							**+**	(+)
no. 2, skin metastasis						**+**							**+**	(+)

* These tumors express the 68,000 mol.wt. neurofilament polypeptide in addition to cytokeratins.

** In 3 other cases of oat cell carcinomas no intermediate filaments have been detected.

pattern has also been found in a lymph node metastasis of a laryngeal squamous cell carcinoma (Fig. 2c). Adenocarcinomas of the lung express the simple epithelial subset, as do most other adenocarcinomas. Large cell carcinomas appear heterogeneous in their CK expression in that some but not all display stratified epithelium-type CKs. The various neuroendocrine carcinomas are characterized by simple epithelium-type patterns with prominent amounts of CKs 8 and 18. While some carcinoids only reveal CK-IFs (16) others apparently co-express CKs and neurofilament proteins.

Malignant pleural mesotheliomas, specifically those of the epithelial type, often are difficult to distinguish from adenocarcinomas metastatic to the pleura. CK analysis (Table 9) shows that, unlike adenocarcinomas, at least some epithelial and biphasic mesotheliomas also express stratified epithelium-type CKs, notably CK 5, in addition to the simple epithelial subset, although these tumors have not shown any evidence of stratification of the tumor cells (G.A. Blobel et al., manuscript in preparation).

Table 9. Cytokeratin patterns of malignant pleural mesotheliomas.

CYTOKERATIN NO.	1	4	5	6	7	8	10/11	13	14	15	16	17	18	19
Epithelial mesothelioma														
no. 1			**+**			**+**			+				**+**	**+**
no. 2			+		+	**+**							**+**	**+**
Biphasic mesothelioma														
no. 1*			**+**		**+**	**+**			**+**			**+**	**+**	**+**
no. 2*					**+**	**+**							**+**	**+**
Fibrous mesothelioma														
no. 1*					+	**+**							**+**	+
no. 2*					+	**+**							**+**	+

* The fusiform tumor cells express vimentin in addition to cytokeratins.

Conclusions.

Results from various laboratories have shown that the IF type of the cell of origin is maintained in the corresponding primary tumors and their metastases (1). The immunohistochemical determination of the IF type is therefore of value for the diagnostic classification of metastatic tumors, notably the poorly differentiated ones. The presence of CKs has become a decisive criterion for the definition of an epithelial cell, and thus the detection of CKs in a tumor can be taken, irrespective of its morphological features, as positive proof of its carcinomatous nature. The additional criterion of desmoplakin-positivity now adds a further, independent marker for the immunocytochemical identification of metastatic carcinomas (13).

Gel electrophoretic analyses of a variety of carcinomas have revealed profound differences between different types of carcinomas. For example, the present data provide evidence that squamous cell carcinomas, including poorly differentiated ones, can be distinguished by their distinct CK patterns from adenocarcinomas of various degrees of differentiation. In our present studies, we have not observed significant differences in the CK patterns between a primary tumor and its metastasis. When the CK patterns expressed in normal epithelia and the corresponding primary and metastatic carcinomas are compared, identical patterns are found in many instances, notably adenocarcinomas. For carcinomas arising from complex epithelia, i.e. epithelia consisting of more than one cell type, an exact relationship between the CK type of the cell of origin and that of the corresponding carcinoma is difficult to establish since individual cells cannot be analyzed by electrophoresis. Many of the differences between the overall CK patterns of normal epithelia and the corresponding tumors may therefore be due to the selection, during tumorigenesis, of a specific cell type maintaining its pattern of CK expression. In other instances, de novo expression of CKs cannot be ruled out. The use of a panel of selective CK antibodies which we are currently developing should answer these questions.

Varying degrees of differentiation in different tumors from the same body site and of same histological type, and also in different cells of the same tumor, may be the reason for certain differences in CK expression within one group of tumors or even within one and the same tumor. It is not clear whether and to what extent such differences reflect intrinsic punctual alterations of programs of expression in some tumor cells or general changes due to changes of the local environment. It is known that CK expression can be influenced by experimental conditions. For example, when cultured cells are exposed to different concentrations of retinoic acids reversible changes in the CK patterns are noted which involve certain CKs whereas others remain constant (27-29). Probably the stability of expression of some CKs is higher than that of others. Apparently certain features of CK patterns are inherent to a given type of carcinoma, independent of its degree of differentiation. Some diagnostic implications of CK typing are already obvious from the present data, such as the differential diagnosis of mesotheliomas and adenocarcinomas. Clearly, more cases have to be analyzed to define the diagnostically relevant features of the CK patterns in metastatic carcinomas.

ACKNOWLEDGEMENT

This study has been supported by the Deutsche Forschungsgemeinschaft (grant Mo 345/3). We thank G.A. Blobel, T. Achtstätter and P. Cowin (Heidelberg) and V.E. Gould (Chicago) for valuable discussions.

REFERENCES

1. Osborn, M. and Weber, K. Lab. Invest. 48:372-394, 1983.
2. Franke, W.W., Schmid, E., Schiller, D.L., Winter, S., Jarasch, E.D., Moll, R., Denk, H., Jackson, B., Illmensee, K. Cold Spring Harbor Symp. Quant. Biol. 46:431-453, 1982.
3. Czernobilsky, B., Moll, R., Levy, R. and Franke, W.W. Eur. J. Cell Biol., in press.
4. Regauer, S., Franke, W.W. and Virtanen, I. J. Cell Biol., in press.
5. Schiller, D.L., Franke, W.W. and Geiger, B. EMBO J. 1, 761-769, 1982.
6. Hanukoglu, I. and Fuchs, E. Cell 33:915-924, 1983.
7. Sun, T.-T., Eichner, R., Schermer, A., Cooper, D., Nelson, W.G., Weiss, R.A. In: Cancer Cells 1, The transformed phenotype (Eds. A.J. Levine, G.F. Vande Woude, W.C. Topp and J.D. Watson), Cold Spring Harbor Laboratory, Cold Spring Harbor, 1984, pp. 169-176.
8. Moll, R., Franke, W.W., Schiller, D.L., Geiger, B. and Krepler, R. Cell 31:11-24, 1982.
9. Crewther, W.G., Dowling, L.M., Steinert, P.M. and Parry, D.A.D. Int. J. Biol. Macromol. 5:267-274, 1983.
10. Franke, W.W., Schiller, D.L., Hatzfeld, M. and Winter, S. Proc. Natl. Acad. Sci. USA 80:7113-7117, 1983.
11. Quinlan, R.A., Cohlberg, J.A., Schiller, D.L., Hatzfeld, M. and Franke, W.W. J. Mol. Biol. 178:365-388, 1984.
12. Mueller, H. and Franke, W.W. J. Mol. Biol. 163:647-671, 1983.
13. Franke, W.W., Moll, R., Müller, H., Schmid, E., Kuhn, C., Krepler, R., Artlieb, U. and Denk, H. Proc. Natl. Acad. Sci. USA 80:543-547.
14. Kartenbeck, J., Schwechheimer, K., Moll, R. and Franke, W.W. J. Cell Biol. 98:1072-1081, 1984.
15. Blobel, G.A., Moll, R., Franke, W.W. and Vogt-Moykopf, I. Virchows Arch Cell Pathol. 45:407-429, 1984.
16. Blobel, G.A., Gould, V.E., Moll, R., Lee, I., Huszar, M., Geiger, B., and Franke, W.W. Lab. Invest. 52:39-51, 1985.
17. Moll, R., Franke, W.W., Volc-Platzer, B. and Krepler, R. J. Cell Biol. 95:285-295, 1982.
18. Moll, R., Krepler, R. and Franke, W.W. Differentiation 23:256-269.
19. Gigi, O., Geiger, B., Eshhar, Z., Moll, R., Schmid, E., Winter, S., Schiller, D.L., and Franke, W.W. EMBO J. 1:1429-1437, 1982.
20. Achtstätter, T., Moll, R., Moore, B., and Franke, W.W. J. Histochem. Cytochem., in press.
21. Moll, R., Levy, R., Czernobilsky, B., Hohlweg-Majert, P., Dallenbach-Hellweg, G. and Franke, W.W. Lab. Invest. 49:599-610, 1983.
22. Debus, E., Weber, K. and Osborn, M. EMBO J. 1:1641-1647, 1982.
23. Fuchs, E. and Green, H. Cell 19:1033-1042, 1980.
24. Moll, R., Moll, I. and Franke, W.W. Arch. Dermatol. Res. 276:349-363, 1984.
25. Ramaekers, F., Huysmans, A., Moesker, O., Kant, A., Jap, P., Herman, C. and Vooijs, P. Lab. Invest. 49, 353-361, 1983.
26. Debus, E., Moll, R., Franke, W.W., Weber, K. and Osborn, M. Am. J. Pathol. 114:121-130, 1984.

27. Fuchs, E. and Green, H. Cell 25:612-625, 1981.
28. Kim, K.H., Schwartz, F. and Fuchs, E. Proc. Natl. Acad. Sci. USA 81: 4280-4284, 1984.
29. Eckert, R.L. and Green, H. Proc. Natl. Acad. Sci. USA 81:4321-4325, 1984.

10

BIOCHEMISTRY AND MOLECULAR BIOLOGY RAW117 LARGE CELL LYMPHOMA

G.L. NICOLSON[1], V. ROTTER[2], D. WOLF[2], T. IRIMURA[1], C.L. READING[1], M. BLICK[3], R. LA BICHE[1], and M. FRAZIER[4]

Departments of [1]Tumor Biology and [3]Clinical Immunology and Biological Therapy, The University of Texas - M.D. Anderson Hospital and Tumor Institute, Houston, Texas 77030, [2]Department of Biophysics, The Weizmann Institute of Science, Rehovot 76100, Israel, and [4]Department of Medicine, Baylor College of Medicine, Houston, Texas 77030

INTRODUCTION

Animal tumor models for studying cellular properties important in metastatic tumor spread have been established by sequential selection _in vivo_ or _in vitro_ to obtain variant cells with enhanced metastatic and organ colonization behaviors (1-4). Using the murine large cell lymphoma line RAW117 we have developed a metastatic system that shows preferential organ colonization (5-7). This tumor system was derived from parental RAW117-P large cell lymphoma or lymphosarcoma, which is an Abelson murine leukemia virus (Ab-MLV) transformed cell line of recent origin that colonizes lungs, liver, spleen and lymph nodes at low frequencies in BALB/c mice (8). Metastatic variant sublines were sequentially selected from RAW117-P for their abilities to colonize liver, and after ten such _in vivo_ selections, subline RAW117-H10 was established (5). The RAW117-H10 variant subline is highly metastatic and rapidly kills its host. For example, H10 cells form more than 200 times as many gross surface liver tumor nodules after intravenous or subcutaneous injection than does the parental RAW117-P line (5, 6, 9, 10).

The biochemical and cell surface properties of RAW117-P and -H10 sublines have been examined, and these data indicate that there are specific changes in the more metastatic cells, including: exposures of cell surface proteins (9) and glycoproteins (7, 10, 11), quantities of viral antigens (10) and lectin-binding sites (7, 11), growth characteristics (12), partitioning behavior in aqueous two-phase systems (13), amounts of liver-adhesion molecules (14), and sensitivities to macrophage-mediated cytolysis and cytostasis (12, 15). In addition, we have examined the RAW117 large cell lymphoma system for differences in gene expression of transformation- (16) and viral-associated (17) components.

MATERIALS AND METHODS

Cells and metastasis assays.

RAW117 parental (RAW117-P) cells and a subline selected ten times for liver colonization (RAW117-H10) were established and grown in Dulbecco-modified Eagle's medium (DME) containing 10% fetal bovine serum (FBS), as described previously (5-7). Cell cultures were used within ten passages from frozen stocks of low passage cells to eliminate possible drift in metastatic and other properties (6). Cultures were tested for the presence of mycoplasma using Hoechst 33258 staining (18) and were found to be negative. RAW117 sublines were assayed for organ colonization by intravenous injection of 5 x 10^3 viable tumor cells in 0.1 ml phosphate buffered saline (PBS) (5, 6). Mice were sacrificed at specific times after injection, and visible surface tumor nodules were counted in all major organs and were confirmed by histologic examination (7, 12).

Cell adhesion assays.

Cells were obtained from radiolabeled tissue of 17-19 day old embryos as described (15). The labeled tissue was rinsed Hepes-buffered medium, and the suspended cells had a final specific activity of 0.1-0.2 cpm/cell (15). RAW117 lymphoma cells were labeled in suspension at a concentration of 1 x 10^7 cells in 1 ml of Hepes-buffered medium containing ^{14}C -labeled amino acids, resulting in a specific activity of 0.1-0.5 cpm/cell (15). Cell adhesion assays were performed by the cell capture technique, as described by Grady and McGuire (19).

Identification of cellular glycoproteins.

Chemical pretreatment of separated RAW117 glycoproteins in polyacrylamide gels and reaction of the glycoproteins with ^{125}I-labeled lectins were performed as described by Irimura and Nicolson (20, 21). The lectins used were: wheat germ agglutinin (WGA), *Ricinus communis* agglutinin I (RCA_I), peanut agglutinin (PNA), *Lens culinarus* hemagglutinin (LCH), and concanavalin A (Con A). All lectins were purified by affinity chromatography and radiolabeled as described previously (20, 21). Standard glycoproteins with known carbohydrate chain structures were separated in adjacent lanes and stained simultaneously with the same ^{125}I-labeled lectins. Binding of ^{125}I-labeled lectins to the RAW117 cellular glycoproteins and to standard glycoproteins was assessed by autoradiography (11, 20, 21).

Analysis of mRNA.

Total RNA was prepared (22, 23) and polyadenylated mRNA(poly A^+) was selected by oligo(dT)-celluose chromatography (24). Aliquots were heated at 60^oC for 10 min and were electrophoresed in a 1% agarose gel containing 6% formaldehyde as described (16, 17). The RNA was transferred onto nitrocellulose sheets (25), and hybridized to nick-translated plasmid inserts containing specific sequences (16, 17, 26).

Immunoprecipitation of proteins.

Tumor cells were labeled with [^{35}S] methionine as described (16, 17). Cell lysates were treated with nonimmune serum and adsorbed with *S. aureus* (27) prior to immunoprecipitation of radiolabeled proteins with various antibody preparations (16, 17). The immunoprecipitates were absorbed with *S. aureus*, separated by sodium dodecyl sulfate (SDS)-polyacrylamide gel electrophoresis and identified by autoradiography (16, 17).

Phosphorylation experiments were conducted with immunoprecipitated complexes from unlabeled lysates. The immunoprecipitates were collected with *S. aureus* and incubated with γ- [^{32}P] ATP as described (17). The complexes were then washed and subjected to SDS-polyacrylamide gel electrophoresis-autoradiography.

Screening a cDNA library.

Double stranded complementary DNA (ds-cDNA) made from RAW117-H10 poly A^+ RNA was cloned into pBR322 according to the method of Gubler and Hoffman (28) except that the *Pst* I site was used and annealing was by homopolymeric tailing (C tailed insert, G tailed vector). Colonies were established and replicated by direct contact as described by Hanahan and Meselson (29). Probes were made from RAW117-P and -H10 poly A^+ RNA by preparing ds-cDNA according to the method of Gubler and Hoffman (28) except that ^{32}P-dCTP was added to the second strand synthesis reaction. The filters were hybridized and washed as described, prior to autoradiography (30).

RESULTS

Biologic properties of RAW117 cells.

Intravenous injection of 1-5 x 10^3 RAW117-P cells into BALB/c mice produced few lung or liver tumor colonies within 15 days (median = 0), whereas intravenous injection of the same number of RAW117-H10 cells produced >250 liver tumor nodules in all animals but few (median = 1) lung tumor nodules (5). RAW117-L17 (lung selected cells formed few liver tumor colonies, but significantly higher numbers of lung tumor colonies (median = 16). Similar results were obtained by subcutaneous injection of 5×10^5 cells (10).

Cell surface glycoproteins of RAW117 cells.

Cellular glycoproteins were identified by ^{125}I-lectin labeling to the isolated components in SDS-polyacrylamide gels (20, 21). Labeling the separated glycoproteins from P and HO cells with ^{125}I-Con A revealed a dramatic decrease in a M_r ~70,000 band and slight increases in a M_r 150,000 band on the SDS gels containing the glycoproteins from the highly metastatic RAW117-H10 subline (7). Using another lectin, ^{125}I-WGA, a major difference in a sialoglycoprotein of M_r ~150,000-200,000 was found. In the highly metastatic H10 cells, the M_r ~150,000-200,000 component bound more ^{125}I-WGA than RAW117-P cells (Fig. 1). Removal of sialic acid by mild acid treatment followed by also indicated an increase in the M_r 150,000-200,000 component (data not shown).

Viral components in RAW117 cells.

The M_r ~70,000 cell surface component has been identified as Mo-MLV gp70, which decreases dramatically with metastatic potential (Table 1). Competition radioimmune assays for Mo-MLV viral components gp70, p30 and p13 indicate that all three are present at significantly lower levels in the highly metastatic RAW117-H10 cells (Table 1).

Table 1. Expression of Mo-MuLV components in RAW117 cells measured by competition radioimmune assay*

Cell sublines	Antigen content (ng/10^6 cells)		
	p12	p30	gp70
RAW117-P	875 ± 100	1400 ± 118	407 ± 10
RAW117-H5	500 ± 30	333 ± 22	120 ± 10
RAW117-H10	23 ± 2	47 ± 7	65 ± 13

*For experimental methods see Reading et al. (10).

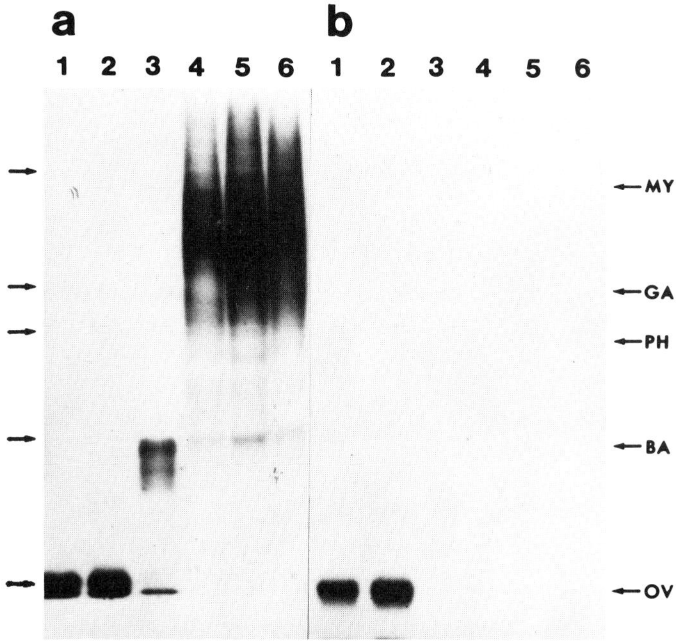

Fig. 1. Autoradiogram of sodium dodecylsulfate polyacrylamide gels (7%) reacted with ^{125}I-WGA without (panel a) or with (panel b) mild acid removal of sialic acid from the glycoproteins. Lanes 1-3 contain the standard glycoproteins: porcine thyroglobulin, bovine fibronectin, human transferrin, bovine fetuin and ovalbumin. Lanes 4-6 contain lysates of RAW117-P, RAW117-H10 and RAW117-L17, respectively. Molecular weight markers are MY, myosin (M_r ~ 200,000), GA, β-galactosidase (M_r ~ 116,000), pH, phosphorylase b (M_r ~ 92,000), BA, bovine serum albumin (M_r ~ 66,000) and OV, ovalbumin (M_r ~ 45,000) (data from ref. 11).

Adhesion components on RAW117 cells.

RAW117 cells express a cell surface antigen that cross-reacts with a fetal liver adhesion molecule that is involved in embryonic liver cell-cell adhesion (6, 15). RAW117-H10 cells express 4-8 times as many of these adhesion molecules than RAW117-P cells (15), and liver colonization of RAW117-H10 cells can be blocked using $F(ab')_2$ or Fab' antibody fragments against the fetal liver adhesion molecule (15). As a control for these experiments, RAW117-H10 cells can be treated with similar amounts of anti-H2 $(Fab')_2$ or Fab' and the degree of liver colonization assessed. In contrast to the $F(ab')_2$ or Fab' fragments against the fetal liver adhesion system, anti-H-2 antibody fragments had no effect on liver colonization (15).

When the adhesion properties of RAW117 cells to liver cells was examined, RAW117-H10 cells showed mored selectivity in binding to

Table 2. Adhesion of RAW117 cells to embryonic mouse cells*

RAW117 subline	Embryonic cell type	Normalized adhesion ratio (mean)	selectivity ratio
P	brain	1.04	-
H10	brain	0.73	0.66
P	liver	0.69	-
H10	liver	2.65	3.63

*Data of McGuire et al. (15).

embryonic liver cells than embryonic brain cells (15). In contrast, no such selectivity was found with RAW117-P cells (Table 2). In this aggregate capture assay RAW117-H10 cells had an approximately 4 times greater rate of selective adherence to embryonic liver cells. Since liver, not brain, is the target organ for colonization of RAW117-H10 cells, these tumor cells were treated as above with antibody fragments against the liver adhesion molecule, and then liver cell adhesive properties were assessed. RAW117-H10 cell adhesion to liver cells was inhibited by the antibody $F(ab')_2$ fragments against the liver adhesion molecule, but not by antibody $F(ab')_2$ fragments against H-2 molecules (15).

Expression of viral- and transformation-associated proteins and mRNA.

RAW117 cells express the transformation-associated molecule p53 that has been found at high levels in a variety of transformed cells compared to normal cells (31). To see if tumor cells express p53 in relation to their metastatic potentials we used the RAW117 system. When highly metastatic RAW117-H10 cells were examined for expression of p53, the amount of p53 protein found was similar to that found in RAW117-P cells (16). In addition, the expression of a 2.0 Kb mRNA containing p53 specific sequences was equivalent in RAW117-P and -H10 cells (16).

Since RAW117 cells were originally transformed by Ab-MLV, we examined the expression of the *abl* oncogene and its phosphoprotein product, p160, in RAW117 cells (17). The *abl* oncogene product p160 can immunoprecipitated from RAW117 cell lysates by antibodies (Ab-T1) against the normal cellular oncogene c-*abl* (32) or by antibodies (Ab-T2) against p160 plus Mo-MLV precursors for the *env* (p80) and *gag* (p65) proteins (33). Equivalent amounts of *abl* encoded p160 was immunoprecipitated by Ab-T1 or Ab-T2 from cell lysates of RAW117-P or -H10 cells. As a positive control for these experiments the *abl* encoded protein p120, synthesized by a different Ab-MLV-transformed cell line 2M3/M, was also immunoprecipitated (Fig. 2).

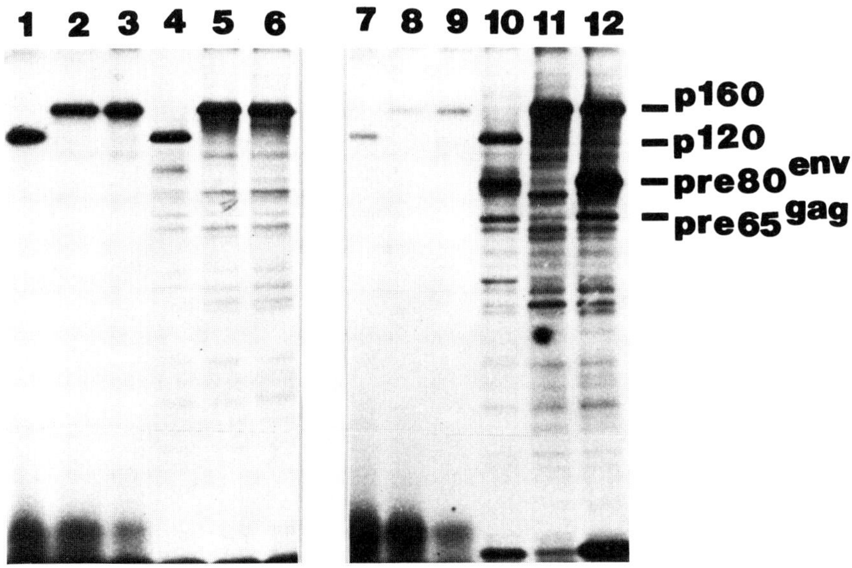

Fig. 2. Synthesis and phosphorylation of *abl*-encoded p160 in RAW117-P and H10 cells and in Ab-MLV-transformed fibroblasts (2M3/M) by immunoprecipitation with Ab-T1 (lanes 1-6) and Ab-T2 (lanes 6-12). Lanes 1, 4, 7, 10 = 2N3/M; lanes 2, 5, 8, 11 = RAW117-P; lanes 3, 6, 9, 12 = RAW117-H10; lanes 1-3, 7-9 = [^{32}P] ATP; lanes 4-6, 10-12 = [^{35}S] methionine (data from ref. 17).

Many oncogene-encoded proteins possess the capacity to be phosphorylated at tyrosine residues (34). We tested whether p160 could be differentially phosphorylated in RAW117 cells of low and high metastatic potential by addition of γ- [^{32}P] ATP to Ab-T1 or Ab-T2 immunoprecipitates (17). When γ - [^{32}P] ATP was added to the p160 immunoprecipitates, similar autophosphorylation occured (Fig. 2). Thus the amounts of p160 and the autophosphorylation activity of p160 were independent of metastatic potential in RAW117 cell (17).

The expression of abl-encoded mRNA in RAW117 cells was examined using poly A$^+$ mRNA and a probe against the entire v-abl sequence. Low and high metastatic potential RAW117 cells expressed abl mRNA in equal amounts (17). When other oncogenes (fos, myc, myb, B-lym) were examined for expression in RAW117 cells, these were also expressed in the same low amounts, if expressed, in RAW117-P and -H10 cells (17).

Differential gene expression in RAW117 cells.

Differential gene expression in RAW117 cells was examined by establishing a pBR322 cDNA library from poly A$^+$ RNA of RAW117-H10 cells (30). The cDNA consisting of 17,600 clones, was then screened using the colony hybridization procedure by replicate exposure of the cDNA clones to ^{32}P-ds-cDNA prepared from RAW117-P and -H10 cells and comparison of the autoradiographs made from such exposures (Fig. 3). We found several differences in ^{32}P-ds-cDNA binding to the cDNA library (Table 3). Most of these differences (76/160) were attributed to Mo-MLV sequences, as found in an independent probing using a ^{32}P-labeled Mo-MLV probe (30). However, some non-Mo-MLV mRNAs were expressed at higher levels in RAW117-P cells, and a few non-Mo-MLV mRNAs were expressed at higher levels in RAW117-H10 cells (Table 3).

Table 3. Differential gene expression in RAW117 cells*

Type	Colonies (no.)	Abundance (%)
Total library	17,600	-
Estimated differentially expressed	160	0.90
Total Mo-MLV	76	0.43
Estimated non-Mo-MLV high in P	63	0.36
Estimated non-Mo-MLV high in H10	21	0.12

*Data of LaBiche et al. (30).

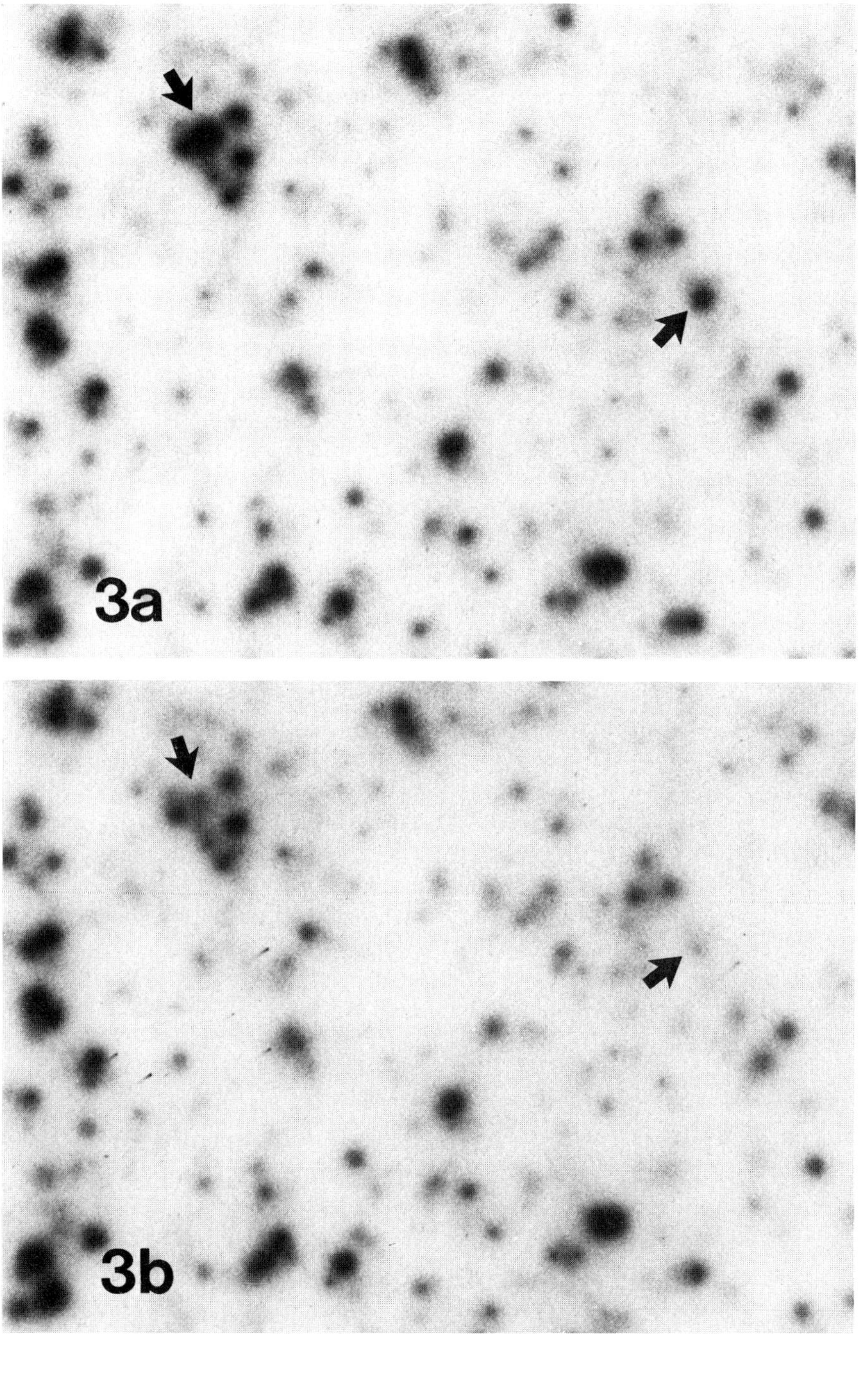
3a
3b

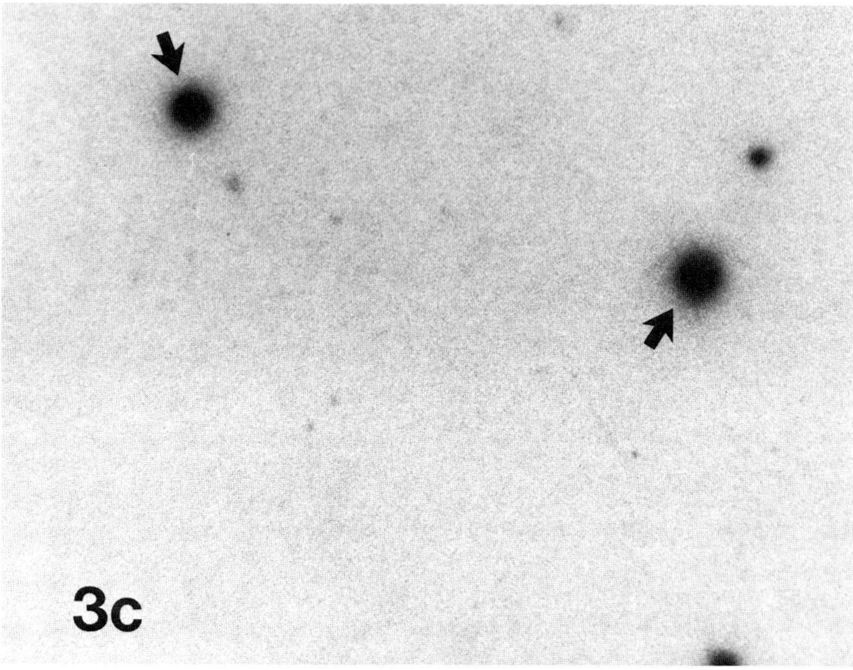

Fig. 3. Differential hybridization with probes made from poly A^+ RNA. a, Autoradiography of a library filter hybridized with ^{32}P-cDNA made from RAW117-P poly A^+ RNA. Arrows indicated colonies with obvious differential hybridization. b, Autoradiography of the same library filter after reprobing with ^{32}P-cDNA made RAW117-H10 poly A^+ RNA. c, The same filter reprobed with ^{32}P-cDNA made from Moloney virus genome, note that the two differentially hybridization colonies in a and b above appear to be viral transcripts (overexposed).

DISCUSSION

The highly metastatic phenotype of RAW117 cells is characterized by a number of properties that may define the quantity and location of metastatic deposits. In our studies we have found that highly metastatic RAW117-H10 cells progressively lose expression of Mo-MLV components, such as gp70, suggesting that host responses may be involved in suppressing RAW117 cells with high amounts of gp70 on their surfaces. Although evidence of selective destruction and/or growth inhibition of RAW117 cells by T-cell, NK-cell or NC-cell-mediated responses could not be demonstrated (12), we did find that RAW117 cells of high metastatic potential were much less susceptible to macrophage-mediated cytolysis and cytostasis (14).

Although host responses may be important in determining the quantities of metastatic deposits, they are probably not that important in determining the locations of tumor colonies (12). Other mechanisms such as preferential adhesion of highly metastatic RAW117 cells to liver cells may account for the colonization of liver over other organs. In addition, we have also found that highly liver metastatic RAW117 cells are less susceptible to cytostasis by liver-released factors which differentially inhibit the growth of RAW117-P cells (G. L. Nicolson, unpublished observation).

The ability of highly metastatic RAW117 cells to successfully colonize distant sits does not appear to be related to the expression of transformation-associated components such as p53 (16) or oncogenes products such as the abl-encoded p160 (17). This latter conclusion has been independently reached by Kris et al. (35), who found that B16 melanoma and UV2237 fibrosarcoma cells of varying metastatic potentials expressed similar amounts of Ki-ras (the major oncogene of these cells) and $p21^{ras}$ in both low and high metastatic potential cells. In the RAW117 system, the Mo-MLV encoded products are, however, differentially expressed, resulting in lowered amounts of gp70, p30, p15 and other Mo-MLV encoded products in the highly metastatic RAW117 cells (10).

In addition to their role in transformation, oncogenes, such as abl, may serve another function in tumor cells. That is, they may set in motion the generation of cellular phenotypic diversification that results in tumor cell heterogeneity (36, 37). Once cells are transformed, oncogene expression may not be essential for achieving the metastatic phenotype (36, 37).

Although oncogenes did not appear to be differentially expressed in highly metastatic RAW117 cells, we did note that several cellular genes were apparently expressed at lower of higher levels in the highly metastatic RAW117-H10 than in the lowly metastatic RAW117-P cells (30). Some of these transcripts undoubtedly code for specific components, such as the liver adhesion molecule important in liver colonization (15), and macrophage recognition structures that are not yet defined (14). Since metastatic cells are know to possess a number of enzymes, receptors and responses that differ from their low metastatic counterparts (1-4), it is not surprising that they show some differential gene expression. In any given metastatic system defining the products of such differentially expressed genes and their roles in maintaining the highly metastatic phenotype remain important goals of our studies.

REFERENCES

1. Nicolson, G.L. Biochim. Biophys. Acta 695:113-176, 1982.
2. Nicolson, G.L. Exp. Cell Res. 150:3-22, 1984.
3. Nicolson, G.L., and Poste, G. Curr. Prob. Cancer 7(6):1-83, 1982.
4. Nicolson, G.L., and Poste, G. Int. Rev. Exp. Pathol. 25:77-181, 1983.
5. Brunson, K.W., and Nicolson, G.L. J. Natl. Cancer Inst. 61:1499-1503, 1978.
6. Nicolson, G.L., Mascali, J.J., and McGuire, E.J. Oncodevelop. Biol. Med. 4:149-159, 1982.
7. Reading, C.L., Belloni, P.N., and Nicolson, G.L. J. Natl. Cancer Inst. 64:1241-1249, 1980.
8. Raschke, W.C., Ralph, P., Watson, J., Sklar, M., and Coon, H. J. Natl. Cancer Inst. 54:1249-1253, 1975.
9. Nicolson, G.L., Reading, C.L., and Brunson, K.W. In: In Tumor Progression (Ed. R. G. Crispen), Elsevier North Holland, Inc., Amsterdam, pp. 31-48, 1980.
10. Reading, C.L., Brunson, K.W., Torriani, M., and Nicolson, G.L. Proc. Natl. Acad. Sci. U.S.A. 77:5943-5947, 1980.
11. Irimura, T., Belloni, P.N., and Nicolson, G.L. Exp. Cell Res. (submitted), 1984.
12. Reading, C.L., Kraemer, P.M., Miner, K.M., and Nicolson, G.L. Clin. Expl. Metastasis 1:135-151, 1983.
13. Miner, K.M., Walter, H., and Nicolson, G.L. Biochemistry 20:6244-6250, 1981.
14. Miner, K.M., and Nicolson, G.L. Cancer Res. 43:2063-2071, 1983.
15. McGuire, E.J., Mascali, J.J., Grady S.R., and Nicolson, G.L. Clin. Expl. Metastasis 2:213-222, 1984.
16. Rotter, V., Wolf, D., and Nicolson, G.L. Clin Expl. Metastasis 2:199-204, 1984.
17. Rotter, V., Wolf, D., Blick, M., and Nicolson, G.L. Clin. Expl. Metastasis (in press), 1985.
18. Chen, T.R. Exp. Cell Res. 104:255-262, 1977.
19. Grady, S.R., and McGuire, E.J. J. Cell Biol. 71:96-106, 1976.
20. Irimura, T., and Nicolson, G.L., Carbohydrate Res. 120:187-195, 1983.
21. Irimura, T., and Nicolson, G.L., Cancer Res. 44:791-798, 1984.
22. Auffray, D., and Rougeon, F. Eur. J. Biochem. 107:303-314, 1980.
23. Glisin, W., Crkvenjakov, R., and Byus, C. Biochemistry 13:2633-2637, 1974.
24. Aviv, H., and Leder, P. Proc. Natl. Acad. Sci. U.S.A. 69:1408-1412, 1972.
25. Thomas, P.S. Proc. Natl. Acad. Sci. U.S.A. 77:5201-5205, 1980.
26. Rigby, P.W., Dieckmann, M., Rhodes, C., and Berg, P. J. Mol. Biol. 113:327-251, 1977.
27. Kessler, S.W. J. Immunol. 115:1617-1624, 1975.
28. Gubler, U., and Hoffman, B.J. Gene 25:263-269, 1983.
29. Hanahan, D., and Meselson, M. Gene 10:63-77. 1980.
30. LaBiche, R., Frazier, M.L., Brock, W.A., and Nicolson, G.L. (submitted for publication), 1985.

31. Rotter, V., Witte, O.N., Coffman, R., and Baltimore, D. J. Virol. 36:547-555, 1980.
32. Witte, O.N., Rosenberg, N., and Baltimore, D. J. Virol. 31:776-784, 1979.
33. Rosenberg, N., and Witte, O.N. J. Virol. 33:340-348, 1980.
34. Collet, M.S., and Erikson, R.L. Proc. Natl. Acad. Sci. U.S.A. 75:2021-2024, 1978.
35. Kris, R.M., Avivi, A, Bar-Eli, M., Alon, Y., Carini, P., Schlessinger, J., and Raz, A. Int. J. Cancer 35:227-230, 1984.
36. Nicolson, G.L. Cancer Metastasis Rev. 3:25-42, 1984.
37. Nicolson, G.L. Clin. Expl. Metastasis 2:85-105, 1984.

11

KARYOTYPIC PROGRESSION AND METASTASIS FORMATION OF HUMAN TUMORS

PETER C. NOWELL and GLORIA BALABAN*

Department of Pathology and Laboratory Medicine, and *Department of Human Genetics, University of Pennsylvania School of Medicine, Philadelphia, PA 19104

INTRODUCTION

In recent years, chromosome studies have contributed significantly to our understanding of fundamental tumor biology in a number or areas. Based on the work from many laboratories, four general statements can be made about karyotypic alterations in neoplasia: 1) most tumors have chromosome abnormalities, which are not present in other cells of the body, 2) in a given tumor, all the neoplastic cells often have the same cytogenetic change, or related changes, 3) although chromosome alterations often differ between tumors, there are nonrandom patterns, and 4) chromosome abnormalities are more extensive in advanced tumors.

The cytogenetic data that support the first two generalizations represent a significant portion of the evidence that somatic genetic changes are important in tumorigenesis. In addition, the second generalization, along with related biochemical and immunoglobulin data, has been the basis for the now generally accepted view that most neoplasms are unicellular in origin (1,2). The fact that in a given tumor all the cells show the same chromosome abnormality (or related abnormalities) suggests the derivation of the tumor from a single altered cell. Presumably, the particular karyotypic change confers on the progenitor cell a selective growth advantage, allowing its progeny to expand as a neoplastic clone (1).

The most significant aspect of the third generalization listed above is that specific alterations in particular chromosomes are associated, with various degrees of consistency, with specific types of tumors or with neoplasia in general (3,4). Recognition of this fact led to the suggestion that these nonrandom karyotypic changes might indicate sites in the genome where particular genes, important in carcinogenesis, are located, and also might provide clues as to how the function of such "oncogenes" might be significantly altered (3-5). Support for this hypothesis has now been provided from the study of specific reciprocal chromosome translocations in various hematopoietic tumors (e.g., Burkitt's lymphoma, chronic granulocytic leukemia), in which alteration in structure and/or function of human proto-oncogenes has been demonstrated (3-6). In addition, chromosomal changes in other tumors, reflecting gain or loss of genetic material, are indicating a critical role for oncogene dosage in some instances of carcinogenesis (3-5,7,8). We will return to these considerations later.

Karyotypic evolution and tumor progression

It is the fourth generalization listed above that provides the basis for the present brief discussion of the role of sequential chromosome changes in the development of metastasis. The observation that more advanced neoplasms typically show more extensive karyotypic alterations has led to the hypothesis that clinical and biological tumor progression, in general, may often reflect the appearance in a neoplastic clone, over time, of subpopulations of cells with further alterations in genetic makeup (1,9). Experimental evidence indicates that neoplastic cells shown increased genetic instability, and are thus more likely than normal cells to generate genetic variants (9,10). Occasionally such a variant cell may have more aggressive biological characteristics, and so its progeny may grow out as the predominant malignant population, providing the basis for clinical tumor progression.

Most of the data supporting this view have been derived from studies of human leukemias and lymphomas, and very little information is available on other malignancies. For instance, this concept has been well documented in chronic granulocytic leukemia, in which the cells in the early indolent stage of the disorder typically show only the Philadelphia chromosome, but the terminal accelerated phase of the disease apparently results from overgrowth of this initial population by one or more subclones having additional karyotypic changes (3).

A similar sequence of events, relating more aggressive tumor growth to additional chromosome changes in the neoplastic cells, has been documented in a number of other human hematopoietic neoplasms, and in a few experimental solid tumors, such as the sarcomas induced in mice and rats by the Rous virus (11,12). It has also been possible, through cytogenetic studies, to demonstrate the coexistence of multiple variant subpopulations within many advanced malignancies, thus providing at least one explanation for the biologically and clinically important phenomenon of heterogeneity in many neoplasms (3,9).

To date, there have been very few attempts, with human malignancies, to compare directly the karyotype of a primary and a metastatic lesion in the same patient, and thus to draw conclusions concerning specific somatic genetic changes that might be important in the acquisition of metastatic capability. In the following sections we will summarize the small amount of information currently available, and suggest some possibilities for future work.

RESULTS

Malignant melanoma. During the past several years, in the course of cytogenetic studies of direct preparations or cell lines derived from melanocytic lesions in all stages of tumor development (13-15), we have had the opportunity to investigate four cases of malignant melanoma in which material was available both from a primary and from one or more metastatic lesions (Table I). In all instances the primary tumor was characterized by a cytogenetically abnormal clone, and the metastases showed at least some of the same cytogenetic alterations as the primary, indicating their derivation from it, as well as additional changes.

In case No. 740, cytogenetic data were available from two metastases and from a complex primary lesion that had areas of low-grade malignancy (radial growth phase) as well as a deeply invasive portion (vertical growth phase) (16). Cells derived from the radial growth phase portion did not show a karyotypically abnormal clone, but the cells of the vertical growth phase had an aberrant near-triploid karyotype characterized by abnormality of chromosome 6 and loss of chromosome 10. Cells from the metastases had the same loss of chromosome 10, new alterations involving chromosomes 1 and 6, and extra copies of chromosome 7 (Table I). The latter findings are of particular interest, since we and others have observed that rearrangements in chromosomes 1 and 6, as well as an extra dose of chromosome 7, occur nonrandomly in a high proportion of advanced malignant melanomas,

Table I

Chromosome abnormalities in primary melanomas and their metastases

Case	Lesion	Chrom. #1	Chrom. #6	Chrom. #7	Other chromosomes
740	Primary				
	RGP*	0	0†	0	0
	VGP	0	6q-	0	-10
	Metastases (2)	1p-	6q-‡	+7	-10
75	Primary	0	0	7q-	9p+, 10q-
	Metastasis	1q-	0	7q-	9p+, 10q-
115	Primary	t(1p;9q)	6p-	+7	11p+, 16q+, M1
	Metastases (5)	t(1p;9q)	6p-	+7	Same, and new markers
983	Primary	1p+, 1q-	6q+, 6q-	+7q-	9p+, 12p-, 15p+, +18, +M1, +M2
	Metastases (2)	1p+, 1q-	6q+, 6q-	+7q-	Same, and new marker (3p+)

*RGP = radial growth phase portion of complex primary; VGP = vertical growth phase portion.

†Non-clonal changes in chromosome #6.

‡Different breakpoint.

perhaps suggesting the site of specific oncogenes of importance in the later stages of clinical and biological progression of these tumors (13-15).

In case No. 75 the primary lesion had an aneuploid karyotype, with abnormalities of chromosomes 7, 9, and 10. A metastasis from this tumor showed these same alterations, as well as an additional change involving chromosome 1 (Table I). The other two melanomas (Nos. 115, 983) in which we have such data were highly aneuploid, with abnormalities, in both cases, involving chromosomes 1, 6, and 7, and the respective metastases showed the same karyotypic changes, as well as one or more new alterations.

Quinn et al. (17) have reported similar findings in one case of malignant melanoma from which they were able to obtain data on both a primary tumor and its metastasis. It is clear, however, that more information is needed before an attempt can be made to relate specific cytogenetic alterations to metastatic potential in melanoma.

Neuroblastoma. Limited cytogenetic data are available on six cases of neuroblastoma in which both a primary and one or more metastases were studied. Four such cases have been investigated in our institution (18,19). In one instance the primary tumor was highly aneuploid with multiple chromosome rearrangements. When the tumor recurred at its original site, with the concurrent appearance of lung and sternal metastases, karyotypic data were obtained on the recurrent primary and the two metastases. In all three specimens, many of the same abnormalities were observed as in the initial primary tumor, but with the loss of several aberrant chromosomes and the addition of two new markers, a 4q- and a 15q-. These findings indicated the derivation of both the recurrence and the metastases from the same subpopulation of the original tumor (18).

In each of the other three cases of neuroblastoma studied locally, the primary tumor and a metastasis shared marker chromosomes, and the metastases had further alterations, including an extra chromosome 1 in one instance (19). Reynolds et al. (20) have reported limited information on two other cases of neuroblastoma, in which metastatic lesions showed markers in common with the primary, and also additional changes in the metastases.

Other tumors. Similar limited data are available on two cases of human ovarian cancer, and on one case each of carcinoma of the kidney, large bowel, and testis (21-25). In each instance the primary tumors were highly aneuploid, and the authors focused on demonstrating in the metastases one or more of the same markers as had been observed in the primary, thus indicating the clonal nature of the process. The data also indicated that the metastases had new abnormalities as well, but there was no attempt to characterize specific additional changes associated with the metastatic state.

DISCUSSION

A number of workers, as indicated elsewhere in this symposium, have begun to extend studies of the biological characteristics of metastasis to the specific molecules and metabolic pathways important in the acquisition of the capacity for successful invasion of adjacent tissue and colonization of distant sites. Primarily in experimental tumors, the relative importance of properties such as various proteolytic enzymes, tumor angiogenic factor, platelet agglutinating capacity, and surface recognition molecules are being actively explored. As the role and importance of each of these factors are clarified, and extended to human systems, the next logical step will be to determine, at the genetic level, just how each of these properties is altered in the malignant cell. These questions can be approached with a variety of molecular genetic techniques, working with cells that differ in specifically defined biological properties.

Recent work with human tumors, particularly of the hematopoietic system, has indicated that chromosome studies represent one useful approach to the identification of genes important in neoplastic processes (3-6). As indicated above, a number of the nonrandom karyotypic abnormalities observed in human leukemias and lymphomas are providing important clues to the involvement of specific genes in human neoplasia, some being known proto-oncogenes and others perhaps representing previously unrecognized oncogenes (3-6).

This approach is just beginning to be extended to solid malignancies, as in the investigation of the relationship between oncogene amplification and the chromosomal homogeneous staining regions (HSR's) and double minutes (DM's) in neuroblastomas and other malignancies (6,7). Brodeur et al. (7) have suggested that in neuroblastoma, amplification units involving the oncogene N-myc may be associated with the more aggressive forms of the disease.

We have also recently observed that the extra dosage of chromosome 7, and specifically the short arm of this chromosome (7p), that is present in many advanced melanomas appears to be associated with expression on the cells of the receptor for epidermal growth factor

(EGFR) (8). Since the human proto-oncogene c-erbB appears to code for a portion of the EGFR (26), this may be another example of a significant alteration in oncogene function reflected in a nonrandom chromosome change in neoplastic cells. In our study, the extra dosage of chromosome 7 appeared to be characteristically associated with advanced stages of melanoma, both primary lesions and metastases. EGFR expression may therefore contribute to a further selective growth advantage in these already malignant cells (8), but may not represent a property specifically important in the acquisition of the capacity for metastasis.

These preliminary findings in neuroblastoma and melanoma indicate both the possibilities and the difficulties of future work. In theory, working initially with experimental tumor lines carefully defined with respect to metastatic properties, and then progressing to human cells, one might be able to relate specific karyotypic changes (and ultimately specific genes) to particular properties necessary for metastasis. The very limited human data, summarized above, indicate that both primary cancers and their metastases will typically be aneuploid, and the metastases will have additional karyotypic changes. It is not certain, however, which, if any, of these additional abnormalities will represent somatic genetic alterations essential for metastasis, since many of the cells in the highly aneuploid primary may already have such capacity, and sorting through the extensive cytogenetic changes in both types of lesions will be extremely difficult.

It is clear, however, that the ultimate understanding of metastasis must be based on identification of the specific genes and gene products involved, and characterization of how they are critically altered. This difficult problem will not be resolved by any single approach, and as important information is provided by biochemical analysis, by transfection experiments, by DNA library subtraction methods, and by other approaches, carefully conceived and executed karyotypic studies, both in human and animal systems, should also contribute usefully to unraveling this complex biological phenomenon.

REFERENCES

1. Nowell, P. Science 194: 23-28, 1976.
2. Fialkow, P. Ann. Rev. Med. 30: 135-143, 1979.
3. Rowley, J.D. and Testa, J.R. Adv. Cancer Res. 36: 103,148, 1982.
4. Yunis, J.J. Science 221: 227-236, 1983.
5. Klein, G. Nature 294: 313-318, 1981.
6. Nowell, P.C., Emanuel, B.S., Finan, J.B., Erikson, J. and Croce, C.M. Microbiological Sciences 1: 223-228, 1984.
7. Brodeur, G.M., Seeger, R., Schwab, M., Varmus, H.E. and Bishop, J.M. Science 224: 1121-1124, 1984.
8. Koprowski, H., Herlyn, M., Balaban, G., Parmiter, A., Ross, A. and Nowell, P.C. Somat. Cell Molec. Genet. 1985, in press.
9. Nowell, P. In: Chromosome Mutation and Neoplasia (ed. J. German), Alan R. Liss, New York, 1983, pp. 413-432.
10. Cifone, M.A. and Fidler, I.J. Proc. Natl. Acad. Sci. USA 78: 6949-6952, 1981.
11. Mark, J. Eur. J. Cancer 5: 307, 1969.
12. Levan, G. and Mitelman, F. Hereditas 84: 1, 1976.
13. Balaban, G., Herlyn, M., Guerry, D., Bartolo, R., Koprowski, H., Clark, W. and Nowell, P.C. Cancer Genet. Cytogenet. 11: 429, 1983.
14. Herlyn, M., Balaban, G., Bennicelli, J., Guerry, D., Halaban, R., Herlyn, D., Elder, D., Maul, G., Steplewski, Z., Nowell, P., Clark, W. and Koprowski, H. J. Natl. Cancer Inst. 1985, in press.
15. Balaban, G., Herlyn, M., Guerry, D., Parmiter, A., Nowell, P. and Clark, W. Jr. Proc. Amer. Assoc. Cancer Res. 1985, in press.
16. Clark, W. Jr., Elder, D., Guerry, D., Epstein, M., Greene, M. and Van Horn, M. Hum. Pathol. 15: 1147-1165, 1984.
17. Quinn, L.A., Woods, L.K., Merrick, S.B., Arabasz, N.M., Moore, G.E. J. Natl. Cancer Inst. 59: 301-307, 1977.
18. Feder, M. and Gilbert, F. J. Natl. Cancer Inst. 70: 1051-1056, 1983.
19. Feder, M.M. Thesis, Department of Human Genetics, University of Pennsylvania, 1983.
20. Reynolds, C.P., Biedler, J.L., Spengler, B.A., Ross, R.A., Reynolds, D.A., Smith, R.G. and Frenkel, E.P. Proc. Am. Assoc. Cancer Res. 23: 11, 1983.
21. Kusyk, C.J., Seski, J.C., Medlin, W.V. and Edwards, C.L. J. Natl. Cancer Inst. 66: 1021-1025, 1981.
22. Kusyk, C.J., Turpening, E.L., Edwards, C.L., Wharton, J.T. and Copeland, L.J. Gynecologic Oncol. 14: 324-338, 1982.
23. Hagemeister, A., Hoehn, W. and Smit E.M.E. Cancer Res. 39: 4662-4667, 1979.
24. Reichmann, A., Martin, P. and Levin, B. Int. J. Cancer 28: 431-440, 1981.
25. Wang, N., Trend, B., Bronson, D.L. and Fraley, E.E. Cancer Res. 40: 796-802, 1980.
26. Lin, C.R., Chen, W.S., Kruiger, W., Stolarsky, L.S., Weber, W., Evans, R.M., Verma, I.M., Gill, G.N. and Rosenfeld, M.G. Science 224: 843-848, 1984.

12

MORPHOLOGICAL AND FUNCTIONAL ALTERATIONS OF OCCLUDENS, ADHERENS, AND GAP JUNCTIONS IN CANCER*

B.U. PAULI AND R.S. WEINSTEIN

Department of Pathology, Rush-Presbyterian-St. Luke's Medical Center, Chicago, Illinois, USA

INTRODUCTION

Cell junctions are defined as structurally specialized domains that are formed at regions of contact between two cells, and to which both cells contribute an equal part. A number of different types of cell junctions, customarily subclassified on the basis of their ultrastructure, have been identified: adherens, occludens, and gap junctions (1,2). These junctions perform the following functions (1-4): (1) They provide strong intercellular adhesion and in turn are the basis for the formation of mechanically coherent tissues; (2) They mediate direct communication between cells by allowing the transfer of ions and small molecules from cell to cell without leakage into the extracellular compartment (metabolic coupling); (3) They act as intercellular conduits for electrochemical impulses (electrical coupling); and (4) They seal cells together into a coherent tissue that can act as a highly selective barrier to diffusion. In cancer, each type of intercellular junction can differ from its normal counterpart with respect to overall size, distribution at the cell surface, level of development, and numerical density (5). These differences may account for or contribute to some properties of malignant tumors, such as abnormalities in cell-to-cell adhesion, growth control, and transepithelial permeability. In this brief review, structural, biochemical, and functional features of the three major classes of cell junctions are summarized, emphasizing information that may be particularly pertinent to the cancer problem.

---

* This work was supported by Public Health Service grant CA25034 from the National Cancer Institute and by funds from the Otho S.A. Sprague Memorial Institute.

CELL JUNCTION STRUCTURE IN NORMAL TISSUES

Classes of cell junctions have been identified on the basis of ultrastructure, function, and, more recently, physico-chemical characteristics of isolated junctions. The three major classes of cell junctions are (1) occludens junctions (tight junctions), (2) adherens junctions, and (3) gap junctions (Fig. 1). These types of cell junctions can be conveniently grouped into two major categories on the basis of the spatial relationships of the junctional membranes (1). In the first category are intercellular junctions at which the surface membrane of neighboring cells come into direct contact (e.g., occludens and gap junctions). The second category includes junctions at which the junctional membranes are separated by a 25-35 nm interspace (e.g., adherens junctions). Cell junctions in both of these general categories are further subclassified on the basis of shape and size (1-4). Occludens and adherens junctions can be in zonula (latin: zone), fascia (latin: sheet), and macula (latin: spot) configuration, while gap junctions are usually maculae (Fig. 1).

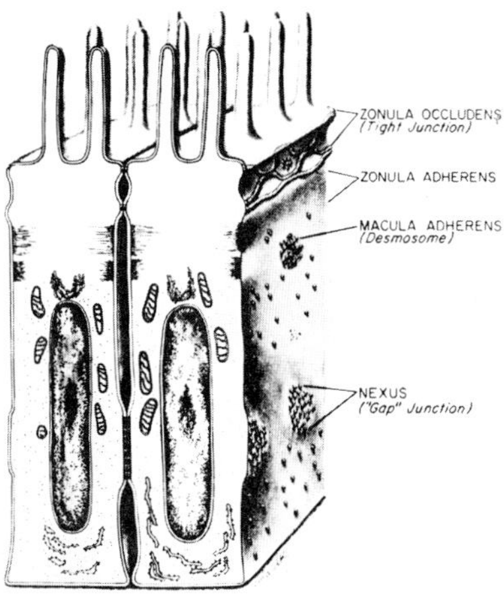

Fig. 1. Diagram of intercellular junctions between two columnar epithelial cells: The right membrane surface shows the major classes of intercellular junctions as viewed in freeze-fracture replicas.

Occludens Junctions

At occludens junctions, membranes of adjacent cells are in intimate contact and the outer membrane leaflets appear to "fuse", reducing the overall thickness of the pair of junctional membranes to less than the combined thickness of the two non-junctional plasma membranes. Membrane splitting by the freeze-fracture process reveals an anastomosing, concertina-like network of fibrils, 6-8 nm in replica diameter and hundreds of nanometers in length (Fig. 2). These fibrils are probably linear protein aggregates (1). Complementary networks of fibrils within the apposing membranes of the junction forming cells associate across the intercellular space, establishing strong cell-to-cell attachment sites and, in doing so, focally obliterating the extracellular space (1,2). In many epithelia (including some endothelia), the occludens junction forms a belt that completely encircles the perimeter of superfical cells near their lumenal surfaces, and is therefore called a zonula occludens (6)(Fig. 2). The occludens junctions in some tissues (e.g., some endothelia) are discontinuous and are thus designated fasciae or maculae occludentes)(7).

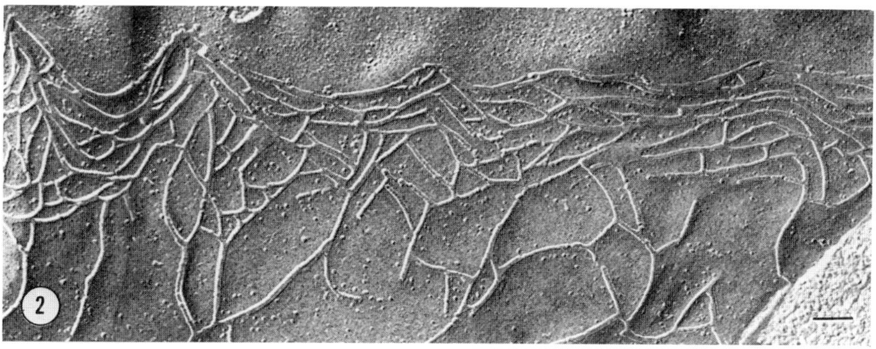

Fig. 2. Freeze-fractured zonula occludens of the rat stomach epithelium appears as a continuous, belt-like network of interconnected, intra-membrane fibrils. Bar: 0.1 µm

Adherens Junctions

The macula adherens (desmosome) is considered to be the prototype for adherens junctions (1,2). Desmosomal junctional membranes are separated by a 25-35 nm interspace containing electron-dense material (Fig. 3). A central dense stratum existing in a plane which is equidistant from the junctional membranes, may be present in the interspace separating the cells. Dense fibrillar plaques subjacent to the junctional membranes within the cytoplasm serve as attachment sites for tonofilaments and possibly other components of the cell cytoskeleton (8). When isolated desmosomal cores consisting of plasma membrane and intercellular components are displayed on polyacrylamide gels under reducing conditions, they have the following composition: an intercellular glycoprotein of $22\text{x}10^3$ M_r, two cytoplasmic plaque proteins of $82\text{x}10^3$ and $86\text{x}10^3$ M_r, an intercellular glycoprotein of $100\text{x}10^3$ M_r, an intercellular glycoprotein doublet of $115\text{x}10^3$ M_r, an intercellular glycoprotein triplet of $150\text{x}10^3$ M_r, and two cytoplasmic plaque proteins of $205\text{x}10^3$ (desmoplakin II) and $230\text{x}10^3$ M_r (desmoplakin I)(9,10). The name "desmogloeins" was assigned to the $150\text{x}10^3$, $115\text{x}10^3$, and $100\text{x}10^3$ M_r glycoproteins to indicate their location in the intercellular space of desmosomes and their involvement in cell-to-cell adhesion (11).

Hemidesmosomes are cell junctions that resemble one-half of a desmosome. They are located at the epithelial-stromal front in normal tissues where they serve as attachment sites of the epithelium to the connective tissue sustratum (basement membrane)(1-4).

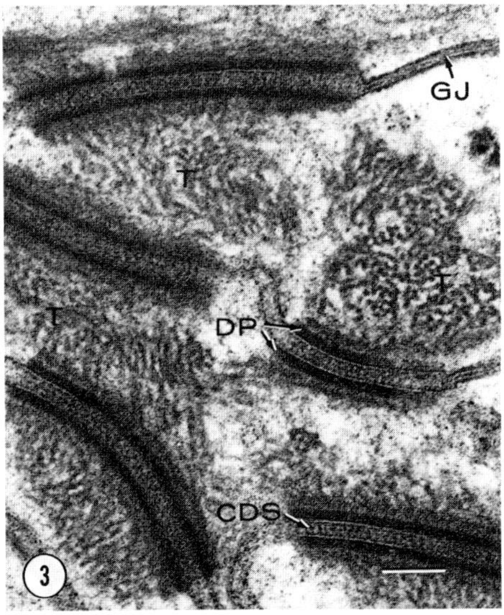

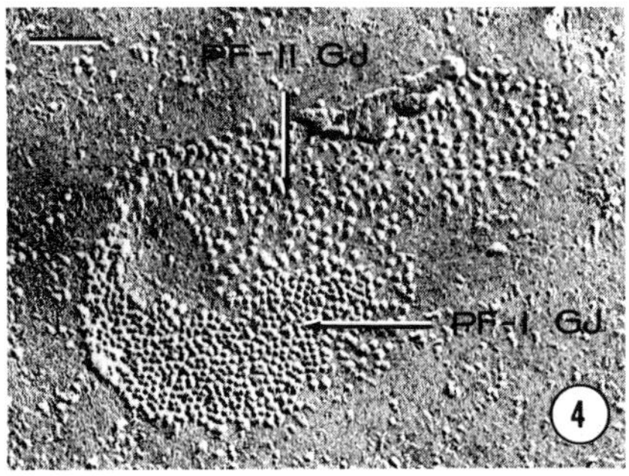

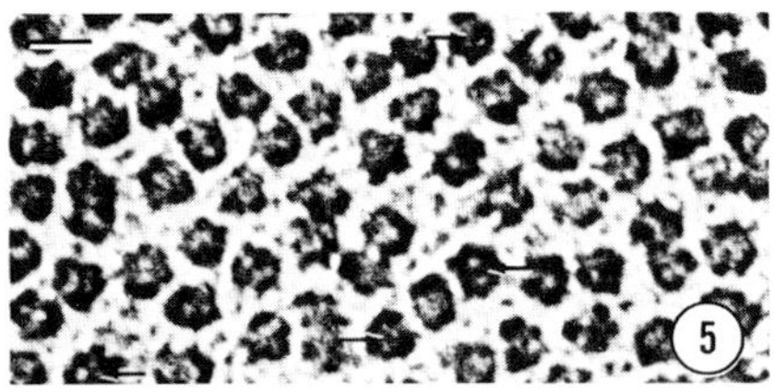

Fig. 3. Desmosomes in rat urinary bladder tumor: DP = cytoplasmic dense plaques; CDS = central dense stratum; T = tonofilaments; GJ = adjacent gap junction. Bar: 0.1 μm

Fig. 4. Freeze-fractured type-I and type-II gap junctions in normal rat urothelium: The type-I gap junction (PF-I GJ) consists of hexagonal array of connexons. The type-II gap junction (PF-II GJ) consists of larger connexons with a center-to-center spacing of 20 nm. Bar: 0.1 μm

Fig. 5. Type-I gap junction connexons of normal rat urothelium consist of 6 subunits that appear in open (arrow) or closed configuration. Bar: 0.01 μm

Gap Junctions

Early electron microscopists believed that tight and gap junctions were identical, until Revel and Karnovsky (12) firmly established that gap junctions are a separate class of intercellular junctions. In the Unwin-Zamphigi model (13), the gap junction is pictured as a bipartite array of 7-nm subunits, called connexons, which recieves equal structural contributions from each partner of the junction forming cell pair (Fig. 4). The connexons are protein oligomers that span the lipid bilayer of the junctional membranes. As seen in freeze-fracture replicas, connexons are often organized in densely packed hexagonal arrays, which are called type-I gap junctions (1-4). Type-I connexons have been described as cylinders composed of 6 subunits which surround a narrow central channel through the junctional membranes. Mapping of connexons from type-I gap junctions by high resolution electron microscopic techniques has shown that the subunits, which tilt around the connexon axis, can exist in 2 configurations: an "open" and a "closed" configuration (13)(Fig. 5). Other types of gap junctions are type-II and type-III gap junctions. Type-II gap junctions are found in close proximity of type-I gap junctions and are distinguished from the latter by the greater size (10-11 nm) and greater spacing (20 nm) of their connexons. Type-III gap junction connexons are arranged in very small rectilinear arrays with a spacing of only 6-8 nm (14).

JUNCTIONAL FUNCTION IN NORMAL TISSUES

The multiplicity of functions mediated at the cell surface and the large number of candidate components that may provide the structural basis of these functions greatly complicates the task of assigning specific functions to specific membrane components. In all likelihood, many individual surface-mediated functions are the responsibility of more than one surface component. In this section, attention will be focused on the three functions which are commonly acknowledged to be the major responsibility of cell junctions: intercellular adhesion, cell-to-cell communication, and control of transepithelial permeability (1-4).

Intercellular Adhesion

Historically, desmosomes were the first class of intercellular junctions to be associated with strong adhesion, therefore, they were designated adherens junctions (6). It is now established that all types of cell junctions provide sites of strong intercellular adhesion. However, the absolute contribution of each of the various classes of cell junctions to adhesion are difficult to measure, especially since several types of junctions are present at the surface of most kinds of epithelial cells (1-4).

Intercellular Communication

The direct exchange of ions and small molecules from cell to cell without leakage into the intercellular space is mediated primarly by gap junctions (1-4,15). Cell-to-cell transfer of substances can be measured as a function of transmembrane electrical resistance or the efficiency of molecule exchange from one cell to another. While it is established that various substances (up to 800 Daltons) can be exchanged at gap junctions under experimental conditions, the physiologic function of these junctions in many organs is a mystery. In the case of excitable tissues such as nerve, electrical coupling is known to be mediated by ion flux through gap junctions. However, the physiologic function of gap junctions in non-exitable tissues such as epithelia is simply not known. Their general importance is indicated by their presence in multicellular systems throughout the animal kingdom, and the fact that they provide the largest hydrophilic channels known to traverse biological membranes. By providing conduits for the transfer of molecules (e.g., potential morphogens, peptide hormones and growth factors) from cell to cell, gap junctions may participate in the regulation of such activities as tissue metabolism, growth, and differentiation.

The Control of Transepithelial Permeability

Zonulae occludentes are the major contributors to the bypass diffusion barrier in many epithelia and endothelia. The tightness of this barrier frequently can be related to junctional morphology (16). Very tight junctions have many points of membrane union in thin sections and numerous intramembrane fibrils in freeze-fracture replicas (Fig.2), whereas relatively leaky junctions have few points of membrane union and few intramembrane fibrils, although there may be exceptions to this rule (17). In their role as a barrier to bypass diffusion, occludens junctions are responsible for maintaining osmotic and electrochemical gradients across epithelia and separating tissue compartments so that they can maintain their compositional integrity and individuality (e.g., the blood-brain barrier, the blood-gas barrier in the lung).

CELL JUNCTIONS IN TUMORS

Description of normal and pathologic intercellular junctions in tumors are numerous, and not without controversy in matters of junction classification. In general, these data show that the tissue-specific classes of cell junctions are maintained in many types of epithelial and mesenchymal tumors, although the occurrence of these junctions often is less frequent than in normal tissues (5). The type of cell junction most frequently reported in solid tumors is the desmosome. However, this overemphasis on desmosomes may be related more to the relative ease with which these junctions are visualized in the electron microscope than to their relative abundance or importance. Systematic quantitative studies on the occurrence and distribution of intercellular junctions have been performed for two tumor systems: squamous cell carcinomas of the human cervix (18-20) and transitional cell carcinomas arising in human urinary bladder (20-22) and after induction with the carcinogen N-[4-(5-nitro-2-furyl)-2-thiazolyl]formamide (FANFT) in rat urinary bladders (21,23-28).

QUANTITATIVE STUDIES OF CELL JUNCTIONS IN TUMORS

Occludens Junctions

Occludens junctions have been examined in detail in human and rat urinary bladders (20,21,23,25,27,28). As is the case in normal human bladder epithelium, zonulae occludentes of low-grade papillary transitional cell carcinomas are confined to the apical regions of the lateral surfaces of superfical tumor cells. In non-neoplastic epithelium, the occludens junction contains 4 to 5 intramembrane fibrils. In low-grade, non-invasive bladder tumors, the width of occludens junctions may be focally attenuated to 1 or 2 parallel fibrils (Fig. 6). In the intervening areas, occludens junctions either appear normal or may be strikingly hyperplastic, consisting of many anastomasing fibrils which may occupy large areas of the cell surface (Fig. 7). In higher-grade, invasive carcinomas of the urinary bladder, occludens junctions are discontinuous, thus forming maculae or fasciae occludentes. In such tumors, they are found on all surfaces of superficial tumor cells and, occasionally, at the surfaces of cells deep within tumors. The presence of occludens junctions in such aberrant locations may be a manifestation of (1) loss of cell polarity, (2) retrograde migration of tumor cells, (3) inappropriate differentiation of tumor cell plasma membranes, and

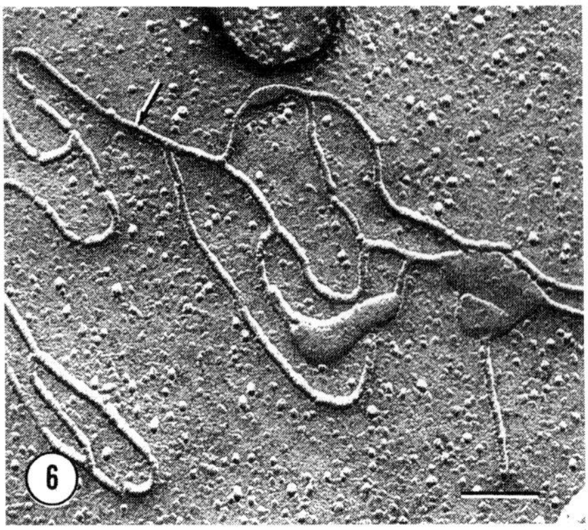

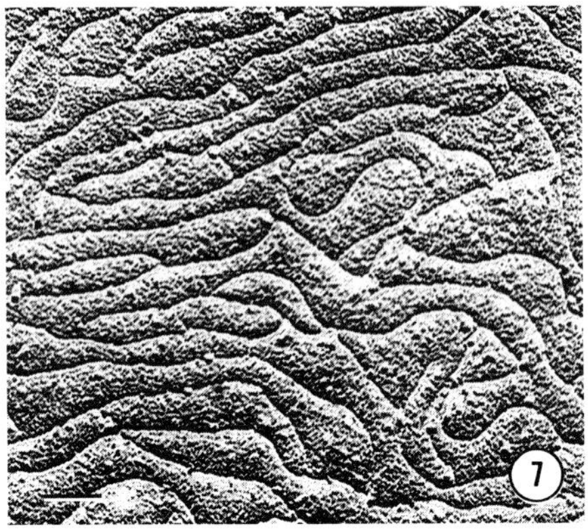

Fig. 6. Occludens junction of the PF membrane in a rat bladder carcinoma cell is focally attenuated to one intramembrane fibril (arrow). Bar: 0.1 µm
Fig. 7. Occludens junction of the EF membrane in a rat bladder carcinoma cell is focally expanded, covering a large area of the plasma membrane. Bar: 0.1 µm

(4) abnormal microenvironment within the tumor (3-5). A similar set of occludens junctional alterations have been described in FANFT-induced rat bladder tumors (21,23,25,27).

It is not known whether abnormalities in occludens junctions in carcinomas are genetically controlled or are a reflection of alterations in the tumor cell microenvironment (e.g., proteolytic enzymes). To test the hypothesis that proteolytic enzymes may be causually related to occludens junctional abnormalities, we have compared the ultrastructure of zonulae occludentes in rat bladder carcinoma cell lines with high surface protease activity and low surface protease activity (27). Freeze-fracture data on their occludens junctions reveal differences in configuration and distribution of intramembrane fibrils between these cell lines. The occludens junctions of cells with low surface protease activity measure 281 (±115) nm in width and have an average of 3.95 (±1.17) intramembrane fibrils. These values are similar to those observed for normal rat bladder epithelium in vivo. In cells with high surface protease activities, the width of occludens junctions is increased to 353 (±224) nm and the number of intramembrane fibrils is increased to 4.60 (±4.16). The average length of intramembrane fibrils is measured as a function of the intramembrane fibril length per micron occludens junction length. These values are found to be statistically similar in the two cell lines: 6.83 (±2.79) µm and 7.84 (±4.97) µm, respectively. Thus, carcinoma cells with high surface protease activity appear to be capable of synthesizing near normal quantities of intramembrane fibrils, but are unable to assemble morphologically normal zonulae occludentes. The alterations in zonulae occludentes which have been induced experimentally by exogenous proteolytic enzymes (29) closely resemble those observed in our tumor cell line with high surface protease activity (27).

Adherens Junctions

Numbers of desmosomes have been estimated for many types of tumors and have been precisely quantitated for two solid tumors, human cervical und urinary bladder carcinomas (5,19,22). In invasive cervical carcinoma, there is a decrease in the size and the total number of desmosomes per cell relative to normal cervical epithelium, yet no significant differences are observed in the number of desmosomes per unit cell membrane length (19). Quantitative electron microscopy studies of desmosomes in human transitional cell carcinoma have provided some circumstantial evidence for an association between abnormalities of desmosomes and malignant behavior (22). Relatively few desmosomes are present in normal urinary bladder epithelium. The mean desmosomal density in normal bladder epithelium is 1.95±0.17 per 100 µm of cell perimeter. The mean desmosomal density in non-invasive transitional cell carcinomas is increased to 3.02±0.31 per 100 µm cell perimeter and in invasive transitional cell carcinomas is decreased to 0.76±0.15 per 100 µm of cell perimeter. These differences exist in human transitional cell carcinomas of similar histopathological grade, but at different stages of tumor progression (stage 0 vs. stage >B). Despite this apparent association between changes in desmosomal numerical density and aggressiveness of human transitional cell carcinomas, it remains to be shown that loss of desmosomes per se is mechanistically related to tumor invasiveness. Arguing against such a cause and effect relationship is data from a small number of cases, showing that the number of desmosomes in invasive human transitional cell carcinomas containing foci of glandular and squamous differentiation is actually greater than the control value (22). Further, numerical densities of cell junctions in FANFT-induced rat urinary bladder tumors do not correlate with invasiveness (24), nor are consistent differences in cell junctions observed in tissue culture lines derived from non-invasive and invasive rat bladder tumors (25). However, this may not be incompatible with the findings in human bladder carcinomas, since the morphology of the rat carcinomas shows evidence of progressive squamous cell differentiation which can influence desmosomal ultrastructure and prevalence (23). The vast majority of human bladder carcinomas are of the transitional cell variety throughout their development.

Loss of hemidesmosomes has been observed in both human basal cell carcinoma (30) and in FANFT-induced rat bladder tumors (23). It has been associated with extracellular proteolytic activities of invasive tumors.

Gap Junctions

The relationship of gap junction deficiences to tumor behavior per se has been examined, using biopsy specimens of human uterine cervix with various preinvasive and invasive lesions (18,20). Cervical squamous epithelium is particularly well suited for quantitative electron microscopy studies because gap junctions are usually abundant in this tissue. The data show that gap junctions are slightly deficient in metaplasia and mild or moderate dysplasia of the cervical epithelium, but are markedly deficient in carcinoma *in situ* and invasive squamous cell carcinomas. This observation led to the conclusion that gap junction deficiencies are not causually related to tumor invasion in the cervix, since progression from a non-invasive (carcinoma *in situ*) to an invasive lesion (invasive squamous cell carcinoma) did not result in a further loss of gap junctions.

Interesting differences in gap junction ultrastructure have been observed between normal rat urothelium and FANFT-induced rat bladder tumors (26). FANFT-induced tumors contain ultrastructurally normal type-I gap junctions (small subunit), but are devoid of type-II gap junctions (large subunit). Depletion of type-II gap junctions which display larger connexons with larger ionic channels than those of type-I gap junctions, may alter metabolic coupling between carcinoma cells.

ABNORMALITIES OF JUNCTION FUNCTION IN TUMORS

Loss of Intercellular Adhesion

A relatively low strength of adhesion between tumor cells may be mechanistically related to tumor invasion and metastasis and may explain certain patterns of tumor growth. Reduced adhesion between tumor cells would facilitate their separation from each other and allow tumor cells with increased motility to invade normal tissues. This concept gained early support from the work of Coman (31), who measured the force that was needed to pull apart a pair of normal or neoplastic cells. He found that cell-to-cell adhesion is weak in tumors and that reduction in adhesion arose at an early stage of malignant transformation. Crude approximations of the strength of adhesion can also be deduced from ultrastructural data, if the assumption is made that strong adhesion occurs exclusively at intercellular junctions (18,20,22,24). For example, basal cells in human uterine cervical epithelium and in rat bladder epithelium have fewer and less well differentiated intercellular junctions than intermediate and superficial cells, suggesting that intercellular adhesion may increase with maturation in these tissues. Further, a junctional complex (e.g., terminal bar), consisting of a zonula occludens, a zonula adherens, and several maculae adherentes (desmosomes) in juxtaposition, forms near the the apex of superficial cells in many normal epithelia and provides a membrane area of very strong intercellular adhesion (1-4)(Fig. 1). These junctional complexes are frequently attenuated, broken up, or completely absent in malignancies. Although it seems likely that decreased intercellular adhesion, particularly within incompletely differentiated superficial cell layers of tumors, contributes to enhanced cell shedding from tumor surfaces, it is less likely that such decreases in adhesiveness are necessary for tumor invasion. Recent quantitative electron microscopic studies in our laboratory demonstrate that adherens junctions are plentiful in nests of invading tumor cells and in metastatic tumors (24,28). However, it is noteworthy that the number and distribution of intercellular junctions at the surfaces of tumor cells may be important determinants of patterns of tumor growth.

Altered Intercellular Communication Between Tumor Cells

Loewenstein and collaborators provided the first evidence of a defect in low-resistance coupling in malignant tumors and proposed that a genetically determined interruption of junctional communication may be one of many causes of cancerous growth (15,32). The same investigators also studied metabolic coupling in malignant cells and found a correlation between defective cell-to-cell transfer of $[^3H]$hypoxanthine-derived substances, and a lack of electric coupling and the capacity to transfer dye. They loaded three cancer cell lines, two derived from Morris 5123 hepatoma and one from an X-irradiated embryonic hamster cell, with [^{3}H]-hypoxanthine, and showed the absence of transfer of the labeled nucleotide to heterotypic IPP^- mutant cells in coculture. These tumor lines also lack junctions that are permeable to either inorganic ions or fluorescein (15,33).

Azarnia and Loewenstein demonstrated a genetic correlation between the simultaneous occurrence of gap junctions, ionic coupling and contact (density-dependent) inhibition of growth (15,34,35). They found evidence that the reinstatement of contact inhibition of growth by cell hybridization is accompanied by a parallel correction of gap junction changes. Contact-inhibited human Lesch-Nyhhan fibroblasts have gap junctions, are electrically coupled, and fluorescein freely diffuses from one cell to another. Non-contact-inhibited mouse Cl-1D cells have a coupling frequency of zero, will not exchange fluorescein with neighboring Cl-1D cells, and lack gap junctions. Hybrids from Lesch-Nyhan and Cl-1D cells contain a nearly complete set of chromosomes, are contact inhibited, have gap junctions, and are electrically and metabolically coupled. This experiment shows that the capacity to form cell association, necessary for coupling, is expressed in hybrid membranes. Azarnia and Loewenstein (34) also showed that as hybrid cells lose human chromosomes with serial passage, clones appear among the segregants which have reverted to the non-coupling and junction-deficient trait of the mouse parent cells. The authors concluded that the human cells may contribute a genetic factor to the hybrids that corrects the junctional deficiency of the mouse cells. The factor could be a junctional component or, alternatively, a component of the general plasma membrane which may be essential for cell-cell recognition or junction assembly (15,34).

Altered Tranepithelial Permeability

Attenuation and disruption of occludens junctions and a concommitant increase in transepithelial permeability is common in carcinomas (5,21). Although the full extent of occludens junctional abnormalities can only be appreciated in anaplastic tumors, attenuations and leakiness of occludens junctions are present at an early stage in tumor progression. This may increase the exposure of basal cell layers to water soluble carcinogens and by this mechanism exert a co-carcinogenic effect.

CONCLUSIONS

Junctional alterations and deficiencies are common in tumors (5). In general, the degree of junctional abnormalities correlates well with the level of tumor anaplasia. In spite of numerous reports to the contrary, there is neither concrete nor compelling circumstantial evidence which supports the popular notion that junctional defects contribute to those properties which are the hallmarks of malignant growth, namely invasiveness and metastasis. Junctional alterations are probably expressions of a transformed genome since cell junctions are thought to contain structural components that are gene products, and since their formation is genetically controlled. Additional alterations in cell junctions may be caused by changes in the microenvironment, and these changes may be determined by metabolic disturbances in the neoplastic tissues, by host reactions against the cancer and its metabolites, or by other extrinsic factors.

REFERENCES

1. McNutt, N.S. and Weinstein, R.S. Progr. Biophys. Molec. Biol. 26:45-101, 1973
2. Staehelin, L.A. Int. Rev. Cytol. 39:191-283, 1974
3. Weinstein, R.S. and Pauli, B.U. In: Advances in Clinical Cytology (Eds. L.G. Koss and D.V. Coleman), Butterworths, London, 1981, pp. 160-200
4. Weinstein, R.S., Alroy, J. and Pauli, B.U. In: Cellular Pathobiology of Human Disease (Eds. B.F. Trump, A. Laufer, and R.T. Jones), Gustav Fischer, New york, 1983, pp. 49-72
5. Weinstein, R.S., Merk, F.B. and Alroy, J. Adv. Cancer Res. 23:23-89,1976
6. Farquar, M.G., Simionescu, M. and Palade, G.E. J. Cell Biol 17:375-412, 1963
7. Simionescu, N., Simionescu, M. and Palade, G.E. J. Cell Biol. 79:27-44, 1978
8. Kelley, D.E. J. Cell Biol. 28:51-72, 1966
9. Mueller, H. and Franke, W.W. J. molec. Biol. 163:647-671, 1983
10. Cowin, P., Mattey, D. and Garrod, D.J. J. Cell Sci. 70:41-60, 1984
11. Gorbsky, G. and Steinberg, M.S. J. Cell Biol. 90:243-248, 1981
12. Revel, J.P. and Karnovsky, M.J. J. Cell Biol. 33:C7-C12, 1967

13. Unwin, P.N.T. and Zampighi, Q. Nature 283:545-549, 1980
14. Staehelin, L.A. Proc. Nat. Acad. Sci. USA 69:1318-1321, 1972
15. Loewenstein, W.R. Physiol. Rev. 61:829-913, 1981
16. Claude, P. and Goodenough, D. J. Cell Biol. 58:390-400, 1973
17. Martinez-Palomo, A. and Erlig, D. Proc. Nat. Acad. Sci. USA 72:4487-4491, 1975
18. McNutt, N.S., Hershberg, R.A. and Weinstein, R.S. J. Cell Biol. 51:805-825, 1971
19. Wiernik, G., Bradbury, S., Plant, M., Cowdell, R.H., Williams, E.A. Brit. J. Cancer 28:488-499, 1973
20. Weinstein, R.S., Zel, C. and Merk, F.B. In: Membrane Transformation in Neoplasia (Eds. J. Schultz and R.E. Block), Academic Press, New York, 1974, pp. 127-146
21. Merk, F.B., Pauli, B.U., Jacobs, J.B., Alroy, J., Friedell, G.H. and Weinstein, R.S. Cancer Res. 37:2843-2853, 1977
22. Alroy, J., Pauli, B.U. and Weinstein, R.S. Cancer 47:104-112, 1981
23. Pauli, B.U., Weinstein, R.S., Alroy, J. and Arai, M. Lab. Invest. 37:609-621, 1977
24. Pauli, B.U., Cohen, S.M., Alroy, J. and Weinstein, R.S. Cancer Res. 38:3275-3285, 1978
25. Pauli, B.U., Kuettner, K.E. and Weinstein, R.S. J. Microscopy 115:271-282, 1979
26. Pauli, B.U. and Weinstein, R.S. Experientia 37:248-250, 1981
27. Pauli, B.U. and Weinstein, R.S. Cancer Res. 42:2289-2297, 1982
28. Pauli, B.U., Alroy, J. and Weinstein, R.S. In: Pathology of Bladder Cancer (Eds. G.T. Bryan and S.M. Cohen), CRC Press, Boca Raton, 1983, pp. 41-140
29. Orci, L., Amherdt, M., Henquin, J.C., Lambert, A.E. and Unger, R.H. Science 180:647-649, 1973
30. McNutt, N.S. Lab. Invest. 35:132-142, 1976
31. Coman, D.R. Cancer Res. 4:625-629, 1944
32. Loewenstein, W.R. and Kanno, Y. J. Cell Biol. 33;225-234, 1967
33. Azarnia, R., Michalke, W. and Loewenstein, W.R. J. Membrane Biol. 10:247-258, 1972
34. Azarnia, R., Larsen, W.J. and Lowenstein, W.R. Proc. Nat. Acad. Sci. USA 71:880-884, 1974
35. Azarnia, R. and Loewenstein, W.R. J. Membrane Biol. 34:1-28, 1977

13

PATTERN OF BASEMENT MEMBRANE DEGRADATION BY METASTATIC TUMOR CELL ENZYMES

K. TRYGGVASON

Department of Biochemistry and Department of Pathology, University of Medicine and Dentistry of New Jersey-Rutgers Medical School, Piscataway, New Jersey 08854

INTRODUCTION

The formation of tumor metastasis occurs through a complex multistep process whereby malignant cells detach from the primary tumor, invade the adjacent tissue, penetrate into the circulatory system, traverse the vessel wall at a distant site, invade the tissue and finally proliferate to initiate the metastatic focus (for reviews, see 1,2,3). Metastasizing tumor cells must penetrate basement membranes (BM) and interstitial connective tissue at several locations during the dissemination process. These extracellular matrices are composed of a variety of specific structural components which have to be removed to allow the passage of invading tumor cells and proliferation of malignant cells.

BMs are ubiquitous, acellular, approximately 50-100 nm thick sheath-like structures which separate epithelial, endothelial and mesenchymal cells from the underlying interstitial connective tissue (4). BMs are of particular interest with respect to metastasis formation since they usually are the first extracellular barrier to be penetrated by disseminating tumor cells (Fig. 1) and they have to be penetrated at least twice more before the cells reach the site of metastatic focus. During tumor invasion the BM matrix must be degraded since it does not contain pores large enough for cells to traverse. A considerable body of data, both histological and biochemical, have indicated that proteolytic enzymes play a significant role in this process. Electron microscopic examinations have demonstrated that the transition from an in situ to invasive carcinoma is characterized by the dissolution of BMs and migration of tumor cells into the underlying interstitial connective tissue stroma (5,6). Many malignant tissues as well as cultured tumor cells contain elevated levels of degradative enzymes as compared with nonmalignant

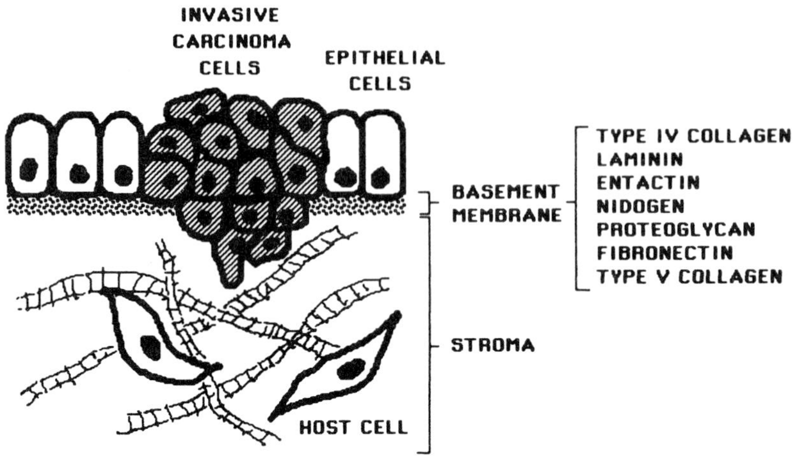

Fig. 1. Invasion of basement membrane by carcinoma cells. During migration of the tumor cells through the subepithelial BM into the stroma the highly crosslinked multicomponent BM matrix must be disintegrated since it does not contain pores large enough for the tumor cells to traverse. The matrix disolution is thought to be accomplished by degradative enzymes primarily secreted by the invading tumor cells.

counterparts, many of these enzymes being capable of degrading matrix molecules (2,3). It appears, therefore, that the breakdown of the BM matrix that takes place during tumor invasion can be accomplished by tumor cell enzymes. Investigation of the enzyme proteins-how they work, why their production is increased in the malignant phenotype and whether this production can be controlled-is one of the major tasks of current metastasis research. Recent advances in the characterization of connective tissue components and their assembly have enabled studies on the biochemical mechanisms of matrix destruction by tumor cells at the molecular level. In this paper current data on basement membrane degradation by malignant cells is reviewed.

STRUCTURE OF BASEMENT MEMBRANES

The molecular composition of the complex BM matrix has been elucidated considerably in recent years and a number of component proteins have been isolated and characterized from BMs. The BM specific type IV collagen is considered to be the major structural component (4). It is a 400 nm long triple-helical molecule with a globular C-terminal end. The molecule is composed of three chains (Fig. 2), presumably two α1

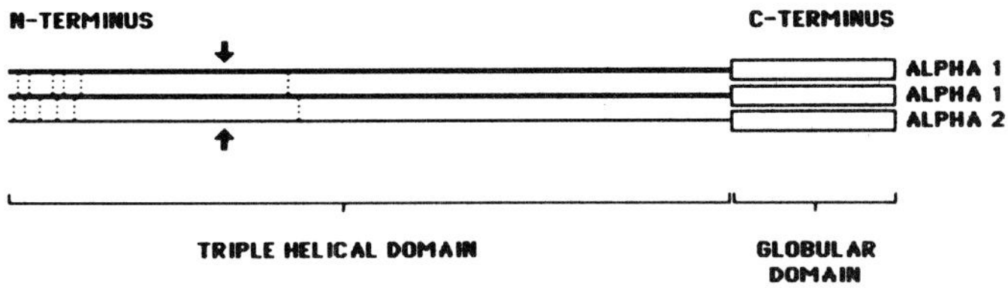

Fig. 2. Schematic picture of the type IV collagen molecule. In the helical domain the three chains are coiled around each other in a triple helical conformation. The chains are bound to each other by disulfide (dashed lines) and hydrogen bonds. The type IV collagenase (see text) cleaves the molecule at a single site at about $\frac{1}{4}$ of the distance from the N-terminal end (arrow).

chains (185 kd) and an α2 chain (170 kd) (7,8). It has been proposed that in the extracellular space four such molecules become linked to each other at the N-termini to form tetramers (see Fig. 3) which are further assembled into a tight network structure (19). Type IV collagen is not degraded by the "classical" mammalian collagenases that specifically cleave fibrillar collagen types (10,11) but it is sensitive to "nonspecific" proteinases such as pepsin, trypsin and elastase (12,13,14). However, plasmin and cathepsin B do not degrade type IV collagen (15). A specific type IV collagen degrading enzyme (type IV collagenase) has been identified in tumor cells (16,17). Laminin is another specific basement membrane component. It is a large glycoprotein of about 900,000 daltons composed of three subunit polypeptides (18,19). Laminin which is considered important for the cell attachment apparently binds to the cell surface through a specific binding protein (20) on one hand and to other BM components on the other. Two different BM proteoglycans, a heparan sulfate and chondroitin sulfate proteoglycan, have been isolated (21,22). These components probably contribute to the filtration of macromolecules through the membrane (23). Entactin (24) and Nidogen (25) are more recently identified components. Fibronectin, present in plasma and interstitial connective tissue, has been visualized in BMs (26) but it is still disputed whether it contributes to its structure or whether it is trapped in the structure from the circulation. Type V collagen has been localized to the BM region but it appears to be located at the interphase between BM and stroma (27). The noncollagenous BM components can be degraded by several proteinases that have broad substrate specificity.

The above described proteins make up the major BM constituents but the matrix probably contains many other minor components (28). The supra-molecular assembly of the BM proteins is poorly understood, but most of the proteins have strong non-covalent interactions stabilized by covalent crosslinks. There seems to be a consensus that the type IV collagen network provides the structural skeleton into which the other proteins are bound in an unknown fashion. With respect to the process of tumor invasion, it is clear that this tough matrix cannot be traversed by tumor or any other cells without its mechanical or chemical disintegration.

ENZYMATIC DEGRADATION OF BASEMENT MEMBRANES BY TUMOR CELLS

Several reports have indicated that the dissolution of the BM at the site of tumor cell penetration occurs through the action of degradative enzymes. An important question is whether the tumor cells themselves produce the enzymes needed or whether they are possibly secreted by host tissue cells such as macrophages or fibroblasts. There is evidence indicating that the tumor cells produce the necessary enzymes since increased levels of degradative enzymes have been reported in many tumor tissues and cultured tumor cells. It has been suggested that these tumor enzymes when diffusing in large enough quantities into the surrounding tissue can be responsible for the disruption of BM and other extracellular matrix structures. Several enzymes which can degrade both the noncollagenous and collagenous BM components have been identified in tumor cells.

Plasminogen activator secretion is increased in most transformed cells and malignant tissues (29,30). A role for the enzyme in the invasion process can be anticipated by the fact that it converts plasminogen to plasmin, since, in addition to being fibrinolytic, plasmin can degrade both laminin and fibronectin (31,32). It does not, however, cleave type IV or V collagens (15). Furthermore, we have demonstrated that plasminogen activator can activate the highly tumor associated type IV collagenase through the activation of plasminogen to plasmin (33). The activity of plasminogen activator does not, however, always correlate well with the metastatic potential of cells (see e.g. 34). The enzyme is also produced by a variety of non-malignant cells and tissues (32). Although the plasminogen activator - plasmin action does not appear to be sufficient for the dissolution of the collagenous skeleton of the basement membrane matrix it is likely to contribute significantly to the process by degrading the noncollagenous components.

Sloane et al., (35) demonstrated a significant correlation with the metastatic potential of *in vivo* grown variants of the B16 murine melanoma and cathepsin B activity. The differential in cathepsin B activity was lost as the cells were sub-cultured. This enzyme, which is active at neutral pH, can degrade laminin, fibronectin and proteoglycans and it can also activate latent collagenases (15,36). Thus, it appears to be able to carry out a similar degradative pattern as the plasminogen activator - plasmin system. However, cathepsin B is an intracellular lysosomal enzyme, and it is not clear whether it functions extracellularly *in vivo* although it can degrade matrix components in *in vitro* experiments. Little evidence has been observed for a positive correlation with metastatic potential and the activity of other lysosomal enzymes such as cathepsin D, β-N-acetylglucosaminidase and β-glucuronidase (35).

In addition to the ability to produce proteinases tumor cells secrete enzymes that degrade the glycosaminoglycan chains of proteoglycans. Nicolson and his collaborators demonstrated that metastatic tumor cells when placed on subendothelial matrix solubilize glycosaminoglycans by the release of an endoglycosidase (3,37). They have also found that the rate of release of heparan sulfate fragments from the matrix is higher in tumor cells with high metastatic potential than in cells of lower invasive potential. This activity was shown to be a heparan sulfate specific endoglycosidase, heparanase. The investigators suggested that this heparanase, which is also secreted by fibroblasts, rat liver cells, human platelets placenta and murine mastocytoma cells, may be important for the remodeling of the BM and its distruction during tumor invasion.

The above described enzymes often associated with the malignant phenotype can bring about the degradation of the non-collagenous components of BM. They have not, however, been shown to degrade type IV collagen which forms the BM structural framework or type V collagen. Proteinases with activity against these collagens do exist though, and they have been identified in tumor cells.

Liotta et al., reported that tissue culture media of a highly metastatic mouse sarcome (PMT) contained type IV collagenolytic activity (16). We have later shown this degradation to be caused by a specific metalloproteinase which we refer to as type IV collagenase because of its specificity for type IV collagen (17,38,39). The secretion of type IV collagenolytic activity has been demonstrated in tissue and cell culture media of numerous malignant tumors. We showed (40) that the enzyme activity in culture media of cells correlated with the metastatic potential of variants of the B16 melanoma (F1, F10, B16, see ref. 1) and PMT cells (40) and was significant in several other cultured tumor cells (33). In contrast, no detectable activity was detected in confluent cultures of fibroblasts, epithelial and endothelial cells (33,40). Starkey et al., (41) detected type IV collagenolytic activity in several tumor cell lines but did not confirm the correlation of enzyme activity with metastatic ability of the F1 and F10 melanoma variants. The reason for the discrepance in the results of the amount of type IV collagenase activity is not clear. We have observed however, that when PMT tumor tissue cells are explanted and sub-cultured *in vitro* their secretion of type IV collagenolytic enzyme decreases gradually with time after several passages (unpublished results). This indicates that the phenotypic changes resulting in type IV collagenase production are not stable in *in vitro* cultures and that tumor-host interactions may be necessary for maintaining the secretion of the enzyme. Accordingly, it may be important to pay attention to the passage number of subcultured tumor tissue cells when assaying type IV collagenolytic activity. Recently, Eisenbach et al., (42) measured type IV collagenolytic activity in conditioned media from tumor cells and tissues of both metastatic and non-metastatic clones of the Lewis lung carcinoma and T10 sarcoma grown in C57Bl/6 mice. The activity was higher in media from the tumor tissue than cultured cells whereas a positive correlation with metastatic potential was not established. However, the tumors were all metastic in immune-deficient host animals suggesting that host-immune reactions can play a role in controlling the dissemination of malignant cells, even though they are capable of producing the enzymes required for BM degradation.

We have isolated the type IV collagen specific type IV collagenase from tumor cells (17,38). We believe that this enzyme is primarily responsible for the type IV collagenolytic activity found in tumor cells described above. A different enzyme that cleaves the BM associated type V collagen has been identified in ovary adenocarcinoma cells (43). These collagenolytic enzymes can provide invading tumor cells with the tools required for the disintegration of the type IV collagen network and the BM associated type V collagen.

TYPE IV COLLAGENASE

The secretion of a metastatic tumor cell associated type IV collagen specific proteinase is of particular interest since it is specific for the major structural component of a significant tumor barrier. In an attempt to learn more about the properties and regulation of this enzyme we have purified and characterized it from two main sources. The enzyme was purified from culture media of metastatic mouse PMT sarcoma (17,38) and partially from a human fibrosarcoma cell line (44). The enzyme which does not degrade fibrillar collagen types, type V collagen, laminin or fibronectin has been referred to as type IV collagenase. The structure of the enzyme protein is not known. It has a molecular weight of about 160,000 when chromatographed on a molecular sieve without detergent but is about 70,000 dalton in the presence of 0.1% Triton X-100 (17). The purified mouse enzyme was resolved into two polypeptides of about 60,000-70,000 daltons on SDS-polyacrylamide gel electrophoresis. The results suggest that the enzyme is either composed of subunit polypeptide chains or is a hydrophobic protein that aggregates in the absence of detergent. Such a variable behavior on molecular sieves has also been observed for "classical" interstitial collagenases that degrade fibrillar collagens (45). The mouse enzyme binds strongly to concanavalin-A indicating its glycoprotein nature. Serine and sulfhydryl proteinase inhibitors such as phenylmethylsulfonylfluoride, aprotinin and N-ethylmalemide do not inhibit type IV collagenase activity. In contrast the metal chelator EDTA reversibly inhibits activity demonstrating that the enzyme is a metalloproteinase. The enzyme activity can be increased in vitro by preincubation with trypsin, plasmin and mercurial compounds. This suggests that the enzyme is produced in a latent form as the "classical" interstitial collagenases (11). The fact that plasmin can activate type IV collagenase is interesting since cells that secrete type IV collagenase activity were also shown to secrete plasminogen activator (33). It is likely that the plasminogen activator-plasmin system functions as the type IV collagenase activator in vivo. Cathepsin B can activate latent interstitial collagenases (36) but it is not known whether it can activate type IV collagenase.

Type IV collagenase cleaves the native type IV collagen molecule at a single site in the helical domain at about $\frac{1}{4}$ of the distance from the N-terminal end (Figs. 2 and 3) (39). The amino acid sequence of the cleavage site has not been determined. However, this finding opens the possibilities for designing substrate analogs that might serve as enzyme inhibitors.

The actual function or presence of type IV collagenase in normal states in vivo is not known. It possibly plays a role in the remodeling of BM during embryogenesis and in repair processes along with other enzymes such as plasminogen activator, plasmin, cathepsins and glycosidases which can degrade the noncollagenous BM constituents. A support for the association of the enzyme with proliferation of nonmalignant cells is our recent observation showing the presence of type IV collagenase in fibroblasts growing in the early log phase in vitro (46).

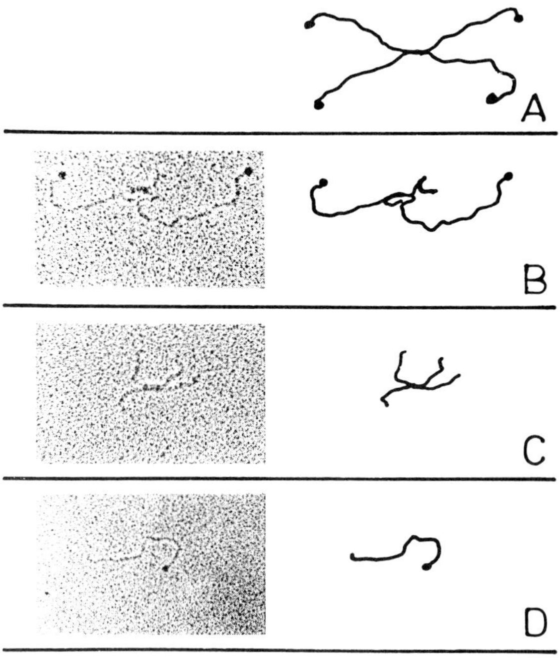

Fig. 3. Rotary shadowing electron microscopy of type IV collagen cleavage products generated by type IV collagenase. The substrate was a tetramer form where four collagen molecules (see Fig. 2) are bound to each other at the N-terminal end, the globular C-terminal ends being free. (A) Scheme of the tetramer substrate. (B) Tetramer partially cleaved by the enzyme; two C-terminal segments have been cleaved off. (C) Completely degraded tetramer, yielding fragments containing four crosslinked N-terminal fragments. (D) Free C-terminal end fragment (Modified from ref. 39).

A key question with respect to tumor invasion is how the production of type IV collagenase and other BM degrading enzymes is controlled. Is type IV collagenase secretion turned on in concomittance with the other degradative enzymes or does it have different control mechanisms? The answer is not yet at hand. We do know, however, that production of type IV collagenase can be induced in cultured fibroblasts with the tumor promoting phorbol ester TPA (46). This phorbol ester can also induce the secretion of other enzymes such as plasminogen activator (47), and interstitial collagenase (48) in cultured cells. These results suggest that at least in some cases the secretion of several proteinases can be induced through the same mechanisms. However, the secretion of these proteinase *in vivo* must also be individually controlled since the resorption needs of their substrates varies depending on the tissue, stage of development, etc.

SUMMARY

The migration of malignant tumor cells through the BM is one of several critical steps in the cascade of metastasis formation. Recent investigations have clearly shown that the tumor cells themselves often possess the proteolytic machinery needed for the dissolution of this highly crosslinked extracellular structure. Based on those studies it

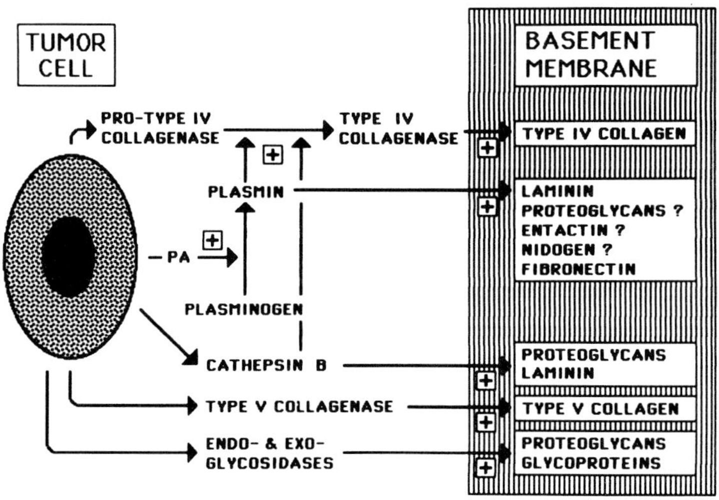

Fig. 4. Secretion of degradative enzymes by tumor cells and their possible role in BM destruction during tumor invasion. The + sign indicates degradative enzyme effect on the respective substrate. Plasminogen can diffuse into the pericellular space from plasma or connective tissue. The degradation process is affected *in vivo* by serum and tissue derived inhibitors.

is probable that enzymes such as type IV collagenase, type V collagenolytic proteinase, plasminogen activator, plasminogen, plasmin, cathepsin B and glycosidases such as heparanase are primarily involved in the disintegration of BM during tumor invasion. The current picture of the pattern of these enzymes in BM degradation is schematically presented in Fig. 4. The picture is obviously a simplification of the process in vivo since the effects of the various degradative enzymes are presumably buffered by both tissue and plasma-derived inhibitors. It is probable that these inhibitors limit the proteolytic effects caused by the tumor cells to the microenvironment adjacent to the invading cells. Furthermore, it is possible that in some cases host tissue cells such as macrophages which are capable of secreting enzymes such as plasminogen activator and cathepsin B might also contribute to the dissolution of BMs.

It is important to realize that none of the highly tumor associated proteinases or glycosidases is specific for the malignant cell. These enzymes are all normal gene products which probably have function in vivo for the remodeling of BMs and other extracellular structures during embryogenesis, growth and regeneration processes. With regards to metastasis the question which remains to be resolved is why the production of these enzymes is turned on in malignant cells and how the individual genes are controlled.

REFERENCES

1. Poste, G. and Fidler, I.J. Nature, 283:139-146, 1980.
2. Liotta, L.A., Garbisa, S. and Tryggvason, K. In: Tumor Invasion and Metastasis (Eds. L.A. Liotta and J.R. Hart), Martinus Nijhoff Publishers, The Hague-Boston-London, 1982, pp. 319-333.
3. Nicolson, G.L., Irimura, T., Nakajima, M. and Estrada, J. In: Cancer Invasion and Metastasis: Biologic and Therapeutic Aspects (Eds. S.L. Nicolson and L. Milas), Raven Press, New York, 1984, 145-167.
4. Kefalides, N.A., Alper, R. and Clark, C. Int. Rev. Cytol., 61:167-228, 1979.
5. Babai, F. J. Ultrastr. Res., 56:287-303, 1978.
6. Frei, J.V. Histopathol. 2:107-115, 1978.
7. Tryggvason, K., Gehron Robey, P. and Martin, G.R. Biochemistry, 19:1284-1289, 1980.
8. Trüb, B., Gröbli, B., Spiess, M., Odermatt, B.F. and Winterhalter, K.H. J. Biol. Chem., 257:5239-5245, 1982.
9. Timpl, R., Wiedemann, H., van Delden, V., Furthmayr, H. and Kühn, K. Eur. J. Biochem., 120:203-211, 1981.
10. Woolley, D.E., Glanville, R.W., Roberts, D.R. and Evanson, J.M. Biochem. J., 169:265-276, 1978.
11. Werb, Z. In: Collagen in Health and Disease (Eds. J.B. Weiss and M.I.V. Jayson), Churchill LIvingstone, Edinburgh, 1983, pp. 121-134.
12. Tryggvason, K. and Kivirikko, K.I. Nephron, 21:230-235, 1978.
13. Mainardi, C.L., Dixit, S.N. and Kang, A.H. J. Biol. Chem., 255:5435-5441, 1980.
14. Tryggvason, K., Pihlajaniemi, T., Salo, T., Liotta, L.A. and Kivirikko, K.I. In: New Trends in Basement Membrane Research (Eds. K. Kühn, S. Schöne and R. Timpl), Raven Press, New York, 1982, pp. 187-192.
15. Liotta, L.A., Thorgeirsson, U.P. and Garbisa, S. Cancer Metastasis Rev., 1:277-288, 1982.
16. Liotta, L.A., Abe, S., Gehron Robey, P. and Martin, G.R. Proc. Natl. Acad. Sci. USA 76:2268-2272, 1979.
17. Salo, T., Liotta, L.A. and Tryggvason, K. J. Biol. Chem., 258:3058-3063, 1983.
18. Timpl, R., Rohde, H., Robey, P., Foidart, J.M., Rennard, S. and Martin, G.R. J. Biol. Chem., 254:9933-9937, 1979.
19. Cooper, A.R., Kurkinen, M., Taylor, A. and Hogan, B.L.M. Eur. J. Biochem., 119:189-197, 1981.
20. Terranova, V.P., Rao, C.N., Kalebic, T., Margulics, J.M. and Liotta, L.A. Proc. Natl. Acad. Sci., 80:444-448, 1983.
21. Hassell, J.R., Robey, P.G., Barrach, H.J., Wilczek, J., Rennard, S. and Martin, G.R. Proc. Natl. Acad. Sci. USA, 77:4494-4498, 1980.
22. Höök, M., Couchman, J., Woods, A., Robinson, J. and Christner, J.E. In: Basement Membranes and Cell Movement, Ciba Foundation Symposium (Eds. R. Porter and J. Whelan), Pitman, London, 1984, pp. 44-50.
23. Kanwar, Y.S., Linker, A. and Farguhar, M. J. Biol. Chem., 86:688-693, 1980.
24. Chung, A.E., Jaffe, R., Freeman, J.L., Vergnes, J.P., Braginski, J.E. and Carlin, B. Cell, 16:277-287, 1979.

25. Timpl, R., Dziadek, M., Fujimara, S., Nowack, H. and Wick, H. Eur. J. Biochem., 137:455-465, 1983.
26. Stenman, S. and Vaheri, A. J. Exp. Med., 147:1054-1064, 1978.
27. Madri, J.A. and Furthmayr, H. Amer. J. Pathol., 94:323-331, 1980.
28. Timpl, R., Fujimara, S., Dziadek, M., Aumailley, M., Weber, S. and Engel, J. In: Basement Membranes and Cell Movement. Ciba Foundation Symposium (Eds. R. Porter and J. Whelan), Pitman, London, 1984, pp. 25-37.
29. Reich, E. In: Biological Markers of Neoplasia: Basic and Applied Aspects (Ed. R.W. Ruddon) Elsevier/North-Holland, Amsterdam, 1978, pp. 491-500.
30. Goldfarb, R.H. and Quigley, J.P. Cancer Res., 38:4601-4609, 1978.
31. Liotta, L.A., Goldfarb, R.H. and Terranova, V.P. Thromb. Res. 21:663-673, 1981.
32. Balian, G., Click, E.M., Crouch, E., Davidson, J.M. and Bornstein, P. J. Biol. Chem., 254:1429-1432, 1979.
33. Salo, T., Liotta, L.A., Keski-Oja, J., Turpeenniemi-Hujanen, T. and Tryggvason, K. Int. J. Cancer, 30:699-673, 1982.
34. Nicolson, G. Biochem. Biophys. Acta, 695:113-176, 1982.
35. Sloane, B.F., Honn, K.V., Sadler, J.G., Turner, W.A., Kimpson, J.J. and Taylor, J.D. Cancer Res., 42:980-986, 1982.
36. Eckhout, Y. and Vaes, G. Biochem. J., 166:21-31, 1977.
37. Kramer, R.H., Vogel, K.G. and Nicolson, G.L. J. Biol. Chem., 257:2678-2686, 1982.
38. Liotta, L.A., Tryggvason, K., Garbisa, S., Gehron Robey, P. and Abe, S. Biochemistry, 20:100-104, 1981.
39. Fessler, L., Duncan, K., Fessler, J.H., Salo, T. and Tryggvason, K. J. Biol. Chem., 259:9783-9789, 1984.
40. Liotta, L.A., Tryggvason, K., Garbisa, S., Hart, I., Foltz, C.M. and Shafie, S. Nature, 284:67-68, 1980.
41. Starkey, J.R., Hosick, H.L., Stanford, D.R. and Liggitt, H.D. Cancer Res., 44:1585-1594, 1984.
42. Eisenbach, L., Segal, S. and Feldman, M. J. Natl. Cancer Inst., 74:87-93, 1985.
43. Liotta, L.A., Lanzer, W.L. and Garbisa, S. Biochem. Biophys. Res. Commun., 98:184-190, 1981.
44. Turpeenniemi-Hujanen, T., Salo, T. and Tryggvason, K. Submitted for publication.
45. Stricklin, G.P., Bauer, E.A., Jeffrey, J.J. and Eisen, A.Z. Biochemistry, 78:1607-1615, 1977.
46. Salo, T., Turpeenniemi-Hujanen, T. and Tryggvason, K. J. Biol. Chem., in press, 1985.
47. Wigler, M. and Weinstein, I.B. Nature, 259:232-233, 1976.
48. Werb, Z., Mainardi, C.L., Vater, C.A. and Harris, E.D. N. Engl. J. Med., 296:1017-1023, 1977.

Part Two

Immunologic Mechanisms

14

GENE PRODUCTS OF THE MAJOR HISTOCOMPATIBILITY COMPLEX CONTROL THE METASTATIC PHENOTYPE OF TUMOR CELLS

L. EISENBACH* S. KATZAV*, G. HÄMMERLING**, S. SEGAL*** AND M. FELDMAN*

*Department of Cell Biology, The Weizmann Institute of Science, Rehovot, Israel; **Institute for Immunology and Genetics, German Cancer Research Center, Im Neuenheimer Feld 280, D6900 Heidelberg, FRG; and ***Department of Microbiology, Ben Gurion University, Beer-Sheva, Israel

INTRODUCTION

To generate metastases, neoplastic cells should be capable of detaching themselves from the local growth, of invading extracellular matrices, of penetrating basement membranes of blood of extravasating while manifesting, in many cases, specific recognition of target organs, and of inducing angiogenesis essential for the growth of metastasis. Since the local tumor cell population was shown to be diverse with respect to the metastatic competence of its individual cells (1), nonmetastatic cells of a metastatic tumor may be impaired with regards to all, a few or any one of the sequential steps culminating in metastatic growth. Recent studies in our laboratories of two metastatic tumors, the 3LL Lewis Lung Carcinoma and the T10 Sarcoma, indicated that metastatic clones and nonmetastatic clones synthesize and secrete similar levels of collagenase IV (2), an enzyme which was previously implicated as part of the enzymatic weaponry for metastatic penetration of basement membranes (3). And although the synthesis of another relevant enzyme, plasminogen activator, was higher in metastatic compared to nonmetastatic clones (4), this did not seem to represent a rate-limiting factor in the initial invasion processes (4).

In fact, cells programmed to generate metastases could be identical to nonmetastatic cells of a given tumor in possessing the entire molecular machinery required to complete the metastatic cycle. In such cases where both the metastatic and the nonmetastatic cells share the cellular mechanisms for invasion and dissemination, a limiting factor for the expression of the metastatic phenotype could be the capacity of the immune system to differentiate between cell surface antigens of metastatic cells, and those of nonmetastatic cells. This need not imply that metastatic cells express different tumor-associated antigens than nonmetastatic cells. Since cytotoxic T-lymphocytes (CTL) recognize cell surface antigens in conjunction with class I antigens of the cell MHC system, metastatic cells and nonmetastatic cells could express the same tumor-associated antigens, yet may differ in the expression of genes coding for the class I molecules. In the mouse, the major relevant genes coding for antigens recognized by CTL are the H-2K and the H-2D genes of the MHC. If each of these, when associated with a tumor cell surface epitope would elicit different immune signals (i.e., immunogenic versus suppressogenic signals), then differences between disseminating tumor cells in the expression of the H-2K and h-2D genes could eliminate one subset of tumor cells, yet leave others to complete the metastatic process. We were therefore tempted to test whether the metastatic phenotype is controlled by the relative expression of the two major class I antigens of the mouse major histocompatibility system (5, 6). This notion gained further support following our observation that our first metastatic model, the 3LL Lewis lung carcinoma, originating in a C57BL ($H-2^b$) mouse, could grow across H-2 barrier; in fact, it grew progressively in mice of any strain tested. Yet, metastases appeared only in syngeneic recipients (7, 8). To test what are the minimal genetic identities between the tumor's strain of origin and the allogeneic recipients, which are required for the development of metastasis, we tested various H-2 recombinants as recipient mice. We discovered that if the recipients expressed the $H-2D^b$ determinant, and the non-$H-2^b$ background, this was sufficient for the generation of spontaneous metastases by the 3LL tumor. The $H-2K^b$ was completely irrelevant. We subsequently revealed that identity at the $H-2K^b$ was unnecessary for the generation of metastases because the 3LL tumor cells hardly expressed the $H-2K^b$ glycoprotein on their cell surface (9). This raised the question of whether the relatively low expression of the $H-2K^b$ determinants, or the low $H-2K^b/H-2D^b$ ratio, is causally related to the metastatic phenotype of the 3LL carcinoma, i.e., to the capacity of these tumor cells to generate spontaneous metastases while growing in syngeneic recipients.

THE METASTATIC PHENOTYPE OF THE 3LL TUMOR CELLS IS CORRELATED WITH THE RELATIVE EXPRESSION OF THE $H-2K^b/H-2D^b$ GENES

To answer this question we cloned in soft agar the 3LL cells and tested the metastatic potency of individual clones. We found that the clones differ with regard to their capacity to generate spontaneous lung metastases, when grown intrafootpad in syngeneic animals. As previously demonstrated for other tumors (1), this tumor cell population was diverse with regards to the metastatic potency of its individual cells. To test whether there is a correlation between the metastatic properties of individual clones and the expression of MHC genes, we used monoclonal antibodies 28-13-3 and 20-8-4, which identify $H-2K^b$ molecules, and antibody 28-14-8, which identifies $H-2D^b$ molecules (5). We analyzed 30 clones by direct radioimmunoassay and by the fluorescence-activated cell sorter. We found that the lower the H-2K^b/H-2D^b ratio, the higher was the metastatic potential of the cloned cells (5, 6).

THE RELATIVE EXPRESSION OF THE H-2K^b/H-2D^b GENES IS CORRELATED WITH THE IMMUNOGENIC COMPETENCE OF THE TUMOR CELLS

To analyze whether the low H-2K/H-2D ratio determines the expression of the metastatic phenotype because it decreases the immunogenic potency of the tumor cells, we analyzed two clones of the 3LL tumor: A low metastatic clone A9 that expresses both the H-2K^b and the H-2D^b glycoproteins, and a high metastatic clone D122 that expresses the H-2^b molecules, but hardly expresses the H-2K^b. Testing the growth of tumor cells of these two clones following transplantation into allogeneic recipients, we found that whereas the D122, similarly to the parental 3LL tumor, did grow, without generating metastases, in mice of BALB/c (H-2^d) and C3H (H-2^k), the metastatic A9 was rejected by the allogeneic mice. Thus, in contrast to the metastatic clone, the nonmetastatic phenotype, manifesting a high H-2K^b/H-2D^b ratio, evoked a strong allograft reaction in allogeneic mice (10). Tests of the growth of these clones in H-2 recombinant mice on a C57BL/10 background, demonstrated that the allograft immune rejection was elicited by the H-2K^b gene products (10).

To examine whether the metastatic phenotype is controlled via the immunogenic properties of the 3LL clones, we analyzed the immune responses evoked by the A9 and the D122 clone in syngeneic mice. C57BL/6J mice were immunized by 3 weekly intraperitoneal injections of 10^7 irradiated A9, D122, or 3LL cells. Ten days later, immunized mice and controls were challenged by A9 or D122 cells. We found that immunization by A9 cells significantly slowed the growth rate of a second A9 tumor but did not affect the growth rate of a graft of the metastatic D122 tumor. Immunization by clone D122 or by the parental 3LL cells did not retard the growth of a second A9 or D122 tumor. Similar results were obtained when mice were intradermally immunized by living A9 or D122 cells (10). Following these in vivo observations, we tested the cytotoxic T lymphocyte (CTL) responses evoked by cells of A9 and D122 clones in syngeneic hosts. A9 induced high levels of cytotoxic activity, which was manifested against A9 cells and to a lower extent against D122 target cells. D122 cells induced a lymphocyte population that manifested low cytotoxic activity against D122 or A9 target cells. Thus, the in vitro interaction of A9 immune lymphocytes with nonmetastatic A9 cells led to the destruction of the tumor cells, whereas lymphocytes interacting with D122 cells are significantly less efficient in destroying the tumor cells (10). It thus appears that the nonmetastatic A9 clone is significantly more immunogenic in syngeneic mice and manifests susceptibility to the lymphocyte cytotoxicity it elicits.

INDUCTION OF ALTERATIONS IN H-2K/H-2D EXPRESSION ALTERS THE METASTATIC PHENOTYPE OF 3LL CLONES

To examine whether the relative expression of class I antigens of the MHC is causally related to its metastatic phenotype, we attempted to induce alterations in the H-2K^b/H-2D^b ratio and subsequently test whether this altered the metastatic potency of the tumor cells. We further focused on the low metastatic A9 clone and the high metastatic D122 clone. In the first set of experiments, we treated in vitro cells of these two clones with either interferon $\alpha + \beta$ (known to stimulate H-2 synthesis), or retinoic acid (5, 6). We observed that interferon caused an increase in both H-2K^b and H-2D^b expression, of A9 and D122 cells, yet the relative increase in H-2D^b expression was significantly higher than that of H-2K^b, resulting in a decrease in the H-2K/H-2D ratio. The interferon treated cells were then tested for the generation of spontaneous metastasis. The result was a significant increase in the metastatic mass produced by both the A9 and the D122 cells (5, 6). Treatment with retinoic acid did not alter H-2K expression but increased significantly H-2D^b production, decreasing even more the H-2K/H-2D ratio. This converted cells of the low metastatic A9 clone to a high metastatic phenotype, and increased the metastatic mass produced by the D122 (5, 6).

We subsequently turned to test whether γ-interferon, as distinct from $\alpha + \beta$, increased the H-2K/H-2D ratio. We found (Fig. 1) that, in contrast to the effect of both retinoic acid and interferon $\alpha + \beta$, γ-interferon caused a dramatic increase in the absolute and relative expression of the H-2K^b molecules, thus increasing the H-2K/H-2D ratio (11). When the treated metastatic D122 cells were then injected intravenously to syngeneic mice, the experimental lung metastases thus produced were of a very low mass (Table 1). Thus, applying this assay, γ-interferon in increasing the H-2K/H-2D ratio seemed to have converted the high metastatic D122, to a low metastatic phenotype. It appeared, however, that the alteration of H-2 expression caused by γ-interferon is of a shorter duration than that induced by interferon $\alpha + \beta$. This we deduced from our observation that when testing the production of spontaneous lung metastases following intra-footpad inoculation of the treated cells of the D122 clone, the reduction in lung metastases was lower than that obtained following intravenous inoculation of the cells.

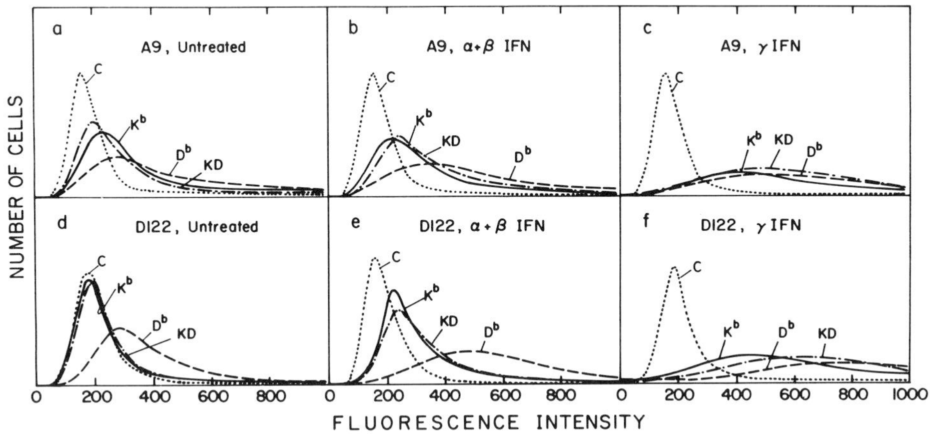

Figure 1. Effect of interferons on cell surface expression of H-2K^b and H-2D^b alloantigens of clones A9 and D122

Two x 10^5 tissue culture propagated cells were transferred to 100-mm petri dishes in 10 ml Dulbecco's modified Eagle's medium supplemented with 10% fetal calf serum, 1% glutamine, 1% sodium pyruvate, 1% nonessential amino acids, and 1% antibiotics. The cells were grown for 5 days in the presence of (a, d) no interferon; (b, e) interferon + 500 U/ml, 10^7 U/mg; (c, f) -interferon 100 U/ml, 12 x 10^7 U/mg. Cells were treated with monoclonal antibody 28-13-3 (anti-(IE7.K^b-1)), 20-8-4 (anti-K^{b}D^b, reacts mainly with (IE7.K^b-1) molecules) or 28-14-8 (anti-D^b) and analyzed by the FACS II.

ABROGATION OF THE METASTATIC PHENOTYPE FOLLOWING TRANSFECTION WITH H-2K GENES

The observations described thus far raised two principal questions: (a) Could the effect of agents such as interferons affecting the metastatic phenotype be attributed to their effect on H-2 gene expression? (b) Is the correlation between the relative expression of class I antigens and the metastatic phenotype confined just to the 3LL tumor, or is it a phenomenon of a more general significance?

Table 1. Average lung weight of low- and high-metastatic clones treated by γ-interferon

Clone	Control	γ-interferon 25 μ/ml	50 μ/ml	100 μ/ml
A9	239±113	258±64	223±35	260±115
D122	736±352	327±95	262±65	322±254

Tumor cells were grown in the presence of γ-interferon, as in Fig. 1 for 5 days. $5x10^5$ cells were injected intravenously into groups of ten C57BL/6J male mice. Thirty days later, mice were sacrificed, lungs were removed and weighed.

To approach the first question we aimed at testing whether transfection of tumor cells by H-2K genes would result in the abrogation of their metastatic phenotype. To address the second question, we turned to investigate a different metastatic tumor the T10 sarcoma which had been induced by methylcholanthrene in mouse of a $(C3HxC57BL)F_1$, $(H\text{-}2^kxH\text{-}2^b)$. Deducing from its original genotype, cells of the T10 sarcoma should have expressed four distinct class I antigens: K^b, D^b, K^k and D^k. Yet, in fact we found that both metastatic and nonmetastatic clones lacked the cell surface expression of K^k and K^b molecules. Nonmetastatic clones expressed only the D^b molecules, whereas metastatic clones expressed both the D^b and the D^k molecules (6, 12, 13) Experiment were carried out to test whether both the D^k and the D^b molecules are required for the expression of the metastatic phenotype. Cells of metastatic clones were transferred in homozygous mice of the parental strains. The result was that metastatic clones (i.e., IE7) which to begin with expressed both the $H\text{-}2D^b$ and the $H\text{-}2D^k$, when serially transplanted in C3H mice $(H\text{-}2^k)$, lost the cell surface expression of the D^b molecule and retained only the D^k molecule. This resulted in an increase of their metastatic competence. Thus products of the $H\text{-}2D^k$ gene determined the metastatic potency of the T10 sarcoma cells (14).

Table 2. H-2 phenotype of transfected IE7 and IC9 tumor clones

Cell line	Transfected genes	H-2 antibody (cpm) Anti-D^b	Anti-K^b	Anti-D^k	Anti-K^k	H-2 phenotype
IE transfected lines						
IE7	-	27,849	580	6 438	697	D^b,D^k
IE7.neo^r-1	neo^r	27,069	270	6,072	294	D^b,D^k
IE7.neo^r-2.	neo^r	20,051	768	7,188	524	D^b,D^k
IE7.K^k-1	neo^r,K^k	16,461	949	4,964	5,785	D^b,D^k,K^k
IE7.K^k-2	neo^r,K^k	21,176	683	5,382	7,413	D^b,D^k,K^k
IE7.K^kK^b-1	neo^r,K^k,K^k	16,160	13,664	5,539	6,992	D^b,D^k,K^b,K^k
IE7.K^b-1	neo^r,K^b	10,447	9,350	3,433	851	D^b,D^k,K^b
IE7.K^b-2	neo^r,K^b	27,780	14,877	6,505	419	D^b,D^k,K^b
IE7.K^b-3	neo^r,K^b	17,350	12 631	4,817	913	D^b,D^k,K^b
IC9 transfected lines						
IC9	-	28,259	453	397	507	D^b
IC9.neo^r	neo^r	16,758	252	187	524	D^b
IC9.K^b-1	neo^r,K^b	10,629	15,944	848	653	D^b,K^b
IC9.K^b-2	neo^r,K^b	18,118	14,496	414	478	D^b,K^b
IC9.K^b-3	neo^r,K^b	21,806	28,420	719	718	D^b,K^b
IC9.K^k-1	neo^r,K^k	17,023	514	745	4,282	D^b,K^k
IC9.K^k-2	neo^r,K^k	19,316	471	516	10,305	D^b,K^k

Legend for Table 2.

The table shows a representative cellular radioimmunoassay. Briefly, 10^6 cells were incubated in microtiter plates with specific anti-H-2 monoclonal antibody (anti-D^b, hybridoma B22-249; anti-K^b, K10-56; anti-D^k, 15-5-5S and anti-K^k, 16-3-22. After washing, ^{125}I-labelled Protein A was added and the amount of bound radioactivity determined. Values represent the mean of triplicate wells. The standard deviation ranged between 5 and 10% (not shown). Background with irrelevant antibodies was always <1,000 cpm (not shown). The cpm obtained with different H-2 antibodies cannot be compared because the affinity and amount of specific antibody in the hybridoma supernatant are probably different.

We subsequently examined whether transfection of clones of the T10 sarcoma with cloned K-gene would alter the malignant properties of the tumor cells (15). A metastatic clone (IE7) and a nonmetastatic clone (IC9) were selected from the two sets of the T10 sarcoma clones. Cells were transfected with H-2K^b or H-2K^k plasmid DNA (16) and plasmid pAG60 containing the neomycin resistant gene (neo^r), then selected in medium containing geneticin. Colonies expressing transfected H-2 genes were identified using specific monoclonal H-2 antibodies. Positive colonies were cloned and the H-2 phenotype was determined using four different assays: (1) cytofluorography, (2) cellular radioimmunoassay, (3) lysis by alloreactive CTL and (4) immunoprecipitation by monoclonal H-2 antibodies. All yielded identical results. A typical cellular radioimmunoassay using monoclonal H-2 antibodies and ^{125}I-labelled protein A is presented (Table 2).

Parental IC9 cells, or IC9 transfected with neo^r alone expressed only the H-2D^b molecule, whereas IC9 cells transfected with the H-2K^b or H-2K^k genes synthesized also the K^b or the K^k antigen. Similarly, the parental IE7 cells or IE7 cells transfected with neo^r alone expressed only the D^b and D^k antigens. On the other hand, K^b, K^k or both K^b and K^k expression was obtained following transfection with the corresponding H-2K genes (Table 2; Ref. 15). Serological analysis with >20 monoclonal antibodies against distinct H-2 determinants and two-dimensional gel electrophoresis showed that the K^b molecules on the transfected cells were identical to those of typical H-2^b cells.

Testing the generation of spontaneous lung metastases by IE7 cells following intrafootpad inoculation of 10^4-10^6 cells into syngeneic $(C3HxC57BL/6)F_1$ mice, we observed that the expression of either K^b or K^k following transfection resulted in the loss of their metastatic competence (Table 3; Ref. 15). Such H-2K transfected IE7 clones did produce lung metastasis when grafted in animals which had been exposed to 550 Rad total body irradiation, a dose which suppress-

Table 3. Metastatic potential of H-2K-transfected IE7 and IC9 cells in normal and irradiated syngeneic (C3HxC57BL/6)F1 mice

Tumor	H-2 phenotype* D^k	K^k	D^b	K^b	Mice inoculated	Weight of lungs (mg)	No. of nodules in lungs
IE7 transfected lines							
IE7	+	-	+	-	Normal	397±70	12, 200, 200, 200, 200
					Irradiated	460±20	200, 200, 200, 200, 200
IE7.neo^r	+	-	+	-	Normal	420±15	200, 200, 200, 200, 200
					Irradiated	395±40	200, 2000, 200, 200
IE7.K^k-1	+	+	+	-	Normal	186±8	0, 0, 0, 0, 2, 6,. 8
					Irradiated	381±15	200, 200, 200, 200, 200
IE7.K^kK^b-1	+	+	+	+	Normal	170±3	0, 0, 2, 2, 4
					Irradiated	401±20	200, 200, 200, 200, 200
IE7.K^b-1	+	-	+	+	Normal	168±8	0, 0, 0, 0, 0
					Irradiated	326±45	200, 200, 200, 200, 200
IE7.K^b-2	+	-	+	+	Normal	182±50	0, 0, 0, 0, 0, 0
					Irradiated	345±50	200, 200, 200, 200, 200
IE7.K^b-3	+	-	+	+	Normal	180±5	0, 0, 0, 0, 0, 3
IC9 transfected lines							
IC9	-	-	+	-	Normal	172±16	0, 0, 1, 3, 9, 200
					Irradiated	255±52	3, 200, 200, 200
IC9.K^b-1	-	-	+	+	Normal	162±6	0, 0, 0, 0, 0, 5
					Irradiated	270±8	200, 200, 200, 200
Ic9.K^b-2	-	-	+	+	Normal	154±4	0, 0, 0, 0, 0
					Irradiated	283±5	200, 200, 200, 200, 200
IC9.K^k-1	-	+	-	+	Normal	171±11	0, 0, 0, 0, 0

Legend to Table 3.

A single-cell suspension 10^5 neo^r- and h-2K-transfected IE7 and IC9 cells were inoculated intrafootpad into intact and sublethally irradiated (550 Rad Co) (C3HxC57BL/6)F1 mice. When the tumor diameter reached 8-10mm, the tumor-bearing leg was amputated using a 3/0 silk tie, which was ligated beyond the iliofemoral joint of the hind leg to prevent bleeding. In each experimental group, mice were killed 21 days post-amputation, and their lungs removed and assessed for metastatic load by weight and by number of nodules counted after fixation in Bouin's solution. Without excision of the primary tumor the number of metastatic nodules is much lower. Corresponding results were obtained with both protocols, but only data generated with the excision protocol are presented. Other independent experiments yielded identical results (not shown).

*H-2 phenotype according to results presented in Table 2.

es the mouse immune reactivity. This suggested that the abrogation of the metastatic phenotype following expression of H-2K antigens, is a function of the acquisition of immunogenic properties conferred on these cells by the $H\text{-}2K^b$ and $H\text{-}2K^k$ molecules.

To further explore the immunogenicity acquired by the IE7 (and IC9 cells) following H-2K transfection, two tests were applied. The first one demonstrated that syngeneic mice transplanted with H-2K transfected IE7 cells, manifested after tumor excision, followed by re-challenge with transfected IE7 cells retarded tumor growth (15). The second test demonstrated that immunization with irradiated cells of the K^b transfected cells of either IE7 or IC9 clones elicited cytotoxic T-cells; which lysed only K^b transfected IE7 and IC9 target cells, but not the parental IE7 or IC9 (Table 4; Ref. 15). Neither did such CTL lyse cells of an unrelated K^b positive EL4 tumor cells. Thus, the CTL elicited by the transfected cells seem to have recognized a T10 associated cell surface antigen shared both by the IE7 and the IC9 cells, restricted by the K^b molecules (Table 4). An immune response against such antigens seems to prevent the generation of metastasis by H-2K transfected IE7 cells.

MOLECULAR NATURE OF MHC GENE PRODUCTS IN CLONES OF THE 3LL CARCINOMA

The MHC control of the metastatic phenotype raises a basic question: What is the molecular basis for the imbalanced expression of the H-2K and H-2D gene products in malignant cells? Analyzing first cells of the 3LL carcinoma, we tested the molecular similarity of the MHC glycoproteins to $H\text{-}2^b$ molecules of C57BL/6J spleens. Immunoprecipitation of ^{125}I cell surface-labeled and ^{35}S-methionine-labeled extracts of A9 and D122 clones showed a strict correla-

Table 4. Generation of CTL specific for transfected tumors in (C3HxC57BL/6)F1 mice

	Target cells (% specific lysis)							
Immunization	IC9	IE7	(IC9.K^b-3)	(IE7.K^b-1)	(1C9.K^b-1)	(IE7.K^k-1)	EL4	L929
IC9	0	0	0	0	0	0	0	0
IE7	0	0	0	0	0	0	0	0
(IC9.K^b-3)	0	0	45	28	0	0	0	0
(IE7.K^b-1)	0	0	40	35	0	0	0	0
(IE7.K^k-1)	0	0	0	ND	ND	59	0	5

(C3HxC57Bl/6)F1 mice were immunized three times at weekly intervals with 10^6 tumor cells irradiated with 3,000 Rad. After 7-10 days, spleen cells were restimulated in vitro with the same tumor cells irradiated with 10,000 Rad. Cytolytic activity was determined at different effector-to-target ratios against ^{51}Cr-labelled target cells including EL4 (H-2^b) and L929 (H-2^k) as controls. Data show percent specific lysis obtained with an effector/target ratio of 50:1. ND, not determined.

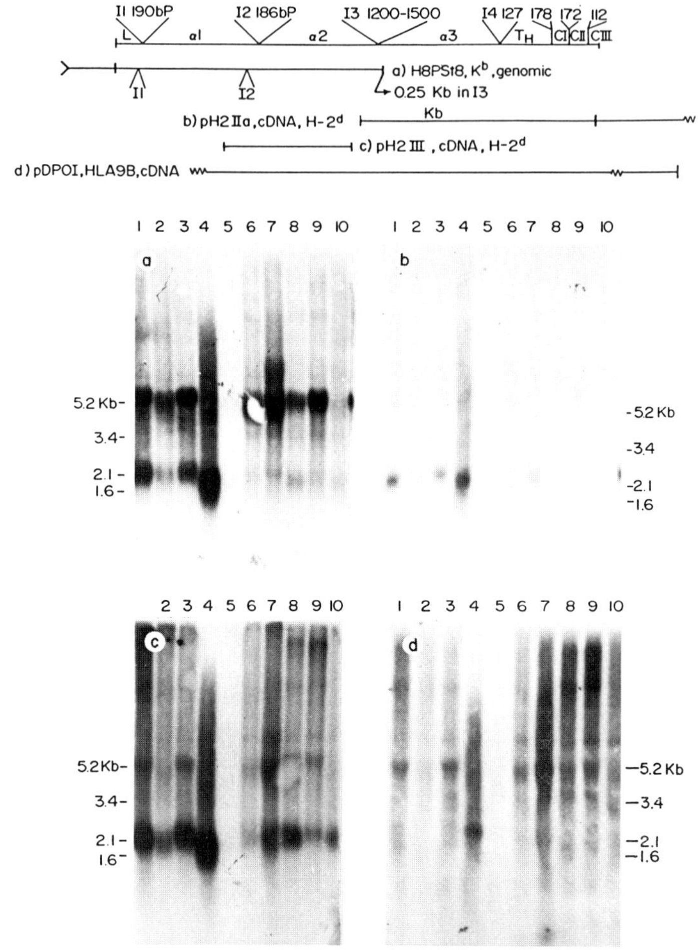

Figure 2. Transcription of MHC class I genes in tumor clones

Northern hybridization performed with poly$(A)^+$ RNAs from 3LL carcinoma and T10 sarcoma. (1) A9, low metastatic, 3LL; (2) D122, high metastatic, 3LL; (3) 3LL, parental; (4) liver; (5) IC9, nonmetastatic, T10; (6) D6 moderate metastatic, T10; (7) IB9, high metastatic, T10; (8) IE7, high metastatic, T10, and (9 and 10) T8, T40, parental T10. Probes used: (a) H8PSt8, (b) H-2IIa, (c) H-2III, and (d) PDPOI.

tion between synthesis and cell surface expression. In D122 and in the parental 3LL cells, but not in A9, synthesis of K^b molecules was suppressed (5). The 45-Kd proteins precipitated by anti-H-2^b serum or monoclonal anti-H-$2K^b$ were similar in migration to molecules precipitated from C57BL/6J splenocytes, and a 12-Kd 2-microglobulin molecule was co-precipitated. Separation on lentil lectin Sepharose showed that most of the H-2^b-encoded proteins were in their glycosylated form (5). To obtain a better understanding of the transcriptional level of the K gene suppression in metastatic clones, we analyzed, by Northern blot hybridization, the mRNA from clones A9, D122, and 3LL as compared with liver mRNA and RNA extracted from metastatic and nonmetastatic clones of the T10 sarcoma. We used four probes: (1) a genomic K^b 5'-region probe, H8Pst8 (8); (2) a cDNA, H-2^d, 3'region probe, pH2IIa (9); (3) a cDNA, H-2^d, 5'-region probe, pH2III (9), and (4) a human HLA.B9 cDNA probe. All these probes hybridize to both K and D end transcripts. Figure 2 shows that the normal H-2 transcript of 2 kb that is expressed in the liver is also expressed at a high level in clone A9 and in the parental 3LL line. A lower level of transcription of the 2 Kb mRNA was observed in clone D122 and in the T10 sarcoma clones as both D122 and all T10 clones lack expression of K end products. An intriguing observation, however, is that besides the normal 2 kb transcripts, an abnormal RNA of 5.5-6 kb was detected in all tumor clones. This transcript hybridized to the three 5'-region probes but not to the 3'-region probe. The origin of this large RNA is not yet known, but it may result from the insertion of a foreign DNA into the H-2K gene of the tumors. In this event the large RNA could represent a transcript containing H-2 sequences plus other sequences and this could explain the inactivation of the H-2K gene.

Southern blot analysis, although very complex, revealed new fragments hybridizing to the H-2 probes when the genomic DNA from tumor clones was digested with EcoRI, BamHI, XbaI, or SStI and compared with liver DNA of the same mouse strains (C57BL for 3LL clones and F_1 for T10 clones). Differences were also observed between clones A9 and D122. The possibility that a mutation or insertion in H-2K genes results in the loss of expression of an H-2 molecule, giving rise to clones insensitive to the host immune system, is investigated.

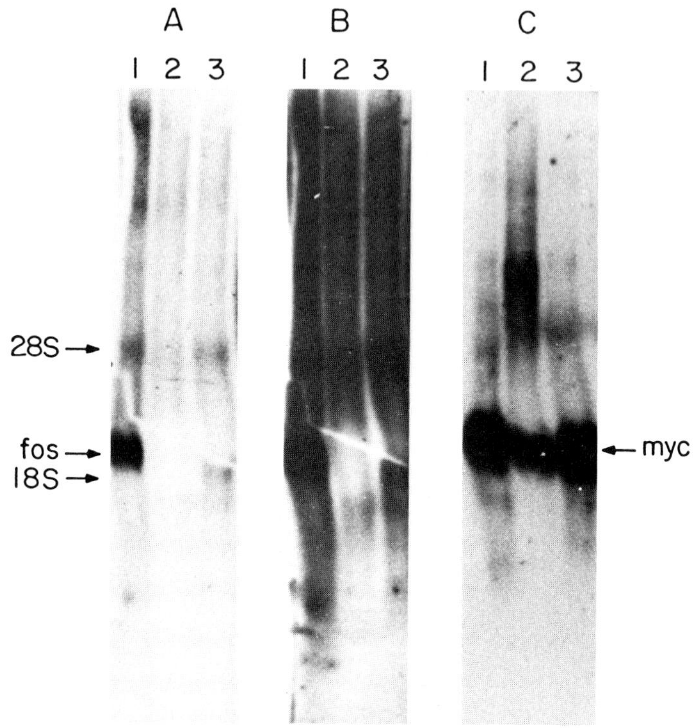

Figure 3. Expression of c-myc and c-fos onc genes in 3LL clones

Polyadenylated mRNA was selected on oligo-dT cellulose and electrophoresed on formaldehyde-agarose gels. RNA was then transferred onto nitrocellulose filters which were then hybridized with a ^{32}P-labelled nick-translated c-myc probe, washed and exposed to x-ray film. The same blot was cleaned of the c-myc probe and rehybridized to a fos probe.

(1) mRNA from low-metastatic A9 clone; (2) mRNA from high-metastatic D122 clone; (3) mRNA from parental 3LL line; (a) hybridization to fos, blots exposed for 1 day; (b) hybridization to fos, blots exposed for 3 days; (c) hybridization to myc, blots exposed for 1 day.

The co-expression of the two class I antigens, which characterizes the low or nonmetastatic clones of 3LL, characterizes also most nucleated normal somatic cells, whereas early embryonal cell, as well as nondifferentiated teratocarcinoma cells, lack expression of H-2 molecules. Do other gene products of the metastatic versus the nonmetastatic phenotypes signify differences in state of differentiation? Of particular interest from this viewpoint is the fos gene. Expression of c-fos, the cellular counterpart of FBJ osteosarcoma viral onc gene was shown to correlate with induction of differentiation in mouse myeloid leukemia line WEHI-3B (17). Transfection of F9 teratocarcinoma cells by a cloned fos gene was shown by Muller and Wagner to induce differentiation of teratocarcinoma cells to endoderm-like cells (18). In a recent study of the organization and expression of onc genes in metastatic and nonmetastatic clones we observed that the c-fos proto onc gene is expressed in the low metastatic clone A9 at high levels while the high metastatic clone D122 does not contain c-fos related mRNA (Fig. 3). A low level of c-fos transcription was also observed in parental 3LL cells. Expansion of these results showed that also other low metastatic 3LL clones expressed c-fos. In addition, we found that the c-myc oncogene was amplified 60-fold in all 3LL clones (Fig. 3). Thus, the level of c-myc mRNA is very high in A9 and D122 clones, as well as in the parental 3LL line. While c-myc is expressed in all three cell types, c-fos is expressed mainly in low metastatic A9 clone (Fig. 3). Since the low metastatic clones of the 3LL tumor resemble normal differentiated cells in their MHC gene expression and in evoking an allograft rejection, it appears that the expression of the c-fos in such clones might be related to the molecular basis of the relative differentiated state of nonmetastatic cells.

CONCLUSIONS

Genes of the major histocompatibility complex (MHC) control the metastatic phenotype of tumor cells in the mouse. This conclusion is based on the analysis of a large number of clones of mouse tumors. Thus, metastatic competence of clones of the Lewis lung carcinoma (3LL) is shown to be correlated with the relative expression of the two class I antigens: the lower the H-2K/H-2D ratio. the higher was the metastatic competence of the cloned cell population. Studies of two selected clones, the nonmetastatic A9 and the metastatic D122, demonstrated that the nonmetastatic clone is significantly more immunogenic in syngeneic animals than the metastatic clone. This was evident from data obtained by measuring both host resistance to secondary tumor grafts and the generation of, and susceptibility to, syngeneic cytotoxic T-cells (CTL). Induced alterations in H-2K/H-2D ratio altered the metastatic competence of tumor clones. Thus, interferon α + β and retinoic acids stimulated the synthesis of H-2D^b gene products significantly more than of H-2K^b, thus decreasing the H-2K/H-2D ratio. This resulted in the conversion of a low metastatic clone A9 to a high metastatic population, and increased the metastatic competence of the D122. Gamma interferon on the other hand induced an increase in the H-2K/H-2D ratio, resulting in a significant decrease in the capacity of the D122 to generate experimental metastases.

To test the applicability of these observations to other tumors, a methylcholanthrene-induced sarcoma, T10, originating in a (H-2^{b}xH-2^k)F1 mouse, was studied. Here, neither the K^b nor the K^k were expressed on cells of this metastatic tumor. Metastatic clones expressed the H-2D^b and the H-2D^k, whereas nonmetastatic clones expressed only the H-2D^b. We studied whether transfection of the metastatic clone IE7 with H-2K^b or H-2K^k gene would result in the abolishment of the metastatic phenotypes. The result was that the induced expression of either the H-2K^b or the H-2K^k led to the abrogation of the metastatic phenotype. Such nonmetastatic H-2K transfected phenotype did produce spontaneous metastases when transplanted into immunosuppressed animals. It seemed, therefore, that the abrogation of the metastatic phenotype was associated with the acquisition of immunogenic capacity. This was substantiated in experiments indicating that the H-2K transfected cells evoked a significant immune response, assayed both by secondary graft rejection and by the generation of CTL manifesting H-2K restricted lysis.

Analyzing the molecular basis for the impaired expression of the H-2K genes in cells of the metastatic clones, we tested the mRNA from clones of the 3LL and the T10 tumors. Probes of the 5' end, detected in all tumor clones but not in normal cells, besides the normal 2 kb mRNA, an abnormal RNA of 5.5-6.0 kb. The relevance of such a messenger to the impaired expression of the H-2K genes remains an inviting possibility.

The nonmetastatic A9 clone manifested immunogenic properties similar to those of normal differentiated cells. We therefore tested the expression of c-fos proto-oncogene, which had been implicated in certain cell systems, with the induction of differentiation. We found that whereas all clones of the 3LL tumor manifested a 60x gene amplification of the c-myc, the c-fos was expressed only in the nonmetastatic phenotype. The possibility of a co-regulated expression of the H-2K and the c-fos is a challenging question.

ACKNOWLEDGMENTS

This work was supported by research grant CA28139 from the National Cancer Institute, National Institutes of Health.

REFERENCES

1. Fidler, I.J. and Kripke, M.L. Science 197: 893-895, 1977.

2. Eisenbach, L., Segal, S. and Feldman, M. J. Natl. Cancer Inst. 74: 87-93, 1985.

3. Liotta, L.A., Tryggvason, K., Garbisa, S., Hart, I.R., Foltz, C.M. and Shafie, S. Nature 284: 67-68, 1980.

4. Eisenbach, L., Segal, S. and Feldman, M. J. Natl. Cancer Inst. 74: 77-85, 1985.

5. Eisenbach, L., Segal, S. and Feldman, M. J. Natl. Cancer Inst. 32: 113-120, 1983.

6. Eisenbach, L., De Baetselier, P., Katzav, S., Segal, S. and Feldman, M. In: Cancer Invasion and Metastasis: Biologic and Therapeutic Aspects, (Eds. G.L. Nicolson, L. Milas), Raven Press, New York, 1984, pp. 101-121

7. Isakov, N., Segal, S. and Feldman, M. J. Natl. Cancer Inst. 66: 919-926, 1981.

8. Isakov, N., Feldman, M. and Segal, S. Inv. Metast. 2: 12-32, 1982.

9. Isakov, N., Katzav, S., Feldman, M. and Segal, S. J. Natl. Cancer Inst. 71: 139-145, 1983.

10. Eisenbach, L., Hollander, N., Greenfeld, L., Yakor, H., Segal, S. and Feldman, M. Int. J. Cancer 34: 567-573, 1984.

11. Eisenbach, L. and Feldman, M. In: Modern Trends in Human Leukemia, (Ed. R. Neth), in press.

12. De Baetselier, P., Katzav, S., Gorelik, E., Feldman, M. and Segal, S. Nature 288: 179-181.

13. Katzav, S., De Baetselier, P., Tartakovsky, B., Gorelik, E., Segal, S. and Feldman, M. In: Biochemical and Biological Markers of Neoplastic Transformation. (Ed. P. Chandra), Plenum Press, New York, 1983, pp. 111-119.

14. Katzav, S., Segal, S. and Feldman, M. Int. J. Cancer 33: 407-415.

15. Wallich, R., Bulbuc, N., Hämmerling, G.J., Katzav, S. Segal, S. and Feldman, M. Nature 315: 301-305, 1985.

16. Mellon, A.L., Golden, L., Weiss, E., Bullman, H., Hurst, I., Simpson, E., James, R.F.L., Townsend, A.R.M., Taylor, P.M., Schmidt, W., Ferluga, J.,

Leben, L., Santamaria, M., Atfield, G., Festenstein, H. and Flavell, R.A. Nature 298: 529-534, 1982.

17. Conda, T.J. and Metcalf, D. Nature 31: 249-251, 1984.

18. Muller, R. and Wagner, G.H. Nature 311: 438-442.

15

GENERATION OF METASTATIC CELLS VIA SOMATIC CELL FUSION : A POSSIBLE MECHANISM FOR TUMOR PROGRESSION IN-VIVO

P. DE BAETSELIER, E. ROOS*, L. BRYS, L. REMELS and M. FELDMAN+

Institute for Molecular Biology, Free University Brussels, Paardenstraat 65, 1640 Sint Genesius-Rhode, Belgium , * Division of Cell Biology, The Netherlands Cancer Institute, 121 Plesmanlaan, 1066 CX Amsterdam, The Netherlands and + The Weizmann Institute of Science, Rehovot, Israel

The question of whether and how properties of tumor cells are altered following somatic hybridisation between neoplastic and normal cells has attracted interest ever since the discovery of the technique of somatic cell fusion. Early studies of somatic hybrids derived from fusion of cells of high malignant lines with cells of low malignant lines suggested that the high malignancy of the parental neoplastic cells was the dominant property (1). Subsequent studies demonstrated, however, that the malignant characteristic could behave like a recessive trait in somatic cell hybrids (2). In such hybrids, it appeared that the neoplastic property was the result of a "deletion" which was complemented by "healthy" chromosomes of the normal partner. The relevance of these observations to the <u>in-vivo</u> processes of tumor progression remained an open question.

Yet, following the initial studies on the outcome of <u>in-vitro</u> somatic cell fusion, reports appeared suggesting that tumor cells grown <u>in-vivo</u> may undergo somatic fusion with the host's normal cells (3,4). In most of these cases, however, the properties acquired by the tumor as a function of somatic hybridization were not studied. Only Goldenberg suggested that the generation of metastases by human malignant cells grown in hamsters was associated with somatic cell fusion with hamster cells (5).

The aim of the present review of our work is twofold: first to summarize our body of evidence indicating that the acquisition of metastatic properties by nonmetastatic tumors after fusion with normal lympho-reticular cells is of general significance and secondly to consider the possible mechanisms whereby tumor-host hybrids formed in-vivo and tumor-normal cell hybrids built up in-vitro may give rise to established, transformed, tumorigenic and metastatic variants.

1. PLASMACYTOMA CELLS ACQUIRE METASTATIC PROPERTIES FOLLOWING FUSION WITH LYMPHO-RETICULAR CELLS.

To test whether, in principle, somatic fusion could confer metastatic properties on non-metastatic tumor cells, we chose to study a model system in which somatic hybridization could readely be achieved in culture and in which the normal non-neoplastic partner was a "circulating cell", and thus possessed some of the disseminating properties of a metastatic cell. We initially investigated the outcome of hybridization of nonmetastatic plasmacytoma cells (NSI, SP2/0) with normal B lymphocytes. The results obtained with these plasmacytoma derived hybridomas (6, 7, 8) can be summarized as followed:

1) Whereas the NSI and SP2/0 plasmacytomas grew locally but did not generate metastases, the B-cell hybridomas were found to do so. In fact, the hybridomas tested could be classified into two distinct categories: hybridomas which generated metastases both in the spleen and the liver and hybridomas which generated metastases only in the liver. It was subsequently tested whether (a) hybrid cells from spleen or liver metastatases were selected for metastatic competence, and whether (b) spleen metastases, as distinct from liver metastases, reflect selection for target organ specificities. The result was that whereas in the primary host 25% of the animals manifested metastases, in the secondary host, grafted with cells from spleen or liver nodules, 100% of the recipients manifested metastases. Thus following one transfer generation, selection for a high metastatic capacity was achieved. Cells from spleen nodules produced metastases both in the spleen and in the liver, and the same was true for liver metastases (Table 1 a, c). Thus, cells from spleen and liver metastases do not reflect the pre-existence of two subpopulations with distinct target organ specificities. Yet, it appears that tumor cells from spleen metastases were more potent in generating spleen and liver metastases, than tumor cells from liver nodules. Cells from liver metastases produced by cells, which metastasized only in the liver, when grafted s.c. to secondary recipients, also produced 100% of metastatic bearing mice. All secondary recipients of this hybridoma generated metastases only in the liver

(Table 1, a).

2) The generation of liver metastases by hybridomas that metastazise to the liver only is controlled by a different mechanism from the generation of liver metastases by spleen- and liver seeking hybridomas. Whereas the latter hybrids do not generate liver metastases when injected into splenectomized recipients, only liver seeking hybrids did produce liver metastases in a splenectomized host. Thus, the spleen exerts a controlling effect on liver metastases of tumors that metastasize to both the spleen and the liver.

3) The normal parental cell partner determines the organ specificity of the metastatic plasmacytoma-derived hybridoma. Whereas B-cell x NSI hybridomas produce spleen and liver metastases a NSI x macrophage hybridoma did not produce spleen or liver metastases. Such cells did, however, produce heavy masses of lung metastases. Thus, fusion between plasmacytoma cells and non B-cells may result in the acquisition of metastatic competence, but the organ specificity of these hybrid cells is different from those of the B-cell hybridomas (Table 1, b).

2. HYBRIDOMAS OF T-CELL LYMPHOMAS ALSO ACQUIRE METASTATIC COMPETENCE.

Thus far, we had emphasized the conversion of nonmetastatic plasmacytoma cells into metastatic hybrids through hybridization with lympho-reticular cells. It was of importance to adress subsequently the question of whether other neoplastic cells, such as thymoma cells, could acquire metastatic properties through somatic hybridization with normal lymphoid cells. As a source of normal lymphoid cells we used activated T lymphocytes since such cells exhibit properties considered to be characteristic for metastatic tumor cells such as invasiveness in cultures of endothelial cells and hepatocytes (9), or the degradation of extracellular matrix components (10). Thus, we studied the metastatic behaviour of T-cell hybridomas produced via somatic fusion between the BW5145 lymphoma and T-cells that had been activated for five days in mixed lymphocyte cultures (MCL activated T-cells). When tested for their capacity to develop organ-specific experimental metastases we found that these cells manifested, upon i.v. inoculation, extensive metastases formation in spleen, liver, kidneys and ovaries. The inherent metastatic capacity of T-cell hybridomas, as compared to BW cells was further confirmed in an in-vitro hepatocyte invasion test system (11). BW and T-cell hybridomas were added to 24h old primary cultures of rat hepatocytes. Adhesion of BW to the hepatocytes was minimal and adherent cells did not infiltrate the monolayer. On the other hand, hybrid T-cells adhered to the hepatocytes and rapidly infiltrated the cultures (Table 2, a). Such in-vitro observations indicate that hybrid T-cells have acquired invasive properties that enable them to diffusely infiltrate liver tissue. The pattern of metastatic development in spleen and liver of T-cell hybrids, as compared to spleen- and liver- seeking B-cell hybridomas, was strikingly different. The T-cell hybrids cells manifested a diffuse pattern of development, as compared to distinct nodular growth of spleen and liver metastases produced by SP2/0-B cell hybridomas. Such differences in metastatic growth patterns may reflect differences in organ-tumor interaction. In fact, we found that SP2/0-B cell hybridomas (11H11) do not adhere to monolayers of hepatocytes, nor do they infiltrate them. Thus, diffuse liver infiltration observed with T-cell hybrids might be due to their capacity to specifically adhere to and infiltrate hepatocyte cultures, while nodular liver infiltration by 11H11 type cells might be due to an organ-specific growth effect.

Apparently for T-cell hybridomas high invasiveness appears to be closely related with metastatic potential as such. Indeed, all of 28 obtained hybridomas that were highly invasive in hepatocyte cultures were highly metastatic. Futhermore, a hybridoma which lost invasiveness in culture, and two low-invasive hybridomas prepared from non-activated T-cells, were not metastatic, suggesting that high invasiveness is indispensable for metastasis formation by this tumor cell type (Table 2, a). Taken together, the present results indicate that invasiveness of T-cell hybridomas is a property which can be contributed by the normal, activated T-cell fusion partner. This notion is strengthened by experiments demonstrating that BW tumor cells acquire similar invasive and metastatic properties through somatic hybridization with an IL-2 dependent continuous T-cell line (i.e. CTL-D). According to their metastatic potential these T-cell hybridomas could be subdivided into three groups i.e. highly metastatic (massive organ infiltration at high incidence, resulting in an increase of organ weight), low or moderately metastatic (occasional metastatic nodules, no significant increase of organ weight) and nonmetastatic (no visible metastatic lesions). Cells derived from all different metastatic lesions manifested upon i.v. inoculation extensive metastases formation in different organs such as spleen, liver and kidneys in virtually all mice. Thus within one transfer generation, selection for a highly metastatic capacity was achieved. It is as yet not clear whether these metastatic cells represent tumor variants present in the original population which were selected in-vivo or alternatively whether metastatic capacity was induced by environmental signals in-vivo . Furthermore a distinct organotropism was observed with certain hybridomas such as organ derived 84-2-14 cells which metastasized mainly or exclusively to the kidneys. All other hybridomas tested, metastasized extensively to liver and spleen and to lesser extent to the kidneys (summarized in Table 2, b).

Invasiveness of some of these T-cell hybridomas was tested in hepatocyte cultures and three metastatic T-cell hybridomas tested, as well as CTL-D cells, were found to be invasive in-vitro , in contrast to one non-metastatic T-cell hybridoma (84-2-22) and BW cells. Interestingly, organ derived T-cell hybridomas (i.e. 84-2-14, kidney and 84-2-19, liver) which are more malignant in-vivo appear to be also more invasive in-vitro (Table 2, b). Yet in-vivo invasiveness in hepatocyte cultures is not completely reflected by the capacity to metastasize extensively to the liver. Indeed 84-2-14 cells which metastasized mainly to the kidney invaded to the same extent hepatocyte monolayers as 84-2-19 cells which metastasized mainly to liver and spleen.

In conclusion, the above described experiments demonstrate that BW tumor cells acquire invasive and metastatic properties through in-vitro somatic hybridization with MLC activated T-cells or a IL-2 dependent continuous T-cell line.

3. IN-VIVO HYBRIDIZATION: A MODEL FOR TUMOR PROGRESSION ?

Fusion in-vivo between tumor cells and host cells has been shown to occur in a variety of animal models. Unfortunately, none of the tumor cell-host cell hybrids thus formed were examined in any detail for changes in their relative tumorigenic or metastatic capabilities. There is, however, a recent study which implies that the in-vivo formation of tumor-host cell hybrids could lead to the emergence of more malignant tumor cell variants as defined by metastatic ability. We refer to the results of Kerbel et al. (12) who, using a nonmetastatic variant (MDW-D4) of the highly metastatic MDAY-D2 DBA/2 (H-2^d) mouse tumor, demonstrated that such variants became metastatic in-vivo after a cellular change took place, such as extinction of recessive lectin or drug sensitivity and acquisition of a higher number of chromosomes. Supportive evidence for a spontaneous cell fusion in-vivo between the MDW4 tumor cells and host cells was provided by the observation that growth of MDW4 tumor cells in either (H-2^k x H-2^d)F1 mice or (H-2^k) ⟶ H-2^d bone marrow radiation chimeras led to the appearance of metastatic MDW4 derived cells which express H-2^k antigens (12). Thus in this experimental tumor model, tumor progression might have arisen as a consequence of tumor-host cell fusion in-vivo .

In view of the results obtained with the metastatic T x T and T x CTL-D hybridoma cells, we were interested in testing whether tranplantation of the AKR derived Thy-1.1 BW lymphoma to semi-allogeneic (CBA x AKR)F1 and allogeneic CBA recipients might lead to metastatic cell fusion products. Thus, CBA and (CBA x AKR)F1 animals were inoculated i.v. with the BW 5147 lymphosarcoma cells and one month after inoculation two animals developed modular growth in the liver. Cells derived from these liver nodules (termed CBA derived BW-Li cells or (CBA x AKR)F1 derived BW-O-Li cells) were grown in culture and re-inoculated i.v. to either CBA or (CBA x AKR)F1 recipients. The animals were killed 2 weeks after tumor inoculation and autopsy revealed diffuse tumor growth in spleen, liver, kidneys and ovaries in 100% of the animals. In particular, the liver was heavely infiltrated in a diffuse manner, distinct from compact nodular growth. Futhermore, these BW liver variants appeared to have acquired inherent invasive properties enabling them to infiltrate in-vitro monolayers of hepatocytes (Table 2, c). Thus inoculation of an AKR T-cell lymphosarcoma to CBA or (CBA x AKR)F1 animals has led within one transplantation cycle to the generation of a tumor cell which manifests the capacity to invade different organs. Futhermore, these BW liver variants were new genetic variants since BW cells were 8-azaguanine drug-resistant tumor cells which died in HAT-medium while BW-Li and BW-O-Li cells were 8-azaguanine-sensitive and HAT-resistant.

We next addressed the question of whether the BW liver variants were real metastatic variants derived from the BW tumor cell. Flow cytofluorographic analysis of cells stained with monoclonal rat anti-mouse and mouse anti-mouse Thy1 antibodies revealed that the BW liver variants expressed the Thy1 marker of the parental BW lymphoma (i.e. Thy1.1) and the Thy1 marker of the CBA host (i.e. Thy1.2), unlike the BW cells which stained completely negative for the membrane expression of Thy1.2. These results imply that the BW liver variants originated from the BW lymphoma and had acquired membrane components of the CBA host such as Thy1.2. Additional staining experiments with monoclonal reagents (i.e. anti-H-2K^k, anti-H-2D^k, anti-Lyt1, anti-Lyt2) revealed that BW liver variants differed from the parental BW cells in the membrane expression of two additional membrane markers, namely Lyt1 and H-2^k alloantigens. Interestingly, the Lyt1 alloantigen appeared to have an AKR origin (Lyt1.2) and not a CBA origin (Lyt1.1). Thus BW liver variants express one host-derived alloantigen (Thy1.2) and two BW tumor derived alloantigens (Thy1.1 and Lyt1.2).

The expression of a mature T-cell marker (i.e. Lyt1) on BW variants raised the question of whether these cells would exhibit T-cell functions such as lymphokine secretion. In fact, liver variants but not BW cells, were found to be inducible to secrete low, yet reproducible levels of IL-2 after stimulation with ConA. Hence the transition from the non-metastatic phenotype of BW cells to the high-metastatic phenotype of BW-Li and BW-O-Li cells was associated with the acquisition or expression of certain differentiation antigens (i.e. Thy1.2, Lyt1.2, H-2^k) and functional characteristics (i.e. IL-2 production). Some of the newly expressed differentiation antigens were unequivocally of host origin (i.e. Thy1.2). Thus, on the basis of phenotypic, functional and drug-sensitive characteristics (summarized in table 3) we propose that BW-Li and BW-O-Li cells have originated from an in-vivo fusion between BW cells and host CBA or (CBA x AKR)F1 T cells (13).

4. POSSIBLE MECHANISMS IMPLICATED IN THE ACQUISITION OF METASTATIC COMPETENCE THROUGH SOMATIC CELL HYBRIDIZATION.

Our results with the T-cell hybridomas derived from fusions either in-vivo or in-vitro between BW and T lymphocytes indicate that normal T-cells can confer metastatic and invasive properties to the hybrid cells. Since we have previously shown that activated T-cells are invasive, in contrast to BW cells it appears that invasiveness is a T-cell property, dominantly expressed in the hybrids. Futhermore, since so far tested, all invasive T-cell hybridomas were highly metastatic, in contrast to the parental lymphoma, to non-invasive revertants and to non-invasive hybrids generated from non-activated T-cells (Table 2), it is tempting to assume that the T-cell derived invasive potential is associated with or possible the actual cause of the high metastatic capacity. Hence, particular properties of normal cells, when introduced into a tumor cell, may apparently give rise to a metastatic behaviour. We reached the same conclusion previously with the B-cell hybridomas (5, 6, 7) although the involved B-cell properties are as yet less clear. In fact, most of the studies on metastatic tumor-host hybrids formed in-vivo do agree in pointing to a cell of bone-marrow origin, possibly a lympho-reticular cell, as the most likely host cell involved in the fusion process (12, 13, 14).

In the context of our data with the BW lymphoma it might, however, be too premature to state that the complete invasive machinery was derived from a normal activated host T-cell. Indeed, the BW variants express the Lyt1.2 alloantigen which belongs genotypically to the BW tumor cells but is not expressed on BW cells, indicating that such cells are arrested in a stage of differentiation. Cell fusion of BW cells with a normal somatic cell might lead to large genotypic changes (chromosome rearrangements, gene amplification,...) resulting in the activation of silent differentiation genes as manifested by Lyt1.2 expression and IL-2 secretion. The resulting hybrid would than express the phenotype of a more differentiated and/or activated T-cell in which invasiveness is constitutively expressed. At present it is difficult to determine whether BW liver variants are T-cell hybrids which derived their metastatic potential from normal T lymphocytes, or alternatively whether cell fusion has resulted in the induction of a repressed invasive phenotype. We tackeled however the question of whether BW cells possess a repressed invasive phenotype and recent experimental evidence points towards this possibility. Indeed, when inoculating BW cells in a hemopoietic inducing microenvironment such as the spleen, a clear-cut induction of metastatic capabilities was observed with BW cells. Intra-splenic inoculation of BW cells to either semi-allogeneic (AKR x BALB/c)F1 animals or syngeneic (AKR) animals resulted in respectively a

30% and 100% incidence of metastatic development in spleen and liver (Table 4). Cells derived from these metastatic lesions behaved as real metastatic variants, generating liver and spleen metastases in 100% of the recipients (Table 5). These variants did not result from a cell fusion process with host cells as evidenced by the presence of the recessive drug resistance marker to 8-azaguanine and the absence of the host membrane markers (i.e. Thy1.2 and H-2^d of BALB/c origin) (Table 5). Futhermore these variants expressed syngeneic membrane antigens, that are absent on BW cells such as H-2^k and Lyt1.2 alloantigens. Thus intra-splenic inoculation of BW cells has resulted in the induction of metastatic variants that resemble phenotypically the BW liver variants generated by cell fusion in-vivo . The phenomenon of shifts in tumor cell phenotype induced by signals from the environment was recently revieuwed (15). It was postulated that signals from certain microenvironments could regulate the expression of various genetic programs, which in turn could activate or repress various cellular activities that might affect metastatic properties of tumor cells. Obviously, the spleen might provide an ideal microenvironment for the generation of differentiation-related inductive signals. Indeed, the capacity of the spleen in producing a "microenvironmental growth milieu" resulting in an enhanced tumorigenicity for B-cell hybridomas has been recently reported (16). Similarly, the growth of a murine leukemia (BCL1) has been reported to be spleen-dependent (17). Finally, as already mentioned, spleen and liver seeking hybridomas, require an inductive signal emitted by the spleen cells in order to generate liver metastases (6, 7, 8). These findings suggest that the spleen plays a growth-controlling role in tumor development, or that certain tumor cells may be induced in the spleen to undergo a differentiation event necessary for their spread. In fact, recent studies indicate that the microenvironment of the spleen induces maturation processes in lympho-reticular cells such as B-cells and macrophages (18, 19). Such processess may promote the expression of properties that control the capacity of leukemias (BCL1), B-cell hybridomas and T lymphomas (BW) to metastasize to other organs or to proliferate in the periphery.

This work was supported by a research grant of A.S.L.K..
P. De Baetselier is a N.F.W.O. fellow and L. Remels an I.W.O.N.L. fellow.

Table 1. Plasmacytomas acquire metastatic properties following fusion with lympho-reticular cells.

Tumor cells	Metastatic capacity (a) (incidence) (b)	(organotropism)
a) NSI x B-cell hybridomas		
NSI	0/10	-
HY1	3/8	spleen, liver
HY2	3/10	liver
HY1 (spleen)	10/10	spleen, liver
HY1 (liver)	6/10	spleen, liver
HY2 (liver)	10/10	liver
b) NSI x macrophage hybridomas		
NSI	0/10	-
HY38	10/10	lungs
c) SP2/0 x B-cell hybridomas		
SP2/0	0/5	-
11H11	2/5	spleen, liver
11H11 (spleen)	5/5	spleen, liver
11H11 (liver)	4/5	spleen, liver

(a) The metastatic potential of the B-cell hybridomas was assessed by injecting 2.10^6 cells i.v. to syngeneic mice. The mice were killed when they looked moribound and examined macroscopically for evidence of metastatic growth in different organs.

(b) Incidence of mice with metastases.

Table 2. Lymphomas acquire metastatic properties following fusion with lympho-reticular cells.

Tumor cells	Invasiveness (a) (Infiltration index)	Metastatic capacity (b) (incidence) (c)	(organotropism)
a) BW x T-cell hybridomas (in-vitro)			
TAM2D2 (BW x MLC primed T-cell)	1.5	4/4	liver, spleen, kidney
TAM4A6 (BW x MLC primed T-cell)	1.3	3/3	liver, spleen, kidney
TAM4D1 (BW x MLC primed T-cell)	0.9	3/3	liver, spleen, kidney
TAM4C4 (BW x MLC primed T-cell)	0.2	0/8	--
TA55C4 (BW x unprimed T-cell)	0.1	0/10	--
TA55C6 (BW x unprimed T-cell)	0.1	0/7	--
b) BW x CTL-D hybridomas (in-vitro)			
BW	0.0	0/3	--
84-2-14	0.6	1/3	kidney, liver
84-2-14 (kidney)	2.3	3/3	kidney
84-2-19	0.8	1/3	liver, spleen
84-2-19 (liver)	2.8	3/3	liver, spleen
84-2-21	0.9	3/3	liver
84-2-22	0.03	0/3	--
CLT-D	0.54	0/3	--
c) BW x T-cell hybridomas (in-vivo)			
BW	0.03	0/3	--
BW-L1	0.9	3/3	liver, spleen, kidney
BW-O-L1	2.7	3/3	liver, spleen, kidney

(a) Invasive capacity was assessed in an in-vitro hepatocyte invasion assay.

(b) The metastatic potential of the B-cell hybridomas was assessed by injecting 2.10^6 cells i.v. to syngeneic mice. The mice were killed when they looked moribound and examined macroscopically for evidence of metastatic growth in different organs.

(c) Incidence of mice with metastases.

Table 3. Characteristics of BW and in-vivo generated metastatic BW variants.

Characteristics		BW	BW-Li	BW-0-Li
Invasiveness		-	+	+
Metastatic capacity		-	+	+
AZA resistance		+	-	-
HAT resistance		-	+	+
Membrane phenotype:	Thy1.2	-	+	+
	Thy1.1	+	+	+
	Lyt1.2	-	+	+
	Lyt2	-	-	-
	$H\text{-}2D^k$	-	+	+
	$H\text{-}2K^k$	-	-	+
IL-2 secretion		-	+	+

Table 4. Incidence of metastases in animals inoculated intra-spleen with BW cells.

Tumor cells (a) (inoculation route)	Metastatic pattern incidence	location
BW-0→(AKR x BALB/c)F1, intra-spleen	2/6	spleen, liver (diffuse)
BW-0→(AKR x BALB/c)F1, intra-footpad	0/6	--
BW-0→ AKR, intra-spleen	10/10	spleen, liver (diffuse)

(a) BW-0 cells were inoculated intra-spleen and intra-footpad to either (AKR x BALB/c) or AKR animals.

Table 5. Metastatic capacity and membrane phenotype of spleen-induced BW variants.

Tumor cells (a)	Membrane phenotype				Metastatic capacity	
origin	Thy1.1	Thy1.2	Lyt1.2	$H\text{-}2^k$	incidence	location
BW-0	++	-	-	-	0/10	-
BW-0 → (AKR x BALB/c)F1 I.S. → spleen	++	-	+	+	10/10	spleen, liver
BW-0 → (AKR x BALB/c)F1 I.S. → liver	++	-	+	+	10/10	spleen, liver
BW-0 → AKR I.S. → spleen	++	-	+	+	10/10	spleen, liver
BW-0 → AKR I.S. → liver	++	-	+	+	10/10	spleen, liver

(a) BW variants were derived from either spleen or liver metastatic lesions from animals inoculated intra-spleen (I.S.) with BW-0 to either (AKR x BALB/c) or AKR animals.

REFERENCES.

1. Barski, G., Sorieul, S. and Cornefert, F. J. Natl. Cancer Inst. 26: 1269-1291, 1961.
2. Harris, H., Mimmer, O.J., Klein, G., Worst, P. and Tachibana, T. Nature 223: 363-368, 1969.
3. Fenyo, E.M., Wiener, F., Klein, G. and Harris, H. J. Natl. Cancer Inst. 51: 1865-1875, 1973.
4. Ber, R., Wiener, F. and Fenyo, E.M. J. Natl. Cancer Inst. 60: 931-933, 1978.
5. Goldenberg, D.M., Pavia, R.A and Tsao, M.C. Nature 250: 649-651,1974
6. De Baetselier, P., Gorelik, E., Eshhar, Z., Ron, Y., Katzav, S., Feldman, M; and Segal, S. J. Natl. Cancer Inst. 67: 1079-1087, 1981
7. De Baetselier, P., Gorelik, E., Eshhar, Z., Ron, Y., Katzav, S., Feldman, M; and Segal, S. In: In-vivo Immunology (Eds. P. Nieuwenhuis, A.A. Van den Broek and M.G. Hanna), Plenum, New York, 1982, pp. 179-185.
8. De Baetselier, P., Roos, E., Brys, L., Remels, L;, Gobert, M., Dekegel, D., Segal, S. and Feldman, M. Cancer Metastasis Revieuws 3: 5-24, 1984.
9. Roos, E. and Van de Pavert,I. Clin. Exp. Metast. 1: 173-180, 1983.
10. Napastek, H., Cohen, I.R., Fuks, Z. and Vlodavsky, I. Nature, 310: 241-244, 1984.
11. Roos, E., Van de Pavert, I. and Middelhoop, O.P. J. Cell Sci. 47: 385-397, 1981.
12. Kerbel, R.S., Lagarde, A.E., Dennis, J.W. and Donaghue, T.P. Mol. Cell. Biol. 3: 523-538, 1983.
13. De Baetselier, P., Roos, E., Brys, L;, Remels, L. and Feldman, M. Int. J. Cancer 34: 731-738, 1984.
14. Larizza, L., Schirrmacher, V., Graf, L. and Braun, M. Int. J. Cancer 34: 699-707, 1984.
15. Schirrmacher, V. Immunobiology 151: 89-98, 1980.
16. Witte, P.L. and Ber, R. J. Natl. Cancer Inst. 70: 575-577, 1983.
17. Slavin, S., Morecki, S. and Weiss, L. J. Immunol. 124: 586-589, 1980
18. Ron, Y., De Baetselier, P. and Segal, S. Eur. J. Immunol. 11: 94-99, 1981.
19. Ron, Y., De Baetselier, P., Feldman, M. and Segal, S. Eur. J. Immunol. 11: 608-611, 1981.

16

THE RECOGNITION AND DESTRUCTION OF METASTATIC CELLS BY TUMORICIDAL MACROPHAGES

I.J. FIDLER

Department of Cell Biology, The University of Texas, M. D. Anderson Hospital and Tumor Institute at Houston, Houston, Texas

The most devastating aspect of cancer is the propensity of malignant cells to spread from a primary site to distant organs, where new lesions develop. Indeed, the majority of deaths from cancer are attributed to the continued proliferation of metastases that are resistant to conventional therapeutics. There are several reasons for the failure of treatment for established metastasis. By the time of diagnosis of primary tumors, metastasis may have already occurred, but the lesions are often too small to be detected. Second, the anatomic location of many metastases may limit the effective dose of therapeutic agents that can be delivered to the lesions without being toxic to normal tissues. The third, and most formidable, problem is the heterogeneous nature of malignant neoplasms, which leads to the rapid biological diversification of tumor cells and to the development of resistance to conventional therapy (1,2).

The morphologic-cellular heterogeneity of neoplasms has been recognized for more than a century, since even early histologic observations of malignant neoplasms characterized them as pleomorphic. More recently, however, cells obtained from individual animal and human neoplasms have been shown to be phenotypically diverse with regard to growth rate, antigenic and/or immunogenic properties, cell-surface receptors and/or products, response to a variety of cytotoxic agents, invasion, and metastatic potential (1,2). Recent data from our laboratory and many others indicate that metastases can arise from the nonrandom spread of specialized subpopulations of cells that preexist within the primary tumor (3), that metastases can be clonal in their origin, and that different metastases can originate from different progenitor cells (4). These data provide an explanation for the clinical observations that even within the same patient, different metastases exhibit different susceptibilities to therapeutic modalities such as chemotherapy.

The issue of biologic heterogeneity is even more complex. Even within a solitary metastasis of proven clonal origin, diversity can develop rapidly (5,6). This process of diversification can result in part from phenotypic instability associated with clonal populations (7) or from a high rate of spontaneous mutation exhibited by cells with metastatic potential (8-10) or both. The implication of the varied response of tumor cells to conventional treatment modalities is that successful therapy of disseminated metastases will have to circumvent the problems of biologic heterogeneity and the development of resistance by tumor cells.

There is now an increasing body of data showing that macrophages activated to the tumoricidal state can fulfill these demanding tasks (11). In this presentation, I shall review some of the evidence to support this hypothesis, as well as sumarize work from our laboratory that deals with the methods to achieve the in situ activation of macrophages and the destruction of established metastases.

Since the 1800's, detailed microscopic studies have allowed investigators to conclude that mononuclear phagocytic cells are associated with processes of tissue turnover. These processes include tissue remodeling during metamorphosis, tissue remodeling during embryogenesis, tissue destruction and repair subsequent to injury, e.g. infection, and tissue renewal, such as the removal of damaged or senescent cells. In short, cells of the macrophage-histiocyte series are important components of the host system responsible for the maintenance of homeostasis (12). A primary function of macrophages in the body is the phagocytosis and disposal of effete cells, such as aged red blood cells (RBCs), cellular debris, and serum protein. The removal of effete RBCs is a continuous process that requires that macrophages distinguish old from young cells, as well as damaged from healthy cells (13). In addition to phagocytosis, macrophages are involved in the controlled metabolism of lipids and iron, in host response to injury (inflammation), and in the defense against microbial infections and parasitic infestations. Macrophages are also an important component of both the afferent and efferent arms of the immune system, and finally these cells are important in the defense against neoplasms (11).

Hibbs (14) suggested that macrophages may provide a surveillance system for the detection and destruction of nascent transformed neoplastic cells. Norbury and Kripke (15) supported this suggestion in studies on skin carcinogenesis induced by ultraviolet (UV) radiation in mice. When the dose of the carcinogen (UV) was not overwhelming, treating mice with a macrophage stimulant (pyran copolymer) prolonged the latent period of tumor development and protected against carcinogenesis. Conversely, treatment of mice with macrophage toxins (carrageenan or silica) shortened the latent period of skin cancer induction. There are also several published reports regarding the efficacy of macrophages in the inhibition of metastasis. Syngeneic mouse macrophages activated in vitro and then injected intravenously reduced the formation of B16 melanoma metastasis (16), and the intravenous injection of nonspecifically activated macrophages prevented the formation of spontaneous fibrosarcoma metastases (17).

INTERACTION OF MACROPHAGES WITH HETEROGENEOUS NEOPLASMS

There are two major ways to achieve macrophage activation in vivo (11). Frequently macrophages are activated as a consequence of their interaction with microorganisms and/or their products, for example, endotoxins, the bacterial cell wall skeleton, and small components of the bacterial cell wall skeleton such as muramyl dipeptide (MDP) (18). In vivo activation of macrophages can also take place after their interaction with soluble mediators released by antigen- or mitogen-sensitized lymphocytes. One soluble lymphokine that induces macrophage activation is referred to as a macrophage-activating factor (MAF). The MAF first binds to a macrophage surface receptor and then is internalized to elicit tumoricidal properties in the macrophages (19).

Recently, a great deal of interest has been devoted to the activation of macrophages by the lymphokine interferon gamma (IFNγ). This lymphokine can synergize with MDP after it binds to a macrophage surface receptor, and then it is internalized to elicit tumoricidal properties in macrophages (20,21).

Regardless of the method of their activation, rodent and human tumoricidal macrophages acquire the ability to recognize and destroy neoplastic cells both in vitro and in vivo while leaving nonneoplastic cells unharmed. The mechanism for this is not known, but appears to be nonimmunologic in nature and to require intimate cell-to-cell contact (22,23). The result of investigations in diverse mammalian systems suggests that macrophage-mediated cytolysis of tumor cells occurs independently of such tumor cell characteristics as metastatic potential, antigenicity and immunogenicity, and drug sensitivity (11).

The major limitations of many cancer therapies are the lack of selectivity and their toxic effect on the patient. The ability of tumoricidal macrophages to distinguish tumorigenic from normal cells presents an attractive possibility for treatment of disseminated cancer. We examined whether human blood monocytes that had been activated to the tumoricidal state by their interaction with liposomes containing immunomodulators can also discriminate _in vitro_ between tumorigenic and nontumorigenic human target cells under cocultivation conditions (24,25). Highly homogeneous preparations of peripheral blood monocytes isolated from normal human donors were activated _in vitro_ by incubation with human MAF encapsulated with liposomes. The cytotoxic properties of these monocytes against several tumorigenic and nontumorigenic allogeneic targets were assessed by an _in vitro_ radioisotope-release assay. Various combinations of tumorigenic and nontumorigenic targets labeled with either [^{3}H]thymidine or [^{14}C]thymidine were mixed and plated onto monolayers of blood monocytes. This experiment was repeated several times with monocytes from different donors. Tumorigenic A375 melanoma, HT-29 colon carcinoma, or NAT-glioblastoma and nontumorigeneic lung cells, dermal fibroblasts, or kidney cells were prelabeled with either [^{3}H]thymidine or [^{14}C]thymidine. Various combinations of tumor and normal cells were cultured alone, with control monocytes, or with tumoricidal monocytes for 72 hr. Control monocytes were not cytotoxic against any of the tumorigenic or nontumorigenic targets. Monocytes preincubated with liposomes containing immunomodulators (MAF or MDP) lysed tumorigenic A375 melanoma, HT-29 colon carcinoma, and glioblastoma cells, but left the lung cells, dermal fibroblasts, and kidney cells unharmed. The ability of tumoricidal monocytes to lyse allogeneic tumor cells was independent of the radiolabel used. In all the combinations, activated monocytes lysed neoplastic cells but not nonneoplastic cells. Moreover, when two different tumorigenic target cells were cocultivated with activated monocytes, release of both ^{3}H and ^{14}C was detected. In contrast, neither radiolabel was released from cocultivated cultures of nontumorigenic target cells (24,25).

Collectively then, the data indicate that the susceptibility of neoplastic cells to lysis by activated macrophages is independent of tumor cell characteristics such as antigenicity-immunogenicity, invasion and metastasis, growth rate, and resistance to lymphocyte- or natural killer (NK) cell-mediated lysis (24-26). Of particular importance for cancer treatment is the finding that tumor cell variants selected for resistance to the drug Adriamycin remain fully susceptible to destruction by tumoricidal macrophages (27). Recently, we attempted to select *in vitro* tumor cell variant lines that exhibit a phenotype of resistance to macrophage-mediated lysis. We used techniques similar to those used previously to successfully select a B16 melanoma tumor cell line resistant to lysis by cytotoxic T-lymphocytes (28) or UV-2237 fibrosarcoma resistant to NK cell-mediated lysis (29). Seven different heterogeneous murine neoplasms (4 fibrosarcomas, a melanoma, a rhabdomyosarcoma, and an osteogenic sarcoma) and one cloned line of a fibrosarcoma were incubated *in vitro* with syngeneic tumoricidal macrophages. Surviving tumor cells were recovered and expanded to undergo subsequent interaction with tumoricidal macrophages. After 6 sequential interactions, all cell lines were examined for their susceptibility to lysis mediated by murine peritoneal exudate macrophages activated with liposomes containing muramyl tripeptide phosphatidylethanolamine. In all 8 systems, no significant differences were detected between the parent tumor cells and cells that survived the sequential interactions. Neither macrophage infiltration into subcutaneous tumors, nor the experimental or spontaneous metastatic potentials of the parental tumors differed from the lines established by cells surviving macrophage-mediated lysis (26). Collectively, the data suggest that tumor cell destruction by activated macrophages is nonselective and does not lead to the development of resistant tumor cells, nor to cells with altered metastatic properties (26).

Taken together, the data generated in rodent and human systems indicate that, at least *in vitro*, tumoricidal macrophages can discriminate between neoplastic and nonneoplastic cells by a process that is independent of transplantation antigens, species-specific antigens, tumor-specific antigens, cell cycle time, or various phenotypes associated with transformation. Although the exact mechanism(s) by which macrophages recognize and lyse tumor cells is still unclear, it is probably regulated by a tumor cell characteristic that is linked with the tumorigenic capacity of tumor cells (30).

IN VIVO ACTIVATION OF MACROPHAGES TO BECOME TUMORICIDAL

In general, attempts to specifically activate macrophages in vivo to enhance host defense against metastases have been unsuccessful. Systemic administration of MAF or IFN-γ is hindered by the lack of purified preparations of MAF and by the fact that lymphokines injected intravenously have a very short half-life, and thus do not activate macrophages in vivo. The systemic activation of macrophages with microorganisms or their products has also suffered from major drawbacks. For instance, administration of whole bacteria or low levels of endotoxins activates various effector cells and is accompanied by serious toxic activity problems such as allergic reaction and granuloma formation (31). The discovery of MDP, a small component of the bacterial cell wall that is capable of activating macrophages (32), has provided an exciting possibility for activating macrophages. However, the use of water-soluble synthetic MDP is limited because by 60 min after parenteral administration, this agent is cleared from the body to be excreted in the urine (33). This brief period is insufficient to activate macrophages, even under ideal in vitro conditions (34).

Recent advances in liposome technology have provided a mechanism for activating macrophages in situ with immunomodulators (35). There are several advantages to using liposome-encapsulated materials to activate cells of the macrophage-histiocyte series in vivo. Liposomes can be used to carry agents to cells of the reticuloendothelial system, since these cells are responsible for the rapid clearance of particulate material from the circulation. Many macrophage-activating agents such as bacterial products or lymphokines can be antigenic, and repeated systemic administrations can lead to adverse reactions. Liposomes consisting of natural phospholipids are nonimmunogenic, and thus elicitation of allergic reactions commonly associated with the systemic administration of other immune adjuvants can be avoided.

Numerous studies have recently shown that immunomodulators entrapped in liposomes are very efficient in activating macrophages to become tumoricidal in vitro (11). Unlike free biological agents, which require binding to a macrophage surface receptor (36), liposome-entrapped immunomodulators enter the cytoplasm via phagocytosis and can activate macrophages that lack receptors for that agent (36). Moreover, for activators such as MAF, IFN-γ and MDP that are ordinarily degraded or cleared from the body too rapidly for effectiveness, encapsulation in liposomes extends their active half-life within the body and enables these agents to activate macrophages *in situ* (37,38).

We examined whether the *in situ* activation of macrophages by intravenous administration of liposomes containing MDP resulted from direct interaction of MDP and macrophages or occurred indirectly by the action of lymphokines released by sensitized T cells. We used three groups of mice with impaired T-cell function: mice exposed to UV radiation, adult mice exposed to thymectomizing doses of X-rays, and athymic nude mice. Mice were given intravenous injections of multilamellar liposomes containing MDP or placebo preparations. Their alveolar macrophages were harvested 24 hr later and assessed for cytotoxic properties against syngeneic tumor targets. Macrophages harvested from all groups of mice became tumoricidal in response to liposome-encapsulated MDP, thus demonstrating that a liposome-MDP activation of macrophages occurred directly and was a thymus-independent process (39).

There are numerous reports indicating that macrophage function can be suppressed in mice bearing tumors (40). If this were the case, then the activation of macrophages in animals with progressing metastasis would not be feasible. We examined this issue by *in vivo* activation of alveolar macrophages in rats with rapidly progressing lung metastases of a syngeneic mammary adenocarcinoma (41). The rats were given i.v. injections of *Nocardia rubra* cell wall skeleton preparations. Even large (>2 mm) metastases did not impair the ability of lung macrophages to respond to the activation stimulus and become tumoricidal (41). Similarly, intravenous administration of liposomes containing immunomodulators can activate fixed liver macrophages (Kupffer cells) to become tumor cytotoxic (42).

ERADICATION OF METASTASES BY THE SYSTEMIC ADMINISTRATION OF LIPOSOMES CONTAINING IMMUNOMODULATORS

These data raise the possibility that macrophages can be activated *in situ* to the tumoricidal state by systemically administered immunomodulators encapsulated in liposomes, and that these could be a potential therapeutic modality for enhancing host destruction of metastases. To test this possibility, we treated mice bearing spontaneous metastases with intravenous injections of multilamellar vesicle liposomes consisting of phosphatidylserine and phosphatidylcholine, and containing entrapped immunomodulators. The B16-BL6 melanoma cell line, which is syngeneic to C57BL/6 mice, was used as the screening model to determine the effectiveness of liposome-encapsulated immunomodulators in the treatment of metastases. After implantation in the footpad, this tumor metastasizes to lymph nodes and the lungs in more than 90% of the mice (43). C57BL/6 mice were each given an intrafootpad injection of melanoma cells. Four weeks later, when the tumors had reached a size of 10 to 12 mm, the leg bearing the tumor, including the popliteal lymph node, was amputated. Three days later, animals were injected intravenously with liposomes containing immunomodulators or placebo preparations. Both test and control groups were treated twice weekly for four weeks. Liposomes were prepared from a chromatographically pure egg phosphatidylcholine and beef brain phosphatidylserine (7:3 mole ratio). We have used these phospholipid liposomes as carriers because they are not toxic at the dose used (44) and because they are arrested efficiently in the lungs as well as in organs of the reticuloendothelial system following intravenous injection (45,46). The internal volume of the multilamellar liposomes is 2.5 ± 0.3 μl/μmole of phospholipid. For the _in vivo_ studies, mice were injected intravenously with 5 μmole of liposomes and thus received 12 μl of entrapped suspension (46).

In experiments designed to evaluate the efficacy of liposome-encapsulated MAF for eradication of spontaneous metastases, control groups of animals were injected intravenously with phosphate-buffered saline solution (PBS), with free MAF (200 μl), or with liposomes that contained PBS but were suspended in 200 μl of free MAF. In experiments to evaluate the efficacy of liposome-encapsulated MDP in eradicating spontaneous metastases, mice were injected with liposomes containing 2.5 μg of MDP. Control groups were injected intravenously with PBS, with free MDP (100 μg per mouse), or with liposomes that contained PBS and that were suspended in PBS containing 2.5 μg MDP per mouse. The mice were monitored daily, and dead or moribund mice were necropsied. Spontaneous metastases in the lungs and lymph nodes were well established at the time liposome treatment began. Many individual metastases could be seen macroscopically. As shown in Table 1, virtually all the mice treated intravenously with PBS, with free MAF, with free MDP, or with liposomes containing PBS were dead by day 90 of the experiment, i.e., 60 days after the amputation of the tumor-bearing leg. In marked contrast, 66% of mice injected intravenously with liposome-encapsulated MAF and 60% of mice injected with liposome-encapsulated MDP were alive when the epxeriments were terminated at 200 days. In this tumor system, at the time of first liposome treatment, the metastases contained an estimated 10^7 cells. Since the median survival time of mice injected with 10 viable B16 cells (admixed with 10^6 dead cells) is 40 to 50 days (17), we speculate the tumor burden in the successfully treated mice (alive on day 200) must have been reduced to less than 10 viable cells. Similar data on the successful treatment of other murine metastatic tumors by the systemic administration of liposomes containing different immunomodulators are now available (48-50). Studies on mice treated with liposome-encapsulated MDP that have residual metastatic disease (albeit reduced compared with untreated control mice) have revealed that the tumor cells present in the lesions are still fully susceptible to destruction by activated macrophages.

This is consistent with the evidence discussed earlier that suggests that tumor cell resistance to killing by activated macrophages is not likely to be a rate-limiting factor in the effectiveness of liposome-encapsulated macrophage activators. The more challenging problem will be the extent of the tumor burden at the time of therapy that can be handled by activated macrophages.

Table 1. Treatment of spontaneous lymph node and lung melanoma metastases by the intravenous injection of liposomes containing MDP.

Treatment group	Survival on day*			
	40	60	90	200
Control mice, saline	30/30	14/30	1/30	1/30
Free MDP, 100 μg	15/15	9/15	1/15	1/15
Liposomes containing PBS suspended in 6 μg of MDP	15/15	7/15	1/15	1/15
Liposomes containing 6 μg of MDP	15/15	9/15	9/15	9/15
Free MAF	14/14	10/14	2/14	1/14
Liposomes containing MAF	15/15	13/15	10/15	10/15

* Survival indicates the number of live mice/total number of mice. The differences in median life span are highly significant ($P<0.001$).

SYNERGISM BETWEEN DIFFERENT IMMUNOMODULATORS ENCAPSULATED IN LIPOSOMES FOR THE TREATMENT OF METASTASES

In an effort to enhance the efficacy of macrophage-mediated destruction of metastasis, we have recently shown that the combination of a lymphokine and MDP at suboptimal concentrations in the same liposome preparations significantly enhances the therapeutic potential against metastasis (51). Mice were injected intravenously with liposomes containing various dilutions of MAF, various dilutions of MDP, combinations of MAF and MDP, or control liposome preparations. Alveolar macrophages were harvested 24 hr later, and their antitumor activity was monitored _in vitro_. Liposomes containing combinations of MAF and MDP were most effective in generating cytotoxic properties in the lung macrophages. Moreover, this therapeutic regimen was very efficacious in the treatment of relatively large tumor burdens (in metastases). The B16 melanoma therapy experiments described previously were carried out again, but this time i.v. administration of liposomes began 7 days after the surgical removal of the subcutaneous (leg) melanoma. In this tumor system, volume doubling time of metastasis is about 3-5 days (47), and thus spontaneous metastases were well established, with some lung metastases being larger than 1 mm in diameter. Here again, liposomes were injected twice weekly for 4 weeks (8 i.v. injections). The survival data shown in Table 2 demonstrate that practically all the mice injected with saline (control) or with liposomes containing saline and free MAF and MDP had died by day 90 of the experiment. Treatment with optimal doses of MDP (6.25 μg) or MAF (1:2 dilution) resulted in the long-term survival (~ 250 days) of about 30% of the mice. Neither MAF at a 1:20 dilution nor MDP at doses lower than 0.6 μg was therapeutically effective. In contrast, treatment of mice with liposome preparations that contained both MAF and MDP at individual suboptimal concentrations resulted in long-term survival (~ 250 days) of 50% of the mice (9/18) (51).

Table 2. Synergism between MAF and MDP encapsulated in liposomes for activation of macrophages and destruction of B16 melanoma metastases.

Treatment group	Alveolar macrophage mediated cytotoxicity*	Survival on day**		
		40	90	250
Control mice, saline	2	20/20	1/20	1/20
Liposome-saline	4	11/18	2/18	2/18
Liposome-MAF (1:2)	45	18/18	5/18	5/18
Liposome-MAF (1:20)	8	18/18	2/18	2/18
Liposome-MDP (6.25 μg)	38	20/20	6/20	6/20
Liposome-MDP (0.3 μg)	9	19/19	2/19	2/19
Liposome MAF (1:20) and MDP (0.3 μg)	42	18/18	9/18	9/18

* Cytotoxicity was determined in vitro.

** Survival indicates the number of live mice/total number of mice. The differences in median life span are highly significnt (P<0.001).

EVIDENCE FOR THE INVOLVEMENT OF MACROPHAGES IN THE DESTRUCTION OF METASTASES.

The regression of established metastases after the systemic administration of liposomes containing immunomodulators required the activation of macrophages to become tumoricidal. Several lines of evidence tend to support this conclusion. First, administration of macrophage-activating agents encapsulated within liposomes that are not retained in the lung fails to activate lung macrophages, and there is no regression of lung metastases (52). Second, the pretreatment of tumor-bearing animals with agents that are toxic for macrophages (silica, carageenan, hyperchlorinated drinking water) before systemic therapy with liposome-encapsulated immunomodulators abrogates the response to liposome therapy, and such animals rapidly die of metastatic disease (52). Third, intravenous injection of macrophages activated _in vitro_ by incubation with liposomes containing MAF effects a reduction in metastatic burden comparable to that achieved by systemic administration of liposome-encapsulated activators (52). Fourth, morphologic and functional analysis of macrophages isolated from pulmonary metastases of mice given intravenous injections of liposomes containing a lipophilic MDP derivative, muramyl tripeptide phosphatidylethanolamine, showed that macrophages containing phagocytosed liposomes infiltrate and localize within pulmonary metastases. Furthermore, once localized, macrophages that have engulfed liposomes containing MDP, but not macrophages that have engulfed liposomes containing a placebo, were tumoricidal against the metastatic tumor cells (53).

SUMMARY

Activated macrophages appear to be able to recognize and destroy neoplastic cells without regard to their phenotype, and macrophage-mediated cytotoxic activity appears invulnerable to the problem of cellular resistance to killing that is routinely encountered in efforts to destroy tumor cells by cytotoxic drugs. Systemic administration of liposomes containing immunomodulators has been shown to activate rodent macrophages _in situ_ and thereby to augment host resistance against spontaneous metastases. The stimulation of host antitumor responses by liposome-encapsulated lymphokines or MDP did not require the participation of functional T lymphocytes or natural killer cells. Rather, the destruction of established metastases after the multiple systemic administration of liposome-encapsulated immunomodulators was mediated by activated macrophages.

Although the results we report are encouraging, it is unlikely that this therapeutic approach could serve for treatment of large tumor burdens. Large tumor burdens, however, can be reduced with a variety of other treatment modalities such as radiotherapy or chemotherapy. Tumoricidal macrophages can then be used to search for and destroy those few tumor cells that resisted conventional therapy.

REFERENCES

1. Poste, G., and Fidler, I.J. Nature 283:139-146, 1979.
2. Fidler, I.J., and Hart, I.R. Science 217:998-1003, 1982.
3. Fidler, I.J., and Kripke, M.L. Science 197:893-895, 1977.
4. Talmadge, J.E., Wolman, S.R., and Fidler, I.J. Science 217:361-363.
5. Poste, G., Tzeng, J., Doll, J., Greig, R., Reiman, D., and Zeidman, I. Proc. Natl. Acad. Sci. USA, 79:6574-6579, 1982.
6. Talmadge, J.E., Benedict, K., Madsen, J., and Fidler, I.J. Cancer Res. 44:3801-3805, 1984.
7. Poste, G., Doll, J., and Fidler, I.J. Proc. Natl. Acad. Sci. USA 78:6226-6230, 1981.
8. Cifone, M.A., and Fidler, I.J. Proc. Natl. Acad. Sci. U.S.A. 78:6949-6952, 1981.
9. Bosslet, K. and Schirrmacher, V. Int. J. Cancer 29:195-202, 1982.
10. Harris, J.F., Chambers, A.F., Hill, R.P., and Ling, V. Proc. Natl. Acad. Sci. U.S.A. 79:5547-5551, 1982.
11. Fidler, I.J., and Poste, G. Springer Semin. Immunopathol. 5:161-174, 1982.
12. Fidler, I.J. Lymphokine Res. 3:51-54, 1984.
13. Tanaka, Y., and Schroit, A. J. Biol. Chem. 258:11335-11343, 1983.
14. Hibbs, J.B. Jr. JNCI 53:1487-1492, 1974.
15. Norbury, K.C., and Kripke, M.L. J. Reticuloendothel. Soc. 26:827-837, 1979.
16. Fidler, I.J. Cancer Res. 34:1074-1078, 1974.
17. Liotta, L.A., Gattozzi, C., Kleinerman, J., et al., Br. J. Cancer 36:639-641, 1977.
18. Chedid, L., Carelli, L., and Audibert, F. Reticuloendothel. Soc. 26:631-641, 1979.
19. Fidler, I.J., and Raz, A. In: Lymphokines (Ed. E. Pick), Academic Press, Inc., New York, Volume 3, 1981, pp. 345-363.
20. Celada, A., Gray, P.W., Rinderknect, E., and Schreiber, R.D. J. Exp. Med. 160:55-62, 1984.
21. Kleinerman, E.S., Zicht, R., Sarin, P.S., Gallo, R.C. and Fidler, I.J. Cancer Res. 44:4470-4475, 1984.
22. Hibbs, J.B., Jr. Science 184:468-471, 1974.
23. Bucana, C., Hoyer, L.L., and Hobbs, B. et al. Cancer Res. 36:4444-4458, 1976.
24. Fidler, I.J., and Kleinerman, E.S. J. Clinical Oncol. 2:937-943, 1984.

25. Kleinerman, E.S., Erickson, K.L., Schroit, A.J., et al. Cancer Res. 43:2010-2014, 1983.
26. Fogler, W.E., and Fidler, I.J. Cancer Res. 45:14-18, 1985.
27. Giavazzi, R., Bucana, C.D., and Hart, I.R. JNCI 73:447-455, 1984.
28. Fidler, I.J., Gersten, D.M., and Budmen, M.B. Cancer Res. 36:3160-3165, 1976.
29. Hanna, N., and Fidler, I.J. JNCI 66:1183-1190, 1981.
30. Fidler, I.J., Roblin, R.O., and Poste, G. Cell. Immunol. 38:131-146, 1978.
31. Allison, A.C. J. Reticuloendothel. Soc. 26:619-630, 1979.
32. Lederer, E. J. Med. Chem. 23:819-825, 1980.
33. Parant, M., Parant, F., and Chedid, L. et al. Int. J. Immunopharmacol. 1:35-41, 1979.
34. Sone, S., and Fidler, I.J. Cell Immunol. 57:42-50, 1981.
35. Schroit, A.J., Hart, I.R., Madsen, J. and Fidler, I.J. J. Biol. Response Modifiers 2:97-100, 1983.
36. Poste, G., Kirsh, R., and Fidler, I.J. Cell Immunol. 44:71-88, 1979.
37. Poste, G., and Kirsh, R. Cancer Res. 39:2582-2590, 1979.
38. Fidler, I.J., and Fogler, W.E. Lymphokine Res. L:73-77, 1982.
39. Fidler, I.J. J. Immunol. 127:1719-1720, 1981.
40. Snyderman, R., Pike, M.C., Blaylock, B.L. and Weinstein, P. J. Immunol. 116:585-589, 1976.
41. Sone, S. and Fidler, I.J. Cancer Res. 41:2401-2406.
42. Xu, Z.L. and Fidler, I.J. Cancer Immunol. Immunother. 18:118-122, 1984.
43. Hart, I.R. Am. J. Pathol. 97:587-600, 1979.
44. Hart, I.R., Fogler, W.E., Poste, G., et al. Cancer Immunol. Immunother. 10:157-166, 1981.
45. Fidler, I.J., Raz, A., Fogler, W.E., et al. Cancer Res. 40:4460-4466, 1980.
46. Schroit, A.J., and Fidler, I.J. Cancer Res. 42:1616-1617, 1982.
47. Griswold, D.P., Jr. Cancer Chemother. Rep. 3:315-323, 1972.
48. Deodhar, S.D., Barna, B.P., Edinger, M., et al. J. Biol. Response Modifiers 1:27-34, 1982.
49. Lopez Berestein, G., Milas, L., Hunter N., Mehta, K., Eppstein, D., Van der Pas, M.A., Mathews, T.R., and Hersh, E.M. Clin. Exp. Met. 2:366-377, 1984.
50. Phillips, N.C., Moras, M.L., Chedid, L., Lefrancier, P., and Bernard, J.M. Cancer Res. 45:128-134, 1985.
51. Fidler, I.J., and Schroit, A.J. J. Immunol. 133:515-519, 1984.
52. Fidler, I.J., Barnes, Z., Fogler, W.E., Kirsh, R., Bugelski, P. and Poste, G. Cancer Res. 42:496-501, 1982.
53. Key, M.E., Talmadge, J.E. Fogler, W.E., Bucana, C., and Fidler, I.J. JNCI 69:1189-1198, 1982.

17

BIOLOGICAL RESPONSE MODIFIERS FOR THE THERAPY OF METASTASES

RONALD B. HERBERMAN, ROBERT H. WILTROUT, RAOUL R. SALUP, JOHN R. ORTALDO AND ELIESER GORELIK

Biological Therapeutics Branch, Biological Response Modifiers Program, Division of Cancer Treatment, National Cancer Institute, Frederick, MD 21701

A wide variety of biological response modifiers, with the ability to affect a diverse range of host resistance mechanisms, have considerable potential for the treatment of tumor metastases (1). Several biological response modifying approaches were discussed in detail in other manuscripts in this proceedings. Therefore, the emphasis here will be on alternative approaches which have yielded promising results.

ROLE OF NATURAL KILLER CELLS IN RESISTANCE AGAINST METASTASES

There are increasing indications that natural killer (NK) cells play an important role in resistance against the development and growth of metastases. An initial indication for this came from studies by Gorelik, et al. (2,3). Tumor cells isolated from metastases were found to be more resistant to NK activity when compared to those isolated from locally growing tumors. Conversely, tumor cells selected in vitro for resistance to NK activity were found to metastasize more readily than the parental, unselected tumor. Subsequently, a series of studies demonstrated that the number of metastases developing from several different tumors was inversely correlated with the level of NK activity in the host (e.g. 4,5). When NK activity was depressed, more metastases were observed, and conversely when NK activity was high or increased, there was more resistance to formation of metastases. The treatments which have been used for depressing NK activity include broadly acting immunosuppressive agents such as cyclophosphamide or estrogen compounds. However, in addition, similar results have been observed when NK activity has been quite selectively depressed (5). For example, administration of an antiserum against the glycolipid asialo GM_1 produces a selective deficit in NK activity. Similarly, the beige mouse has a selective deficit in NK activity, with most of the other parameters of the immune response remaining intact.

Most of the available evidence has pointed toward a key role of NK cells in rapid elimination of circulating tumor cells, before they can extravasate and establish metastases. An initial indication for the important involvement of NK cells in these early phases of the metastatic process came from studies by Riccardi, et al. (6), in which radiolabelled tumor cells were injected intravenously. It was found that very rapid elimination of the tumor cells from the lungs and other organs correlated closely with levels of NK activity in the recipients. This elimination occurred within a few hours after injection of the tumor cells, prior to the time of extravasation into the parenchyma. This ability of NK cells to eliminate tumor cells from the circulation has subsequently been demonstrated to contribute to the resistance to development of detectable metastases (4,5). In addition to observing such a correlation after intravenous inoculation of tumor cells, it has also been possible to demonstrate the contribution of NK cells to resistance to the development of spontaneously developing metastases after surgical removal of primary tumors growing in the footpad (5). Performance of such studies with the B16 melanoma in beige mice or in mice treated with anti-asialo GM_1 serum has demonstrated a substantial increase in the number of metastases which developed in the lungs of the NKdepressed mice. In addition, metastases have also been observed in the liver and other organs of such mice, even though these are not usual sites for the development of metastases by this tumor. Thus, the failure of the B16 melanoma to metastasize to the liver and other organs appears to be related to the NK cell-mediated resistance to establishment of metastases in these organs rather than to an inherent inability of the tumor cells to localize at these sites.

To more definitively demonstrate that NK cells play a critical role in resistance to the formation of metastases, Barlozzari, et al. (7) established the conditions for selective reconstitution of NK activity, in rats which had been pretreated with anti-asialo GM_1 antiserum, by intravenous transfer of a purified population of large granular lymphocytes (LGL). This small subset of cells, which account for most if not all NK activity, largely restored NK activity in rats pretreated with anti-asialo GM_1. In contrast, a comparable infusion of T cells did not restore NK activity. This system was then utilized for examining the development of metastases from a syngeneic mammary adenocarcinoma (8). Intravenous inoculation of tumor cells produced metastases in the lungs. When NK activity was depressed by pretreatment with anti-asialo GM_1, there was a substantial increase in the number of pulmonary metastases. In contrast, rats infused with purified LGL after pretreatment with anti-asialo GM_1 developed very few metastases. In contrast, following transfer of T cells into the asialo GM_1-treated rats, a considerable number of metastases still developed in the lungs.

An issue which is frequently raised regarding the potential importance of NK cells in resistance against metastases is whether the available information is applicable only to established tumor cell line or whether it may also be generalized to primary autochthonous tumors. There is increasing evidence that fresh tumor cells, harvested directly from autochthonous tumors, are susceptible to lysis by NK cells, even by NK cells from the autochthonous tumor-bearing host (e.g. 9-11). Highly purified populations of both human and mouse LGL have been shown to have substantial cytotoxic reactivity against a variety of freshly harvested tumors, including some carcinomas as well as some lymphomas and leukemia cells. The levels of reactivity against fresh tumor cells become particularly impressive if the assay is performed over an extended, overnight period of incubation rather than the usual four hour period. In addition, the NK reactivity against fresh tumor cells can be substantially augmented by pretreatment of the effector cells by various stimuli like interferon or interleukin 2.

TREATMENT OF METASTASIS BY AUGMENTATION OF NK ACTIVITY

With the substantial evidence for an important role of NK cells in resistance against metastases, and for reactivity against autochthonous tumors as well as tumor cell lines, strategies for the therapy of metastases might be focused on the augmentation of NK activity, particularly in the organs most susceptible to the development of metastases. Recent studies by Wiltrout, et al. (12) have demonstrated that NK cells are not limited to the blood or spleen but also can accumulate in some of the major organs, particularly the liver and lungs. Pretreatment of mice with various biological response modifiers e.g. the pyran copolymer MVE-2, polyI:polyC or C.parvum, has been shown to cause a major increase in the number of LGL in these organs, which is associated with very high levels of NK activity. These and other biological response modifers have pleiotropic actions, affecting the functions of a variety of effector cells in addition to NK cells. Therefore an important issue is whether the protective effects of these BRM's on tumor growth or metastases are mediated by stimulation of NK activity or by stimulation of the activity of other effector cells, e.g. macrophages. To address this question, mice have been treated not only with a BRM but also with anti-asialo GM_1 or cyclophosphamide to selectively depress NK activity (13). In the initial experiments, it appeared that NK cells were not involved in the strong resistance to metastases induced by MVE-2, since pretreatment with anti-asialo GM_1, at a concentration sufficient to almost completely ablate NK activity in the blood and spleen and to almost completely inhibit the rapid tumor cell clearance in the vasculature, did not interfere with the resistance against metastases. However, it was then found that this treatment was not sufficient to eliminate the high levels of NK activity in the liver or lungs. By using more vigorous treatment with the antiserum, sufficient to completely ablate NK activity in each of the compartments, i.e. in the tissues as well as the vasculature, the protective effect of MVE-2 was abrogated. These results indicate the important role of NK cells in the resistance to metastases induced by MVE-2 and also demonstrate that NK cells can play an important role in the establishment of metastases in the major organs, beyond the point of extravasation from the vasculature.

COMBINED ANTIMETASTATIC EFFECTS OF ANTICOAGULANT AND BIOLOGICAL RESPONSE MODIFIERS

Various anticoagulants have been demonstrated to have antimetastatic effects, for some human tumors as well as for tumors in experimental animals (e.g. 14). This has been mainly attributed to the ability of the anticoagulants to interfere with the attachment of circulating tumor cells to the vascular endothelium, thereby interfering with extravasation into the parenchyma of the major organs. An alternative hypothesis is that tumor cells may be coated with clotting factors and thus protected from interactions with effector cells of the host. To explore the possibility that anticoagulants facilitate interactions between potentially metastatic tumor cells and NK cells, studies were performed in which anticoagulants were given alone or in combination with NK-augmenting or NK-depressing treatments (15). Treatment of mice with heparin, warfarin or prostacyclin plus theophylline as anticoagulants had a significant protective effect on the development of metastases from B16 melanoma (Table 1) or from 3LL and M109 lung carcinomas. Treatment with a biological response modifier, e.g. poly ICLC, in addition to one of these anticoagulants led to increased antimetastatic effects. Conversely, when NK activity was selectively blocked by treatment of mice with anti-asialo GM_1, a large number of metastases developed in the lungs and there were virtually no long term survivors. These results suggested that the antimetastatic effects of the anticoagulants were dependent on the presence of NK cells. To more rigorously test this hypothesis, experiments were performed in which NK activity was selectively restored in NK-depressed mice, to determine whether this would reconstitute in parallel the antimetastatic efficacy of the anticoagulants. In these experiments, NK activity was depressed by cyclophosphamide and this resulted in a large number of metastases. As with the anti-asialo GM_1 treatment, the treatment with cyclophosphamide blocked the ability of anticoagulants to have an antimetastatic effect. However, when NK activity was restored by adoptive transfer of spleen cells, the antimetastatic effect of the anticoagulant was restored in parallel. In contrast, when the spleen cells were pretreated with anti-asialo GM_1 and complement, to remove NK activity but not affect macrophages or most other cell types, the transferred cells did not reconstitute the antimetastatic effects of the anticoagulant. This further supports the concept that the antimetastatic effect of anticoagulants is dependent on the involvement of NK cells.

The practical implication from this conclusion is that improved therapeutic efficacy against metastases might be achieved by combined treatment with an anticoagulant and a biological response modifier with the ability to induce in vivo augmentation of NK activity. Such additive or synergistic antimetastatic effects have been observed when treatment with heparin or other anticoagulants was combined with poly ICLC or other NK-augmenting biological response modifiers. The logistical limitation to this approach is that it appears to be effective mainly while tumor cells are in the circulation. Upon pretreatment of mice with these agents prior to, or simultaneous with, intravenous challenge with tumor cells, there was a marked antimetastatic effect. In contrast, when the treatments were given more than four hours after challenge with tumor cells, the

Table 1

Antimetastatic Effect of Anticoagulants in Mice with Augmented or Suppressed NK Cell Function

Treatment	Theophylline + PGI_2		Heparin		Warfarin	
	Median no. of lung metastases	Percent of metastasis-free mice	Median no. of lung metastases	Percent of metastasis-free mice	Median no. of lung metastases	Percent of metastasis-free mice
None	83	0	100	0	60	0
Poly I:C	3	0	30	0	5	20
Anticoagulant	35	0	11	0	1	30
Poly I:C + Anticoagulant	1	25	1	30	0	70
Anti-asGM_1	147	0	>300	0	142	0
Anti-asGM_1 + Anticoagulant	151	0	>300	0	105	0

antimetastatic effect was no longer observed. Therefore, it appears that the clinical analogy of this strategy would be to attempt to interfere with the establishment of metastases during a period in which tumor cells are likely to be in the circulation. A focus on such treatments during the perioperative period, when circulating tumor cells frequently have been observed, would seem warranted.

THERAPY OF METASTASES BY ADOPTIVE TRANSFER OF EFFECTOR CELLS PLUS CYTOKINES

The major limitation with current approaches to therapy with biological response modifiers is that there is insufficient information regarding the protocols for optimal augmentation of the activity of effector cells, particularly at the site of tumor growth. Until such optimal biological response modifying protocols are developed, an interim approach would be to directly transfer large numbers of activated effector cells into the tumor-bearing host. An example of the successful application of this approach has come from studies with a transplantable renal cell carcinoma in mice, in which the course of tumor growth closely parallels the clinical situation (16). Inoculation of the tumor cells under the kidney capsule results initially in local tumor growth and then regional invasion, followed by distant metastases to the liver and lungs. Therapeutic attempts were initiated at one week after transplantation, at a time when metastases appear to already have occurred (Table 2) (17). Removal of the kidney at this time did not result in long term survival of the mice. Chemotherapy at this time with doxorubicin hydrochloride or with cyclophosphamide had only modest antitumor effects but did not result in cures of a substantial proportion of the mice. The main immunologic approach to treatment was with nonadherent spleen cells incubated for one day in the presence of interleukin 2. Although these cells had high levels of cytotoxic activity, with a specificity pattern consistent with NK activity, the transfer of such cells together with interleukin 2 had no detectable protective effect on mice with established metastases. However, when the mice were first treated with doxorubicin hydrochloride or cyclophosphamide to reduce the tumor burden and then given the IL-2-stimulated spleen cells together with interleukin 2, the majority of the mice survived for greater than twelve weeks and virtually all of these long term survivors could be shown to be cured of detectable disease. More recent studies have

Table 2. Treatment of 1 Week Murine Renal Adenocarcinoma by Chemotherapy and Adoptive Immunotherapy[1]

	% Survivors Post Treatment	
Treatment[2]	45 days	90 days[3]
None	10	0
Dox	100	15
CLC + IL-2	30	0
Dox + CLC + IL-2	100	67

[1]BALB/c mice were injected with 1 x 10^5 Renca cells under the capsule of the left kidney 7 days prior to initiation of treatment.

[2]Doxorubicin hydrochloride (Dox) was administered on day 7 ip at a dose of 8 mg/kg body weight. For immunotherapy, normal spleen lymphocytes/ml were incubated with 50 units/ml human recombinant (hr) IL-2 (Biogen) for 24 hours (cultured lymphoid cells, CLC). Three daily IV injections of 3.5 x 10^7 cells were given, beginning on day 8. Immediately following each injection of CLC, 10,000 u of hrIL-2 was also administered IV.

[3]Mice surviving 90 days are regarded as cures since no tumor was detectable upon autopsy.

demonstrated that this chemoimmunotherapeutic approach is also applicable for treatment of mice at 2-3 weeks after transplantation of tumor cells, when histologically detectable metastases are readily detectable in various organs. These results support the concept that treatment by immunologic approaches or by biological response modifiers in general is most likely to be effective against micrometastatic disease. It seems likely that the chemotherapeutic treatment sufficiently reduced the tumor burden to permit effective elimination of residual tumor cells by the transferred effector cells. It seems likely that biological response modifiers will be most effective for treatment of metastases when utilized in combination with conventional treatment, when the overall tumor burden of the host is low.

Therapeutic studies with adoptively transferred effector cells have also been performed in a model of human ovarian carcinoma growing in the peritoneal cavity of nude mice. This tumor system closely resembles the frequent clinical problem of ovarian carcinoma, i.e. local growth and metastasis of the tumor in the peritoneal cavity. The emphasis in these studies was to develop effective therapy for the regional disease. Human effector cells from normal donors were transferred into the peritoneal cavity one day after tumor challenge, to determine the potential of various cell types for antitumor effects (18). The transfer of purified human LGL, T cells or monocytes alone had no detectable protective effects. However, the transfer of LGL together with human recombinant interleukin 2 or interferon α, to stimulate their NK activity and to possibly prolong their in vivo survival, resulted in significant prolongation of survival of the tumor-bearing mice. Thus, in two different tumor modals, the adoptive transfer of effector cells with NK activity together with cytokines has been shown to have substantial potential for the therapy of regional or systemic metastases.

TREATMENT OF METASTASES BY MONOCLONAL ANTIBODIES

This approach to therapy was discussed in detail elsewhere in this volume by Koprowski, et al. However, it seems worthwhile to emphasize a few points regarding the potential of this biological approach to the therapy of metastases. The main strategy for therapeutic efforts with monoclonal antibodies has been to combine the antibodies with toxic moieties, either radionuclides, toxins, or drugs, with the goal of having the antibodies selectively carry the toxic agents to the tumor site. However, in addition to the immunoconjugate approach, some antibodies by themselves appear to have promising therapeutic effects. This might be attributed to a variety of mechanisms, including complement-dependent lysis of tumor cells, interactions with effector cells for antibody-dependent cellular cytotoxicity, or immunoregulatory effects, such as the induction of anti-idiotypic responses as discussed by Koprowski, et al. An example of the therapeutic effects of a monoclonal antibody by itself has come from studies with antibodies to the GD_3 ganglioside in human malignant melanoma. Several monoclonal antibodies of the IgG_3 subclass have been developed against this antigen and have been shown to be able to mediate both complement-dependent cytotoxicity and antibodydependent cellular cytotoxicity. A preliminary phase I clinical trial with one of these antibodies has given encouraging results, with partial regression of tumor in three of twelve patients with advanced melanoma (19).

These suggestions of clinical efficacy led us to investigate the mechanism for antibody-dependent cellular cytotoxicity (ADCC). It has not been clear from previous studies whether this or other mouse monoclonal antibodies collaborate predominantly with monocytes or macrophages or with LGL, the two subpopulations of human leukocytes known to be able to mediate ADCC with tumor target cells. Our studies have clearly indicated that LGL from all normal human donors tested gave high levels of ADCC when incubated with melanoma target cells in the presence of the anti-GD_3 antibody (20). In contrast, monocytes gave no detectable ADCC. Since human NK activity has been shown to be augmented substantially by pretreatment of purified LGL effector cells with interferon or IL-2, it was of interest to determine whether the ADCC by monoclonal antibodies also could be augmented by such pretreatment. We have found that pretreatment of LGL from normal donors with either recombinant interferon α or recombinant IL-2 induced substantial augmentation of ADCC activity with the anti-TD_3 antibody. In contrast, interferon γ had no detectable ability to augment the ADCC induced by this antibody.

A further issue in considering the possible role of ADCC in therapeutic effects of monoclonal antibodies in patients with cancer is the functional capability of the patients effector cells. We have, therefore, begun studies to evaluate the ability of peripheral blood mononuclear cells from patients with melanoma to mediate ADCC in cooperation with the anti-GD_3 monoclonal antibody. The results thus far have indicated that melanoma patients have lower but detectable ADCC against melanoma target cells and that such reactivity can be augmented substantially by pretreatment of the effector cells with interferon α for IL-2. Such results are encouraging since they indicate that therapeutic approaches with monoclonal antibodies with the capability of mediating ADCC, perhaps in combination with in vivo interferon α or IL-2 treatment, may be useful in malignant melanoma or other types of human cancer.

CONCLUDING REMARKS

The results presented above and many of the other findings presented at this conference provide considerable indications for the potential value of biological response modifiers for the treatment of metastatic disease. It seems likely that such approaches will be most effective for the treatment of micrometastatic disease. Biological response modifier therapy may be most useful when administered in an adjuvant setting, after primary cytoreductive therapy. A quite encouraging example of the efficacy of this approach has come from the studies of the combined use of chemotherapy and adoptive transfer of IL-2-stimulated lymphocytes for the treatment of metastatic renal carcinoma. Another important therapeutic strategy which deserves increased emphasis is the use of biological response modifiers during the perioperative period, to eliminate tumor cells which are in the circulation and which have not as yet become established as metastases.

REFERENCES

1. Herberman, R.B. In: Rationale of Biological Response Modifiers in Cancer Treatment (Ed. S. Sone), Tokyo, in press.
2. Gorelik, E., Fogel, M., Feldman, M. and Segal, S. JNCI 63: 1397-1404, 1979.
3. Gorelik, E., Segal, S. and Feldman, M. Cancer Immunol. Immunother. 12: 105-109, 1982.
4. Hanna, N. and Fidler, I. JNCI 66: 1183-1190, 1981.
5. Gorelik, E., Wiltrout, R.H., Okumura, K., Habu, S. and Herberman, R.B. Int. J. Cancer 30: 107-112, 1982.
6. Riccardi, C., Santoni, A., Barlozzari, T., Puccetti, P. and Herberman, R.B. Int. J. Cancer 25: 475-486, 1980.
7. Barlozzari, T., Reynolds, C.W. and Herberman, R.B. J. Immunol. 131: 1024-1027, 1983.
8. Barlozzari, T., Reynolds, C.W., Wiltrout, R. and Herberman, R.B. J. Immunol., in press.
9. Serrate, S.A. and Herberman, R.B. In: NK Cells and Other Natural Effector Cells (Ed. R.B. Herberman), Academic Press, New York, 1982, pp. 1069-1076.
10. Serrate, S.A., Vose, B.M., Timonen, T., Ortaldo, J.R. and Herberman, R.B. In: NK Cells and Other Natural Effector Cells (Ed. R.B. Herberman), Academic Press, New York, 1982, pp. 1055-1060.
11. Uchida, A. and Micksche, M. Int. J. Cancer 32 : 34-44, 1984.

12. Wiltrout, R.H., Mathieson, B.J., Talmadge, J.E., Reynolds, C.W., Zhang, S.R., Herberman, R.B. and Ortaldo, J.R. J. Exp. Med. 160: 1431-1440, 1984.
13. Wiltrout, R.H., Herberman, R.B., Chirigos, M.A., Ortaldo, J.R., Green, K. and Talmadge, J.E.: J. Immunol., in press.
14. Honn, K., Sloane, B.F., Crissman, J. and Powers, W.E. (Eds.) Mechanisms of Metastasis: Potential Therapeutic Implications. Martinus Nijhoff, New York, in press.
15. Gorelik, E., Bere, W.W. and Herberman, R.B. Int. J. Cancer 33: 87-94, 1984.
16. Salup, R.R., Herberman, R.B. and Wiltrout, R.H. Submitted for publication.
17. Salup, R.R. and Wiltrout, R.H. Submitted for publication.
18. Ortaldo, J.R., Porter, H.R., Miller, P., Stevenson, H.C., Ozols, R.F. and Hamilton, T.C. Submitted for publication.
19. Houghton, A.N., Mintzer, D., Cordon-Cardo, C., Welt, S., Fleigel, B., Vanhan, S., Carswell, E., Melamed, M.R., Oettgen, H.F. and Old, L.J. Proc. Natl. Acad. Sci., in press.
20. Herberman, R.B., Morgan, A.C, Reisfeld, R., Cheresh, D.A. and Ortaldo, J.R.: In: Monoclonal Antibodies and Cancer Therapy: UCLA Symposia, Vol 27. New York, Alan R. Liss, Inc., in press.

18

DIFFERENCES IN CELL SURFACE CHARACTERISTICS OF POORLY AND HIGHLY METASTATIC LEWIS LUNG TUMOR VARIANTS

K. LAPIS*, J. TIMÁR, F. TIMÁR, K. PÁL, L. KOPPER

I. Institute of Pathology and Experimental Cancer Research, Semmelweis Medical University, 1085 Budapest, Hungary

INTRODUCTION

The acquisition of invasive and metastatic phenotypes in a transformed cell population is the key event in tumor development and progression. The animal and human neoplasms are characterized by a wide diversity in their phenotypes (1), which is the basis for the selection of subpopulations with invasive characteristics (2-4).

Although in humans the liver is one of the most frequently involved organs in the metastatic process, very few experimental tumors have been selected for liver colonization (see ref. 5). A simple in vivo liver metastasis model in mice was described recently (6); Lewis lung tumor cells (LLT) were inoculated intrasplenically and after the development of primary spleen tumor, regular liver metastases were formed. Using in vivo selection procedure (tumor cells from metastatic foci in the liver were repeatedly injected into the spleen) a highly metastatic tumor line (LLT-HH) was established (7). The higher metastatic capacity was not the result of the increased proliferation rate of the tumor cells in the metastases or in the primary tumors. A series of biological differences appeared indicating altered host-tumor relationship with increased resistance to nonspecific host effector cells (Table I) (8, 9).

Cell surface properties have great importance in cell-cell and cell-extracellular matrix (ECM) interactions, and as a consequence in the invasion and metastatic cascade (4, 10). Therefore, we performed a comparative study on the cell surface glucosaminoglycan (GAG) of the two tumor cell populations with different metastatic capacity. Furthermore, we also studied the interaction of these tumor cells with the main ECM

---

*Present address: National Institutes of Health, Building 16, Room 306, Fogarty International Center, Bethesda, MD 20205

components as well as with host cells (heterotypic interactions), which might be involved in liver metastasis formation.

Table 1. Biological differences between LLT and LLT-HH cells in vivo

	LLT	LLT-HH
Minimal cell number injected i.s. to get liver metastases	10^4	10^5
Tumor growth rate (cell kinetics)	no difference	
Effect of macrophage and NK cell blocking	increase in number of metastases	no change in number of metastases
Localization of metastatic foci in liver	lobular dominance	portal dominance
Ultrastructure of tumor cell-host cell (endothelial cells, macrophages) interactions	transitional junctions	no junctional structures

MATERIALS AND METHODS

Tumor cells.

Lewis lung tumor cell lines with "low" (LLT) and "high" (LLT-HH) metastatic capacity were transplanted intramuscularly (i.m.) or intrasplenically (i.s.) into C57BL mice.

Primary culture of LLT and LLT-HH.

Single cell suspensions of tumor cells isolated mechanically from i.m. tumors were adhered to plastic for 2 hr in RPMI 1640 medium completed with 10% fetal calf serum (FCS) at 37°C in 7.5% CO_2 atmosphere. This adherent cell population formed morphologically stabile primary culture up to 7 days.

Morphology and cytochemistry.

Tumor cell suspensions of i.m. tumor or primary cultures on the second day were fixed and processed routinely for light microscopic (LM), transmission (TEM), and scanning electron microscopic (SEM) studies. The surface negative charges were demonstrated on fixed cells [0.5% glutaraldehyde (GA) for 10 min] using 1 mg/ml cationized ferritin (Miles) in TC 199 medium for 5 min at room temperature. After multiple washings, the samples were processed for TEM routinely. The total GAG content and its highly sulfated components were measured on the cell surface with flow cytometric method using acridine orange staining on fixed cells (11). Concanavalin A (Con A; Sigma) (1 mg/ml in distilled water) was used to detect the surface α-D-mannoside residues on fixed cells [1% GA in phosphate buffered saline (PBS) for 30 min]. In controls, the labeling was performed in the presence of specific sugar (α-D-mannopyranoside). The surface bound lectin was labeled by a HRP-5 nm colloidal gold conjugate at a dilution of 1:5 in PBS for 30 min. Wheat germ agglutinin (WGA; Sigma) (1 mg/ml in distilled water) was used to detect surface N-acetyl-glucosaminyl residues on fixed cells. The controls contained D-glucosamine in the labeling solution. After 30 min the bound WGA was revealed using an ovomucoid-12 nm colloidal gold conjugate at a dilution of 1:5 in PBS for 30 min. RCA I (chromatographically purified) (12) was bound to fixed cells at a concentration of 10^{-7} M in PBS for 30 min. The controls contained no lectin in the first step. The bound RCA I was labeled with a rabbit-anti RCA I IgG at a dilution of 1:25 in PBS. In the third step GARG-10 (Janssen) colloidal gold conjugate diluted 1:10 in PBS was used to visualize the complex.

Immunoelectronmicroscopy of surface fibronectin (FN).

The cells were washed and kept at 4°C and were labeled with an anti-FN IgG (sheep) (Miles) diluted 1:20 in PBS for 30 min, then were labeled with an anti-sheep IgG (rabbit) (Miles) 1:20 in PBS for 30 min, and made visible in TEM using a ferritin-anti-rabbit IgG (goat) (Miles) 1:20 in PBS for 30 min. In the controls the first step was omitted or an inappropriate antibody was used. After this the cells were processed routinely for TEM.

Immunoelectronmicroscopy of mouse MHC determinants.

The cells were kept at 4°C and were labeled with a mouse monoclonal IgG against KbDb and H_2D6 determinants (Serotec) at a dilution of 1:100 in PBS for 30 min. The bound antibody was made visible in TEM using GAMG-5 colloidal gold conjugate (Janssen) at a dilution of 1:10 in PBS for 30 min. The controls were labeled with the second antibody or with an inappropriate monoclonal antibody.

Confrontation cultures.

BAEC-802/5 bovine aortic endothelial cell line was kindly provided by Prof. H. Jellinek (Budapest, Hungary). DF2 primary culture of human skin fibroblasts were derived from a Down syndrome patient. Hepatocytes from adult rat liver were prepared according to Seglen (13) and were cultured routinely for 24 hr. LLT and LLT-HH cell suspensions from i.m. tumors were confronted with these monolayers. After 2 hr the nonadherent cells were discharged and the cells were cultured for 24 to 48 hr. C57BL/6 mouse spleen cells (nonfractionated) were confronted with LLT and LLT-HH primary cultures for 24 hr.

Morphological assay for cell attachment to ECM components.

Plastic tissue culture dishes (Falcon) were covered by rat tail collagen (14), 0.5 mg/ml human FN (Serotec), 1 mg/ml hyaluronic acid (Sigma), 1 mg/ml chondroitin sulfate (Sigma) and 1 mg/ml heparan sulfate (Sigma) - all dissolved in PBS. When the substrates were dried, the LLT-HH cells from i.m. tumors were adhered to covered plastic surface for 2 hr at 37°C, washed and cultured for 12 to 48 hr, and studied morphologically as described above.

RESULTS

Morphological and cytochemical characterization of the cell surface.

Although there were no LM or TEM differences between the LLT and LLT-HH (in metastases), SEM revealed a more complicated cell surface for LLT-HH cells due to the much higher number of microvilli present on them (Fig. 1a,b). This difference was also detected on cells maintained in vitro (Fig. 2a,b).

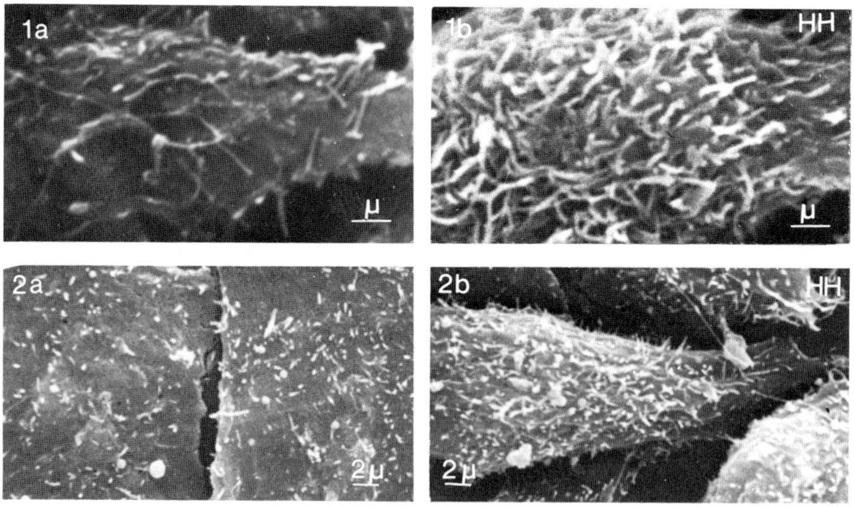

Fig. 1. a) LLT in vivo. b) LLT-HH in vivo.
Fig. 2. a) LLT in vitro. b) LLT-HH in vitro.

Table 2. Membrane cytochemical characteristics of LLT and LLT-HH cells (electronmicroscopy)

	LLT	LLT-HH
Cationized ferritin binding (net negative charge)	+	++
GAG (highly sulfated KS, DS, HS)*	+	++
Con A binding	++	++
WGA binding	±	±
RCA I binding	+++	+(++)
Fibronectin	±	++
H_2D6	-	+
KbDb	-	++

*Measured by flow cytometry; KS - keratan sulfate, DS - dermatan sulfate, HS - heparan sulfate.

Table 2 summarizes the cell surface characteristics found by EM-cytochemistry. More cationized ferritin binding sites were observed on LLT-HH cells indicating an increased number of negative charges. One of the main contributors to the surface net negative charges are the GAGs, particularly the highly sulfated ones. The increased highly sulfated GAG content of LLT-HH cell membrane was demonstrated by flow cytometry (11), which would be consistent with the higher negative charges. Analysis of the terminal sugar composition on the cell surface of the two lines revealed no difference in Con A and WGA binding, but marked difference was found in RCA I binding. This indicates high gal/gal-Nac expression on LLT cells. An increased amount of FN was observed on the surface of LLT-HH cells compared to that seen on LLT cells. Two mouse determinants (H_2D6 and KbDb) were studied. Both were expressed strongly on the surface of LLT-HH cells, but were not detectable on LLT cells.

Morphology of tumor cell interactions with ECM components.

It is now well known that there are specific cell-membrane receptors/ binding sites to interact with various ECM components (collagen, FN, GAG components). Marked differences appeared in the behavior of LLT and LLT-HH cells on various ECM substrates (Table 3), especially on collagen and FN (Figs. 3, 4). There was a much higher affinity of LLT-HH to these

Table 3. Adherence and 48 hr culture on ECM components

ECM	LLT		LLT-HH	
Collagen	+	R	++	S
Fibronectin	+	R	++	S
Hyaluronic acid	++	R/S	++	S
Chondroitin sulfate	++	R/S	++	S
Heparan sulfate	+++	R/S	++++	S

+ = effectivity of adherence; R or S = morphology of adherence (R = round, S = spread).

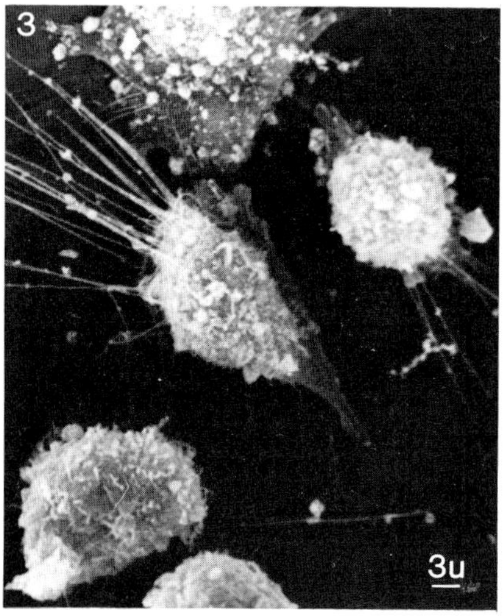

Fig. 3. Rounded LLT cells on collagen substrate.

Fig. 4. Spread LLT-HH cells on collagen substrate.

substrates accompanied by altered morphological appearance of tumor cells, i.e., the spreading of LLT-HH and the maintenance of round morphology of LLT. Interestingly, a slight difference was detected in the attachment and culture on GAG substrates, suggesting relatively even GAG binding sites on both cell types.

Heterotypic interactions in confrontation cultures.

In vitro confrontation assay systems were used to study the heterotypic interactions occurring during liver invasion. Endothelial cells, fibroblasts, adult rat liver cells and spleen cells were used as confrontation partners with LLT as well as LLT-HH cells. The two lines showed marked differences in the binding capacity only to the spleen cells, although the tumor cells behaved differently on each host cell type. The LLT cells were relatively inactive on endothelial (Fig. 5) and fibroblastic (Fig. 7) monolayers, producing microinvasion only on rat liver cells, which was started from the apical membrane (Fig. 9). The LLT-HH cells showed a much higher level of activity in a quite uniform way, i.e., they undermined the endothelial (Fig. 6) and the fibroblastic cells, lifted them off, and proliferated on their substrate very intensively (Fig. 8). Even the hepatic cell monolayer was undermined (Fig. 10). The difference in the interaction of tumor cells with spleen cells was remarkable. The LLT cells did not bind the spleen cells well, while the LLT-HH coexisted nicely with them (Fig. 12). However, the bound spleen cells showed morphological signs of activity (spreading) on LLT cells (Fig. 11), while in the case of LLT-HH there was no change in their morphology. These results are summarized in Table 4.

Table 4. Confrontation culture characteristics of LLT and LLT-HH cells

	LLT	LLT-HH
BAEC-802/5 bovine endothelial cells	inefficient binding inactivity	efficient binding microinvasion proliferation
DF/2 primary culture of human fibroblastic cells	binding spreading	efficient binding spreading microinvasion proliferation colonization cytopathic for fibroblasts
Primary culture of adult rat liver cells	efficient binding spreading microinvasion "melting" type	inefficient binding microinvasion "bottom" type
Adult mouse (syngeneic) spleen cells	inefficient binding spleen cell spreading over the tumor cell	coexistence spleen cell inactivity

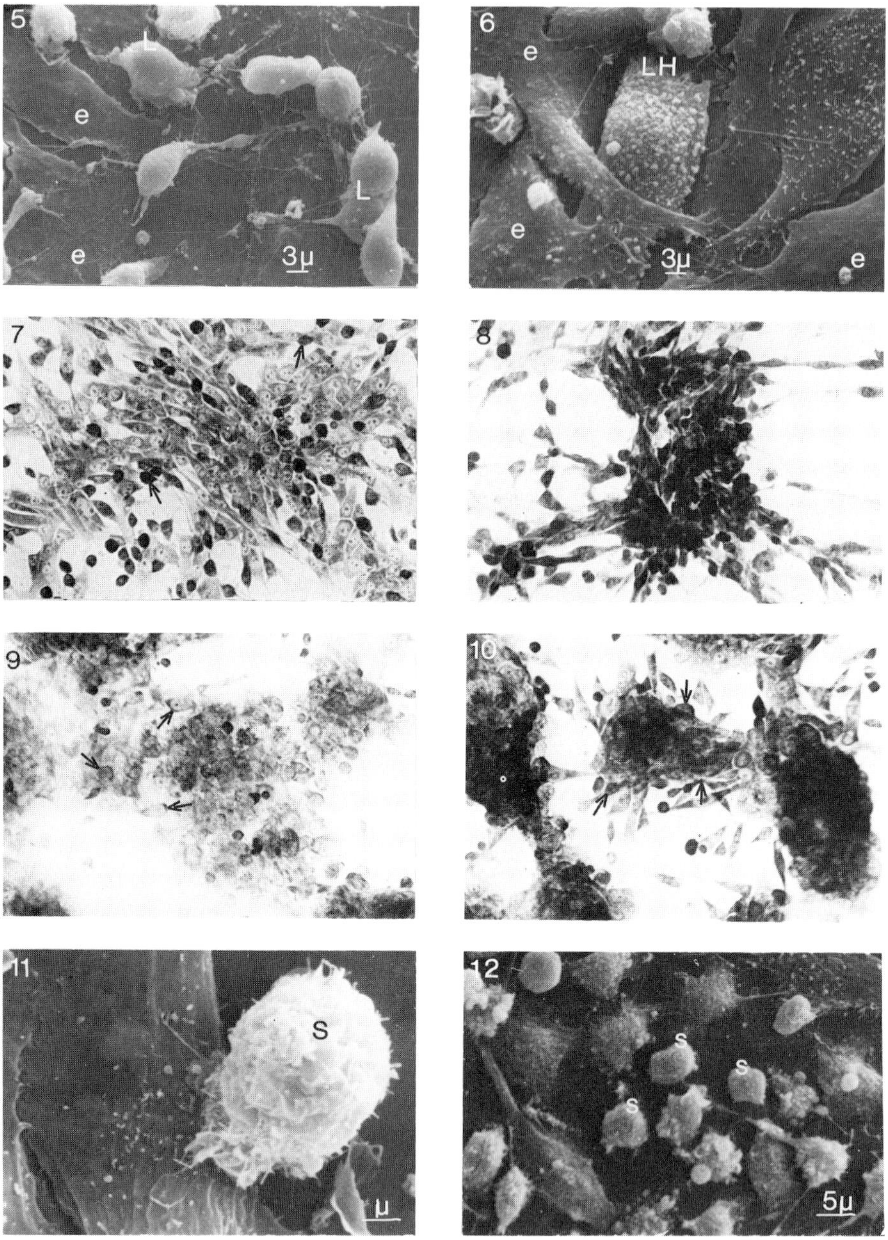

Figs. 5 through 12. LLT with endothelial cells (5), fibroblasts (7), liver cells (9), and spleen cells (11). LLT-HH with endothelial cells (6), fibroblasts (8), liver cells (10), and spleen cells (12).

DISCUSSION

The invasion of the liver by tumor cells is a complex multistep process with numerous interactions between tumor cells and host cells as well as tumor cells and ECM. We attempted to find correlations between cell membrane structures and the biological behavior of metastatic cells in the liver. Heterotypic interactions have been studied in vitro by several investigators (15-18). Based on their studies we also aimed to model the basic heterotypic interactions occurring in vivo during liver invasion.

Our previous morphological and biological studies on liver metastases in vivo revealed altered heterotypic interaction of LLT-HH cells with endothelial, stromal, and parenchymal constituents of the liver (8) compared to LLT cells. In the present experiments complex changes were described at the level of plasma membrane glycoproteins. The differences in the terminal sugar architecture between LLT and LLT-HH correlate well with the interaction of these cells with liver cells both in vivo and in vitro. The tumor cell-liver cell interactions are mediated by galactosyl receptors of hepatic cells, and the decrease in the terminally expressed gal/gal-Nac residues altered the binding capacity and microinvasive characteristic of LLT-HH cells. The increased content of highly sulfated GAG components in the membrane of LLT-HH cells reflects the increased biosynthesis of heparan sulfate, as was demonstrated biochemically by Móczár et al. (19). The increased heparan sulfate content in the cell membrane of LLT-HH correlates with the increased in vitro adherence and spreading on ECM substrates. The FN molecule has specific binding sites for ECM components, namely for heparin, collagen, fibrin (20). The increased FN could explain the intensive interaction of LLT-HH with the portal vasculature and stromal structures in vivo (8) and with the collagen, GAG substrates in vitro. The interaction of LLT-HH with endothelial and fibroblastic cells also showed a much higher affinity and invasive capacity to the matrix produced by these cells. Consequently, the LLT-HH cells disintegrated these monolayers, while the LLT did not, reflecting the in vivo observed differences.

The "melting" type microinvasion (16) of liver cells was a characteristic feature of LLT cells and was probably mediated by the terminal galactosyl residues/liver cell galactose receptors described by Chengsong-Popov et al. (21). In this respect the LLT-HH cells showed higher affinity to the matrix produced by the liver cells and a decreased binding to the liver cells themselves. Confrontation of the LLT and LLT-HH cells with spleen cells demonstrated striking differences between the two lines. The lower binding to the LLT cells is in agreement with the almost complete absence or masking of MHC determinants H_2D6 and KbDb, while the coexistence with LLT-HH probably reflects the high expression of these determinants on their membrane. It was demonstrated on LLT cells that the high Kb/high Db expression is the characteristic of highly metastatic variants (22). It was also observed that spleen selected tumor cells are resistant to the T cytotoxic cells (23). It is possible that the acquisition of resistance against cytotoxic cells (8) and the high expression of MHC determinants are closely correlated. Our recent results strongly support the importance of immunologic mechanisms participating in the selection process, resulting in a highly metastatic subpopulation from a heterogenous tumor. The results also suggest that the tumor-ECM interactions have equal importance in the metastatic cascade. We have demonstrated that the change in molecular-biological conditions of the plasma membrane resulting in decreased interactions with the parenchymal cells of an invaded organ does not necessarily mean a decrease in organ invasion. The enhanced affinity to ECM components of the vasculature and stromal tissues, or to basement membrane (24), could effectively overbalance the former and could lead to the acquisition of "highly" metastatic phenotype.

These observations point out the importance of heterotypic interactions in the metastatic process, which could be the new targets for selective antimetastasis therapy.

REFERENCES

1. Hart, I.R. and Fidler, I.J. Biochem. Biophys. Acta 651: 37-50, 1981.
2. Harlos, J.P. and Weiss, L. Int. J. Cancer 32: 745-750, 1983.
3. Fidler, I.J. Cancer Treat. Rep. 68: 193-198, 1984.
4. Nicolson, G.L. Exp. Cell Res. 150: 3-22, 1984.
5. Nicolson, G.L. and Poste, G. Int. Rev. Exp. Pathol. 25: 77-181, 1983.
6. Kopper, L., Hahn, T.V. and Lapis, K.J. Cancer Res. Clin. Oncol. 103: 31-38, 1982.
7. Pál, K., Kopper, L. and Lapis, K. Invasion Metastasis 3: 174-182, 1983.
8. Pál, K., Kopper, L., Timár, J., Rajnai, J. and Lapis, K. Invasion Metastasis, 1985, 5: 159-169, 1985.
9. Timár, J., Kopper, L., Pál, K. and Lapis, K. Pathol. Res. Pract. 177: 47-59, 1983.
10. Liotta, L.A., Rao, C.N. and Barsky, S.H. Lab. Invest. 49: 636-649, 1983.
11. Timár, J., Boldog, F., Kopper, L. and Lapis, K. Histochem. J. 17: 1985, accepted.
12. McIntosh, D.P., Edwards, D.C. and Davies, A.J.S. Toxicon 22: 293-299, 1984.

13. Seglen, P.O. Meth. Cell Biol. 13: 29-83, 1976.
14. Michelopoulos, G. and Pitot, H.C. Exp. Cell Res. 94: 70-78, 1975.
15. Nicolson, G.L. J. Histochem. Cytochem. 30: 214-220, 1982.
16. Roos, E., Pavert, V. de and Middelkoop, O.P. J. Cell Sci. 47: 385-397, 1981.
17. Benke, R.M., Werling, H.O. and Paweletz, N. Anticancer Res. 4: 241-246, 1984.
18. Vlodavsky, I., Schirrmacher, V., Ariav, Y. and Fuks, Z. Invasion Metastasis 3: 81-97, 1983.
19. Móczár, E., Timár, J., Timár, F., Pál, K., Kopper, L. and Lapis, K. Invasion Metastasis, 1985, submitted.
20. Yamada, K.M., Hayashi, M. and Akiyama, S.K. In: Extracellular Matrix (Eds. S. Hawkes and J.L. Wang), Chapter I. Academic Press, New York, London, 1982, pp. 25-35.
21. Chengsong-Popov, R., Robinson, P., Altevogt, P. and Schirrmacher, V. Int. J. Cancer 32: 359-366, 1983.
22. Eisenbach, L., Segal, S. and Feldman, M. Int. J. Cancer 32: 113-120, 1983.
23. Bosslet, K. and Schirrmacher, V. J. Exp. Med. 154: 557, 1981.
24. Terranova, V.P., Liotta, L.A., Russo, R.G. and Martin, R.G. Cancer Res. 42: 2265-2269, 1982.

19

CHARACTERISTICS OF LL_2 AND ITS LECTIN-RESISTANT NOT METASTASIZING VARIANTS

C. RADZIKOWSKI, D. DUŚ, A. OPOLSKI and L. STRZĄDAŁA

Department of Tumor Immunology, The Ludwik Hirszfeld Institute of Immunology and Experimental Therapy, Polish Academy of Sciences, Wrocław, Poland

Considering the metastasis as a highly complexed process both on the level of heterogeneity of tumor cell population and host environmental factors involved, it is very difficult to select a satisfactory experimental model and an approach to be applied to the given research program. In our motivation to study an immunobiological aspects of metastasis which could be relevant to improved, more rational designing of combined experimental therapy, we have considered several points, to mention just two: 1. to have the tumor variants which differ utmostly in their metastatic properties, but remain similar in their other traits, 2. to possess the tumor cells variants accessible for biochemical and serological analysis of their surface characteristics.

Lectins have been used for selection in vitro tumor cell variants which differed in surface carbohydrates, usually lectin-resistant variants are known to be glycosylation defective (1, 2). Wheat germ agglutinin WGA selected tumor cells usually have lowered metastatic phenotypes and in some cases have even lost of their metastatic ability (3,4).

The maintained in vitro LL_2 line, which has preserved its tumorigenicity and metastasing potential on s.c. inoculation to F_1 recipients, and two WGA-Resistant WGA-R LL_2 variants were subjected to further studies. The Lewis lung carcinoma, 3 LL, a C57BL/6 ($H-2^b$) - originated spontaneous lung tumor is known to progress at the site of transplantation in various mouse strains. However, this tumor produce lung metastases exclusively in mice of those strains, which shared the H-2D region and the non H-2 genetic background with C57BL/6 mice (5, 6) . In our present studies we attempted to correlate the altered metastasizing capacity of WGA-R variants with their: in vivo growth kinetics, the quantitative expression of MHC encoded antigens, and total cellular glycopeptides differential distribution.

The quantitative differences in the expression of H-2 antigenic specificities inclined us to test the immunogenicity of WGA-R variants in comparison with parental LL_2 tumor line in normal and immunosuppressed by pretreatment with CY recipients.

MATERIALS AND METHODS

Mice

Inbred mice were supplied by the Inbred Mice Center of the Institute of Immunology and Experimental Therapy Wrocław, Poland. The mouse strains and their H-2 haplotypes are as follows: C57/BL/6/J Boy/IiW ($H-2^b$) , BALB/c/IiW ($H-2^d$) DBA/2/Cr/IiW ($H-2^d$) , AKR/JBoy/IiW ($H-2^k$) , C57BL/6 ($H-2^k$) and F_1 hybrids of BALB/c x DBA/2 BDF1 and C57/BL/6 x C3H B6C3F1.

Male and female mice, aged from 6 to 12 wk were used in these studies.

Tumor

The Lewis lung carcinoma , 3LL , was maintained by serial passage in the syngeneic strain, as described elsewhere (7). The tumor cells were adapted to growth in tissue culture at 37°C with 5% CO_2, 95% air atmosphere. The growth medium was Eagle's minimal essential medium with Earle's salts (EMEM) supplemented with 10% of fetal calf serum and antibiotics. The procedures applied to maintain LL_2 cell line in vitro and its characteristics are described elsewhere (8).

Selection of WGA-resistant variants.

For selection, toxic concentration of WGA was added to LL_2 cells (48 passage in vitro) and incubated for 7 - 10 days, until only fewer than several cells survived. The WGA-containing medium was then replaced with fresh growth medium which allow the surviving cells to recover and growth. Cells from these cultures were again subjected to another cycle of lectin treatment. The selection procedure was repeated 5 times to produce $LL_2$5 variant resistant to 20 µg/ml of WGA and 8 times to produce $LL_2$8 variant, resistant to 100 µg/ml of WGA (9). For in vivo assays the cells were removed from tissue culture vessels by mechanical means, washed and inoculated into syngeneic or semisyngeneic recipient mice (9). The LL_2 cell line and its WGA-R variants were newer propagated in vitro for period longer than 2 mo during which the cellular material was harwested for immunobiological analysis. For following cycles of experiments the frozen material has been recovered (8, 9).

Assays for tumorigenicity.

Cells in numbers of 1 x 10^6 per mouse were inoculated s.c.. The growth rates of individual tumors were estimated by measuring their length and width and by calculation of tumor masses as described elsewhere (7).

Assays for metastasis.

Two different routes of tumor cells inoculation were applied. For natural metastases 1 x 10^6 or 2 x 10^6 cells were inoculated s.c. and 21 or 28 days later the mice were sacrified, their lungs were removed and pulmonary metastases evaluated following the method of Wexler (10).

For "artificial" metastases 3 x 10^5 cells were inoculated i.v. and 21 days later the number of metastatic lung nodules counted, following the same procedure as after s.c. inoculation.

Pretreatment with Cyclophosphamide (CY), (Germed, GDR)

To test an enhancing effect of CY on metastatic properties of the LL_2 and WGA-R variants, the earlier selected single dose of 200 mg/kg of CY was administered i.p. to recipient mice 24 hr before i.v. inoculation of tested cells (11). The procedure followed those for "artificial" metastases assay.

Antisera.

Anti H-2 alloantisera anti $H\text{-}2^b$, $H\text{-}2^d$ and $H\text{-}2^k$ were obtained in our Department according to the uniform immunization protocol following the method of Ivanyi et al (12) as described elsewhere (13). To remove the cytotoxic activity against public common H-2 specificities the appropriate absorptions of these alloantisera were performed. To determine H-2 specificities on tumor cells the preabsorbed antisera, after verification and titration in cytotoxic assay against referential lymphnode lymphocytes, were applied (13).

Quantitative absorption assay.

The increasing numbers of analysed tumor cells and control cells, referential for the given H-2 antigenic haplotype, were suspended in diluted antiserum one dilution before the last cytotoxic to 100% of lymphocytes of the given H-2 haplotype. After 1 hr incubation at 4°C with intermittent shaking the absorbing cells were removed by centrifugation and the recovered antiserum tested for residual cytotoxic activity to referential lymphocytes in cytotoxic assay as described elsewhere (14, 15).

Analysis of surface carbohydrates.

The data and procedure applied in these studies are described in details elsewhere (16). These studies are being carried in the collaboration with Dr. H. Debray, the Laboratory of Biological Chemistry, University of Lille, France.

The LL_2 cells and its WGA-R variants were labeled metabolically with D-$[6\text{-}^3H]$- or $[1\text{-}^{14}C]$-glucosamine hydrochloride. Differentially labeled materials from two lines examined LL_2 and LL_25 or LL_2 and LL_28 were mixed and cofractionated on Ultrogel ACA 202 column.

After the fractionation the first fraction containing proteoglycans and large glycopeptides with O-glycosidically linked glycans was taken separately. The following 3 fractions were pooled and refractionated by affinity chromatography on ConA-Sepharose and subsequently on LCA-Sepharose.

RESULTS

Resistance to WGA toxicity.

$LL_2 5$ and $LL_2 8$ variant cells were selected through 5 or 8 cycles of WGA treatment. Resistance to lectin and some of in vitro characteristics were tested and described in details elsewhere (9). The summarized results are presented below.

Table 1. Characteristics of LL_2 and WGA-R variant cells in vitro.

Analysed properties	LL_2	$LL_2 5$	$LL_2 8$
Population doubling time, hr	22	22	21
Saturation density x 10^5/ml	5,2	6,3	8,3
Cloning efficiency in semisolid medium[x]	6,5%	0,55%	0,84%
WGA-resistance[xx], µg/ml	10	20	100
WGA concentration required for cytoagglutination, µg/ml	25	50	100

[x] 1 x 10^3 cells in 0,3% agar medium in 50 mmØ Petri plate. Clones which contained more than 30-50 cells were scored on 10th - 14th day.

[xx] 5 x 10^3 cells/ml in the presence of 0, 10, 20, 50 and 100 µg/ml WGA. Cells were considered as resistant when more than 50% survived as compared to control. Means of at least triplicate experiments.

The analysis of kariotypes performed on LL_2 cells after 24th and 50th passage in vitro revealed hypotetraploidity with 72 modal chromosome number (about 20% of cells) and variations in range 68-94 at 24th passage and 66-88 at 50 passage (8). It appeared that LL_2 cell line is an established cell line, but to avoid any possible alterations which could occur during long term propagation in vitro, cells from 48th passage were stored in liquid nitrogen. Similarly $LL_2 5$ and $LL_2 8$ cells were stored at early passages in liquid nitrogen. For experimental purpose, the cells were propagated in vitro during a period never exceeding two months and then were renewed from frozen stock samples.

The main difference between LL_2 and WGA-R variant cells in addition to different sensitivity and agglutinability appeared to be in cloning efficiency (Table 1). The altered lectin binding properties are often found in lectin selected mutants, which were showed to be glycosylation defective and exerting altered carbohydrates structures (2).

Tumorigenicity and metastasibility.

LL_2 cells appeared to be tumorigenic in syngeneic or semisyngeneic recipients, causing 100% of tumor takes after s.c. inoculation of 1 x 10^6 cells. The growth rate, as measured by tumor wt on day 16th, was lower for LL_2 and WGA-R variants than for 3LL in vivo maintained tumor line (Table 2). In all mice inoculated s.c. with 3LL and LL_2 tumor cells on day 30th the lung metastases were present, however there were distinct differences in their number. $LL_2 5$ and $LL_2 8$ cells did not cause any metastases (Table 2) .

Table 2. Tumorigenicity and metastasibility.

Cell line	Average tumor wt in mg[x]	No of mice with pulmonary metastases[xx]/Total No	Lung weight	No of metastatic foci
3LL	3308 ± 768	12/12	588 mg	
LL_2	842 ± 325	14/14	130 mg	6,85
$LL_2 5$	530 ± 205	0/12	-	0
$LL_2 8$	785 ± 340	0/12	-	0

[x] tumor wt on day 16th following s.c. implantation of 1 x 10^6 cells into BDF_1 mice

[xx] score on day 30th

Accord. to Duś et al. (9)

The experiments were repeated few times to check the metastatic potential of LL_2 cells and WGA-R variants. It appeared that this trait of resistant variants was maintained, however in some cases single metastatic foci could be detected (Table 3).

Table 3. Lung metastatic foci on day 21 st .

Cell line	Mean No.	± SD	Range	No. of mice with metastases/Total[x]	%
LL_2	6,2	4,6	0-12	8/9	88
$LL_2 5$	0,2	0,4	0- 1	2/13	15
$LL_2 8$	0,1	0,3	0- 1	1/13	7

x BDF_1 mice inoculated s.c. with 2 x 10^6 cells; results are averages from triplicate experiments.
accord. to Duś et al. (9)

<u>Analysis of H-2 expression.</u>

The method of quantitative absorption of anti $H-2^b$, $H-2^d$ and $H-2^k$ antisera was applied (15). Normal lymphocytes from C57BL/6 mice and leukemia EL-4 of C57BL/6 mice origin were applied in these comparative studies as referential cells with known $H-2^b$ antigenic haplotype.

Using the quantitative absorption of cytotoxic antibodies from anti H-2 antiserum the presence of $H-2^b$ antigens could be demonstrated on all cells analysed, with the highest expression on $LL_2 5$ cells. The cytotoxicity could be absorbed with the lowest No. of these cells (Fig. 1a).

Applying anti $H-2^k$ alloantiserum, 1 x 10^7 $LL_2 5$ cells could removed 50% of cytotoxic activity from anti $H-2^k$ serum. Lower expression of this "alien" antigenic specificity was detectable on $LL_2 8$ and LL_2 cells as well; 5 times higher No. of cells were required to absorb 50% cytotoxic activity of this alloantiserum LL_2 cells or to absorb 40% of this activity $LL_2 8$ cells (Fig. 1b).

Using similar procedure, the presence of "alien" $H-2^d$ antigenic specificities could be demonstrated on $LL_2 5$ cells: 1 x 10^7 cells removed 70% of cytotoxic activity from anti

$H\text{-}2^d$ alloantiserum. Relatively lower absorption capacity was demonstrated for LL_2 cells: 1 x 10^8 cells removed 55% of anti $H\text{-}2^d$ cytotoxic activity, and for LL_28 cells: 5 x 10^7 cells removed 45% of this activity (Fig. 1c).

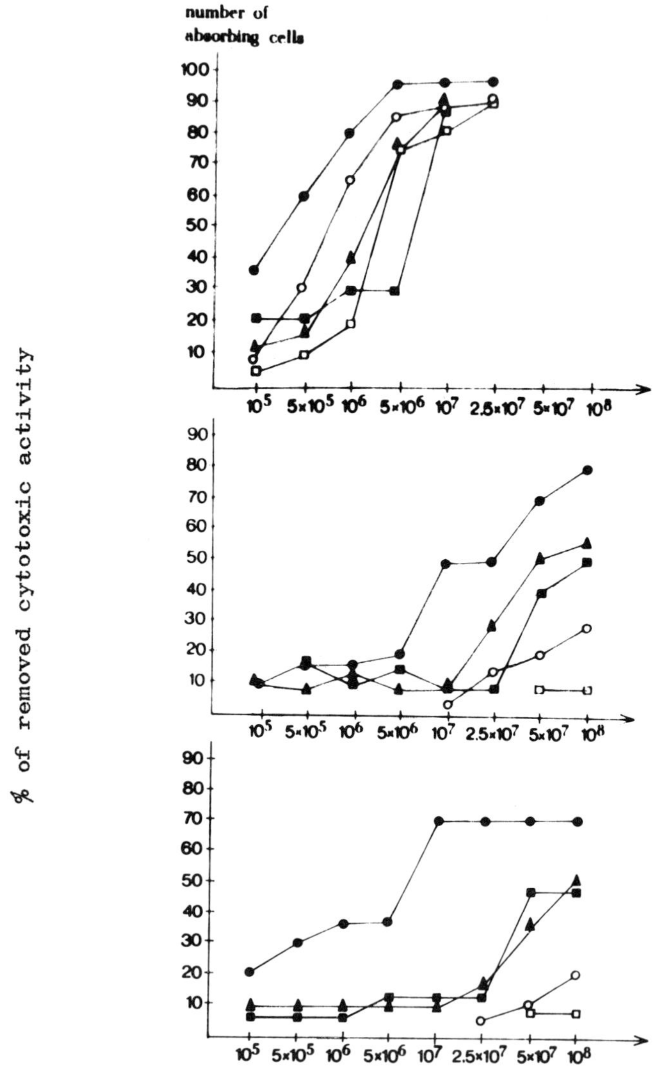

Fig. 1. Absorption in vitro of anti-H-2 allosera with LL_2 and its WGA-R variants: a) anti-$H\text{-}2^b$, b) anti-$H\text{-}2^k$, c) anti-$H\text{-}2^d$

●——● LL_25 ▲——▲ LL_2 □——□ C57BL/6 lymphocytes

■——■ LL_28 ○——○ leukemia EL-4

Considering the observed quantitative and qualitative differences in antigenic expression of MHC gene products on the surface of all cell lines under investigation, we performed some preliminary studies on a possible role of immunogenic factors which could be involved in metastases formation control.

Metastasibility in immunosuppressed recipients (CY-pretreated)

For this purpose the cells of all cell lines were inoculated i.v.. The results of these experiments have shown that the cells are able to form metastatic foci in lung, however there are quantitative differences between LL_2 cells which showed the highest No. of lung foci almost in all mice, as compared to WGA-R cells (see Table 3).

Table 4. Lung metastatic foci on day 21st following i.v. inoculation of tumor cells

Cell line	Mean No.	± SD	Range	No of mice with metastases/Total[x]	%
LL_2	13,8	11	0-14	37/38	97
LL_25	1,2	1,4	0- 5	28/39	73
LL_28	0,8	1,1	0- 4	18/41	43

x BDF_1 mice inoculated i.v. with 3×10^5 cells/mouse. accord. to Duś et al. (9)

In our previous studies in which the approach described by Van Putten et al. was applied (17), it has been demonstrated that treatment of mice prior to LL_2 cells enhance the formation of lung metastases in a dose-dependent manner (11). This effect was time dependent and it persisted up to 7 days after CY administration. The similar method was applied to investigate the metastatic potential of all cell lines inoculated i.v. to CY pretreated recipients.

It appeared that the number of lung metastases increased, 3.2 times for LL_2 cells inoculated to CY pretreated mice. In the case of LL_25 and LL_28 cells the increase was even higher, however still not in all mice the lung metastases have occured (see Table 5). In elucidation of the possible mechanisms involved in lung metastases enhancing effect, several host factors could be considered, to mention a cytotoxic effect of CY on lymphoid tissue and a direct toxic effect for lung tissue compartments.

Table 5. Lung metastatic foci on day 21st in CY pretreated recipients

Cell line	CY 200 mg/kg	Mean No.	± SD	Range	No. of mice metastases/Total[x]
LL_2	-	17,5	11,0	1-40	27/27
	+	56,0	19,2	16-85	17/17
LL_25	-	1,4	1,5	0- 5	17/29
	+	12,0	12,0	0-33	20/25
LL_28	-	0,9	1,2	0- 4	15/30
	+	6,7	5,0	0-22	25/26

[x] BDF_1 mice were injected i.p. with CY 20 hr before i.v. inoculation of 3×10^5 tumor cells. The results are means of triplicate experiments.

accord. to Duś et al. (9)

Surface carbohydrates

The main differences found were among N-acetyllactosaminic glycans. The LL_2 cells showed relative higher % of tri- and tetraantennary glycans with secondary lover % of biantennary glycans. No marked difference was observed between the amount of the oligomannosidic type glycopeptides in all cell lines examined (see Table 6).

Table 6. Total cellular glycopeptides distribution in % in LL_2 and its WGA-R variants

Cell line	Tetra- and Tri-antennary	Biantennary	Oligomannosidic type
LL_2	70,2	19,7	10,0
LL_25	56,0	34,8	9,1
LL_28	60,3	29,3	10,4

accord. to Debray et al. (18)

DISCUSSION

Lectin-resistant cell lines have been often used as a tool for the study of the role of surface oligosacharides in membrane structure and function. With such variants exhibiting different in vitro and in vivo characteristics, specific properties could be analysed for their relevance in determining the ability of cells to metastasize. The rationale for these study was to determine if a correlation might also exist between cell membrane's carbohydrate composition of the given cell lines with their metastasibility and immunogenicity.

Following this type of approach we have selected in vitro from LL_2 tumor cell line, exposed to treatment with increasing doses of WGA, two resistant variants which have lost their metastasing ability following s.c. inoculation into F_1 recipients. Both WGA-R variants were different in their cloning efficiency in semisolid medium, sensitivity to various doses of WGA and its concentration required for cytoagglutination from parental LL_2 cell line (see Table 1).

The comparative biochemical analysis of total cellular glycopeptides revealed that WGA-R variants showed marked decrease in tetra- and triantennary glycans with parallel increase in biantennary glycans, as shown by sequential affinity chromatography on immobilized Con A and LCA (see Table 6) . Similar results indicating on the increase in highly branched oligosaccharides in cells which differ quantitatively in their tumorigenic capacities have been reported by other authors (for review see 2) .

Whether there exists any direct relationship between quantitative differences in highly branched oligosaccharides and lost of metastasing properties of WGA-R variants remains obscure. It seems to be an accompanying trait, like the quantitative alterations in the expression of MHC encoded antigens. Considering the limitations related to anti-H-2 alloantisera applied and quantitative absorption assay used in these studies, the quantitative differences in the expression of expected $H-2^b$ and alien $H-2^d$ and $H-2^k$ antigenic specificities were found, mostly on LL_25 cells. In a series of elegant studies, the group from the Department of Cell Biology, The Weizmann Institute of Science in Revohot (5 + 6, 20) has demonstrated that metastases-derived line from the 3LL tumor in vivo was antigenically different from the primary local tumor. Namely it differed in the phenotypic expression of $H-2^b$ molecules. The cells of these different sublines exerted lower density of $H-2^b$ antigen on their surface (19) . The same authors have recently found that highly metastatic clonal subpopulation of 3LL tumor have a decreased expression of the $H-2D^b$ gene product on their surface compared with poorly metastatic clones.

The WGA-R variant cells appeared to be similar to parental LL_2 cells in growth rate in vitro. The lowest growth rate in vivo of LL_25 cells following s.c. inoculation can be related to the highest expression of MHC encoded antigens, including the presence of "alien" specificities on these cells. Host immune response against the most immunogenic variant cells could be in part responsible for the lack of metastatic capacity as well. This explanation is not valid for LL_28 cell line, which in spite of the lack of metastasing capacity following s.c. inoculation is similar to parent LL_2 line in the expression of MHC encoded antigens and in the rate of growth as a s.c. tumor. The results of experiments in which tumor cells were inoculated i.v. into CY pretreated recipients partially support such a possibility. In immunosuppressed mice nonmetastasing WGA-R variant cells have developed metastatic foci, however, their number and incidence were lower than for parental LL_2 tumor cells.

In conclusion, the role of H-2 antigenic components in determining metastatic phenotype should be further elucidated as well as the role of various cellular compartments of immune system in the interplay with immunogenic tumor cells. Both cellular and host factors seems to be involved in determining local tumor growth and metastasis.

REFERENCES

1. Briles, E.B. and Kornfeld, S. Trends in Biochem. Sci. 3; 223-227, 1978.
2. Nicolson, G.L., Exper. Cell Res. 150; 3-22, 1984.
3. Tao, T.W., and Burger, M.M. Int. J. Cancer 29; 425-430, 1982.
4. Kerbel, R.S. Amer. J. Path. 97; 609-622, 1979.
5. Isakov, N., Katzav, S., Feldman, M. and Segal, S.,J. N.C.I. 71; 139-145, 1983.
6. Isakov, N., Feldman, M. and Segal, S. Transplant. Proc. 13; 778-782, 1981.
7. Budzyński, W. Arch. Imm. Ther. Exp. 30; 363-372, 1982.
8. Duś, D., Budzyński, W. and Radzikowski C. Arch. Imm. Ther. Exp. 1985 (in press) .
9. Duś, D.,Debray, H., Opolski, A., Strządała, L., Rak, J., Kuśnierczyk H., Montreuil, J. and Radzikowski C. (manuscript in preparation).
10. Wexler, H. J. Natl. Cancer Inst. 36; 641-645, 1966.
11. Strządała, L., Rak, J. and Kuśnierczyk, H. Arch. Imm. Ther. Exp. 1985 (in press) .
12. Ivanyi, P., van Mourik, P., Breuning, M. and Melief, C. J. M. Immunogenetics 10; 319-332, 1980.
13. Opolski, A. and Radzikowski C. Arch. Imm. Ther. Exp. 1985 (in press) .
14. Opolski, A., Duś, D. and Radzikowski C. Arch. Imm. Ther. Exp. 1985 (in press) .
15. Strządała, L., Opolski, A., and Kisielow, P. Arch. Imm. Ther. Exp. 29; 439-442, 1981.
16. Debray, H., Delannoy, P., Debeire, P. and Montreuil, J. Biochim. Biophys. Acta(Submitted for publication).
17. van Putten, L. M., Kram, L. K., van Dieredonck, H.H.C., Smink, Tand Füzy, M. Int. J. Cancer 15; 588-595, 1975.
18. Debray, H., Hueso, P., Delannoy, P., Duś, D., Radzikowski, C. and Montreuil, J. (manuscript in preparation).
19. Brodt, P., Feldman, M. and Segal, S. Cancer Immunol. Immunother. 16; 109-113, 1983.

20

A ROLE FOR CELL SURFACE SIALIC ACID IN LIBERATING METASTATIC TUMOR CELLS FROM HOST CONTROL

V. SCHIRRMACHER, J. DENNIS*, C. A. WALLER, and P. ALTEVOGT

Institut für Immunologie und Genetik, Deutsches Krebsforschungszentrum, D-6900 Heidelberg, FRG
*Department of Pathology, Richardson Laboratory, Queen's University, Kingston, Canada

Recent studies have indicated that cell surface carbohydrates can severly affect the metastatic potential of cancer cells (1-4). It was found that changes in the expression of carbohydrates can influence cancer cells' bloodborne arrest and survival properties (1), their adhesive characteristics towards plastic (4), endothelial cells (1) and subendothelial matrix components (3) as well as their immunological properties (5, 6).

In this article we summarize our recent results on the role of surface carbohydrates in metastasis which were obtained in two different tumor systems, each consisting of several related tumor lines, of either high or low metastatic capacity. The origin and characteristics of these tumor lines are listed in Table 1.

Table 1. Characteristics of tumor lines studies

Tumor line	Host strain	etiology	metastatic capacity	Ref.
1. Eb T lymphoma	DBA/2 mouse	chem. ind.	low	7
2. ESb	"	spont. variant of 1	high	7
3. ESb-M	"	plastic adherent variant of 2	low	4
4. MDAY-D2 (undiff.)	"	from MDAY	high	8
5. MDW40	"	WGA^R lectin resistant variant of 4	low	9
6. MDW40-24a	"	WGA^S "revertant" of 5	high	9

Sialic acid content, cell surface charge and electrophoretic mobility

It has been reported (10) that the metastatic capacity of a range of murine tumor lines was positively correlated with the degree of sialylation of available galactose (Gal) and N-acetylgalactosamine (GalNac) residues on cell surface carbohydrate molecules. We first determined the total amount of neuraminidase-releasable sialic acid (SA) for all the lines studied. The wheat-germ agglutinin resistant variant MDW40 of the high metastatic line MDAY-D2 showed a 3-fold reduction in SA content, while the high metastatic revertant (Nr. 6 in Table 1) showed values similar to those of the parental line (11). In case of the Eb/ESb/ESb-M tumor cell lines, the values for neuraminidase accessible SA differed only slightly (1.1-1.4 nmol SA/mg cell protein). There was thus no correlation in this tumor system between metastatic potential and net amount of neuraminidase-releasable SA.

It has been suggested (12) that cell surface charge in general may be important for metastasis and that metastatic cells may be more negatively charged than nonmetastatic cells. Table 2 shows results of a determination of overall cell surface charge by carrier free flow electrophoresis.

Table 2. Effect of neuraminic acid on cell surface charge and electrophoretic mobility.

Neuraminidase Pretreatment	Relative free flow electrophoretic mobility[1)] of Eb	ESb	MDAY-D2
-	87	71	99
+	53	41	59
Reduction of charge	34	30	40

1) % B_{EIZ} = Peak values of electrophoretic mobility relative to that of fixed sheep erythrocytes (13).

There was no correlation between relative metastatic potential and cell surface charge, neither before nor after neuraminidase treatment. The extent of charge reduction seen after neuraminidase treatment corresponded with the quantitative determination of neuraminidase-releasable SA.

Lectin binding and cell surface sialylation on related high and low metastatic tumor lines

The expression of cell surface carbohydrates was further analyzed by flow cytofluorography of the cells after staining with fluorescein isothiocyanate-conjugated lectins. All low metastatic lines expressed receptor sites for the lectins soybean agglutinin (SBA) and Vicia villosa (VV) whereas the metastatic lines had these binding sites mostly blocked by SA (11). It is known that VV and SBA have a similar sugar specificity, namely for D-GalNac or D-GalNac-1-3-DGal.

Since the amount of neuraminidase-accessible SA was similar for all cell types (except MDW40) we interprete the above results as indications of differences between high and low metastatic lines in the positioning of SA on cell surface glycoconjugates. We assume that the differences in positioning of SA on the cell surface may lead to masking or unmasking of terminal sugars thereby influencing the metastatic potential of the tumor cells.

Recognition of surface carbohydrates by lectin receptors of hepatocytes.

This hypothesis of a difference in positioning of SA was supported by tumor cell-hepatocyte binding studies. The spontaneous high metastatic variant ESb which shows heavy liver involvement was found to form spontaneous rosettes in vitro with isolated autologous hepatocytes, whilst the low metastatic parental line Eb did not. Table 3 shows quantitative data on this differential hepatocyte binding. As can be seen, pretreatment of Eb tumor cells with neuraminidase uncovered cryptic sites for hepatocyte binding and led to a high percentage of hepatocyte-tumor cell rosettes. On ESb cells such sites were already exposed but neuraminidase treatment increased the binding even further.

Table 3. Differences in recognition of Eb and ESb cells by syngeneic hepatocytes

Tumor line	pretreatment	% Rosette forming hepatocytes [1)]
Eb	-	5
	neuraminidase	85
	ß-galactosidase	0
ESb	-	74
	neuraminidase	92
	ß-galactosidase	26

[1)] for details see (14).

Since we expected the ß-galactosyl specific lectin-like receptors of hepatocytes (15) to be involved in this tumor cell binding, we tested the effect of ß-galactosidase treatment. The rosette-forming capability was affected after ß-Gal removal and reduced in the case of ESb cells from 74 to 26%.

Further information about the role of SA in tumor cell recognition by hepatocytes was obtained from rosette inhibition experiments with defined glycoproteins and their desialyzed derivatives (Table 4).

Table 4. Effect of neuraminic acid on inhibitor capacity of glycoproteins in the hepatocyte-ESb rosette assay.

Inhibitor of cell adhesion[1)]	Approximate mass D	SA %	Concentration required for 50% inhibition of tumor-hepatocyte rosettes ug/ml
Orosomucoid	44 000	12.4	1000
Asialo-Orosomucoid	39 000	0.4	1
O,MN glycoprotein	600 000	23.8	800
Asialo-O, MN (T-antigen)	555 000	1.2	3

[1)] For details see (16).

Asialo-glycoproteins such as asialo-orosomucoid or the T antigen of human erythrocytes were powerful inhibitors at concentrations of a few ug/ml while their sialylated forms (the % SA content is given in Table 4) inhibited only at a 300-1000 fold higher concentration (16). These results demonstrate the strong protective function of SA in a cellular tumor cell binding assay and thus indicate the potential impact in cell-cell interaction systems of a change in the degree of sialylation of cell surface glycoproteins.

When defined mono- or disaccharides instead of glycoproteins were used as inhibitors much higher concentrations were required to achieve rosette inhibition. The results in Table 5 indicate a fine specificity for Galß1-3GalNac, which is the immunodeterminant of the T antigen.

Table 5. Inhibitory capacity of mono- and disaccharides in the hepatocyte-ESb rosette assay

	% inhibition of ESb-hepatocyte rosettes 20 mg/ml	1 mg/ml
Gal	47	0
GalNac	59	0
Glc	0	0
Galß1-3 GalNac	ND	25
Galß1-3 GlcNac	ND	0
Galß1-3 GalNac α-benzyl	ND	0

ND = not done

A molecular analysis revealed that the hepatocytes recognized ESb tumor cells via receptors with molecular weights of 52,56 and 110 Kd and specificity for Gal and GalNac residues. More than ten different surface glycoproteins of ESb tumor cells and none of Eb type tumor cells served as ligands in the hepatocyte interaction (14).

Although lectin-carbohydrate interactions have been found to play a crucial role in many intercellular recognition processes, this apparently is the first molecular description of a selective lectin-carbohydrate interaction between organ-derived normal parenchymal cells and metastatic tumor cells. The possible relevance of such an interaction for cancer metastasis is suggested by the finding that spleen-selected ESb sublines differed from liver selected ones in their organotropism as well as in their ability to form hepatocyte rosettes.

<u>Differences in recognition of Eb and ESb cells by macrophages and effects of pretreatment with neuraminidase.</u>

Previous histological studies (7) and cell separation procedures (17) had revealed a striking difference in the degree of infiltration of primary tumors from either Eb or

ESb lines by host-derived mononuclear cells, particularly by histiocytes. While ESb tumors had only few of the cells, the low metastatic tumors of Eb were heavily infiltrated.

In order to elucidate mechanisms of macrophage-tumor cell interactions we have investigated in vitro tumor cell binding properties of normal host macrophages and the effect of macrophages on tumor cell growth. In contrast to hepatocytes which reacted preferentially with ESb type cells, macrophages were found to preferentially react with Eb type cells (Table 6). This was found to be true in three different binding assays, at 4°C and at 37°C, and at various macrophage to tumor cell ratios (18).

Pretreatment of tumor cells with neuraminidase increased the numbers of macrophages binding Eb or ESb tumor cells (Table 6). These and other findings suggested that carbohydrates, especially SA, may also play a role in tumor cell recognition by macrophages. A difference in positioning of SA on Eb and ESb cells could possibly explain the differences observed in lectin binding, macrophage binding and hepatocyte binding. In all three binding systems, a low binding by either Eb or ESb cells could be converted into a high binding upon removal of SA.

Table 6. Differences in recognition of Eb and ESb cells by macrophages and effect of neuraminidase pretreatment.

Tumor line	Neuraminidase pretreatment	% Rosette forming macrophages[1)] Exp. 1	Exp. 2
Eb	-	15 ± 2	12 ± 1
	+	28 ± 4	30 ± 2
ESb	-	2	1
	+	10 ± 2	13 ± 1

1) % of normal syngeneic peritoneal macrophages binding ≥ 4 tumor cells (18).

Upon coculture with macrophages a tumor cytostatic effect was consistently observed with Eb but not with ESb tumor cells. This growth inhibitory effect was found to be exerted via direct tumor cell contact. It is possible that ESb cells have been selected for escape from such growth inhibitory effects by macrophages. This could have been achieved by a change in positioning of SA such that cell surface binding sites for macrophages would have become blocked. SA could thus liberate metastatic tumor cells from growth controlling influences of macrophages.

Role of cell surface carbohydrates in tumor cell attachment to fibronectin, laminin or collagen IV.

A growth inhibitory effect could also be exerted via tumor cell adhesion to extracellular matrix (ECM) components. Many of the isolated structural components of ECM have been shown to mediate cell attachment in vitro, and these include collagen types I-V, fibronectin, laminin, chondronectin, entactin and proteoglycans. Studies on mutant cell lines displaying altered attachment to substrate and changes in cell-surface carbohydrate composition provided evidence for the involvement of glycoconjugates in cell attachment (19,20).

We have performed comparative attachment studies with our paired tumor lines using a sensitive drop assay in Terasaki Microtest plates. The high metastatic cells (ESb, MDAY-D2 and MDW40-24a, Table 1) were found to be generally less adhesive to the substrates tested than the corresponding low metastatic lines. Representative results for the Eb/ESb/ESb-M tumor system are given in Table 7. The data obtained with neuraminidase treated cells furthermore revealed a role of cell surface sialic acid in the reduced attachment of the tumor lines to fibronectin.

Table 7. Effects of neuraminidase pretratment on tumor cell attachment top fibronectin and collagen

Adhesion Substrate	Neuraminidase Pretreatment of tumor cells	% attached cells [1] Eb	ESb	ESb-M
Fibronectin	-	91±15	16±7	39±18
	+	92±12	52±7	62± 8
Collagen IV	-	13± 3	5±4	21± 8
	+	24±11	5±3	26±12

1) mean ± SD from three exp. with triplicates each; similar results were obtained previously with MDW4 cells (9).

Similarly, surface SA had been found previously to be responsible for the reduced attachment of metastatic MDAY-D2 tumor cells to fibronectin and collagen type IV (9). The increased adhesiveness of the lectin resistant mutant MDW40 appeared to be closely related to a lesion in cell surface carbohydrate structures and reduced expression of cell surface SA. In an attempt to identify the carbohydrates involved in cell attachment, glycopeptides were isolated from mutant and wild-type cells as well as from purified glycoproteins and tested for their ability to inhibit the attachment of MDW40 cells to plastic surfaces coated with fibronectin, laminin or collagen type IV. Table 8 shows a summary of the results. Sialylated poly-N-acetyllactosamine-containing glycopeptides (inhibitor SSS) isolated from metastatic MDAY-D2 cells but absent from MDW40 cells, inhibited MDW40 attachment to laminin but not to collagen IV or fibronectin. Mannose-terminating glycopeptides of various sources on the other hand inhbited MDW40 cell attachment to collagen IV but not to laminin or fibronectin.

Table 8. Selective inhibition of MDW4 tumor cell attachment to adhesion substrates by tumor derived Asp-linked glycopeptides

Inhibitor[1] (Triantennary glycopeptides)	Terminal sugars on glycopeptides	%Inhibition of attachment to		
		Fibronectin	Laminin	Coll.IV
1. SSS	Neu-Ac.	10	96	15
2. GGG	Gal	-10	85	11
3. GnGnGn	GalcNAc	- 7	0	22
4. Man_{5-9}	Man	-10	12	50
5. Man_5	Man	5	28	92

1) for details see (22).

Tumor cell attachment to and detachment from the subendothelial extracellular matrix proteins laminin and collagen IV are thought to be important events in the metastatic process. Our results indicate that tumor cell attachment to these proteins may be partially modulated by the expression of specific oligosaccharide structures associated with the cell surface.

This conclusion was further supported by the effects on adhesion observed with the inhibitor tunicamycin (21) or with the enzyme peptide N-glycosidase (22). Both reagents affect the expression of asparagine-linked oligosaccharides.

The more completed and sialylated the complex oligosaccharides at the cell surface, the more autonomous from host control and the more metastatic the tumor may become. A number of under-sialylated mutants isolated from the B16 melanoma (23), the RAW-117-H10 lymphosarcoma (24) or the MDAY-D2 tumor (25) have been found to be less metastatic than the respective parental cell line. Furthermore, we suggest an important role of the site-specific positioning of sialic acid, because this negatively charged carbohydrate is a powerful residue which can modulate (mask or unmask) the expression of specific cellular substrate adhesion sites (see Table 7), lectin binding sites, cell-cell-recognition sites (see Table 3) and antigenic determinants. It is also possible that natural tumor defense mechanisms mediated by macrophages or by natural killer (NK) cells involve recognition of tumor cell surface carbohydrates. Interestingly, in both tumor systems investigated the low metastatic lines were found to be more sensitive to NK cell lysis than their metastatic variants.

REFERENCES

1. Irimura, T., Gonzales, R. and Nicolson, G.L. Cancer Res. 41:3411-3418, 1981.
2. Yogeeswaran, G. and Salk, P.L. Science 212:1514-1516, 1981.
3. Dennis, J., Waller, C., Timpl, R. and Schirrmacher, V. Nature 300:274-276, 1982.
4. Fogel, M., Altevogt, P. and Schirrmacher, V. J.Exp. Med. 157:371-376, 1983.
5. Springer, G. F., Desai, P. R., Murthy, M. S., Tegtmeyer, H. and Scanlon, E. F. Progr. Allergy 26:42-96, 1979.
6. Coon, I. S., Winstein, R.S., and Summers, I. L. Amer. J. Clin. Path. 77:692-699, 1982.
7. Schirrmacher, V., Shantz, G., Clauer, K., Komitowski, D., Zimmermann, H.-P. and Lohmann-Mathes, M. T. Int. J. Cancer 23:233-244, 1979.
8. Kerbel, R. S., Florian, M., Man, M. S., Dennis, J. and McKenzie, I. F. C. J. Natl. Cancer Inst. 64:1221-1230, 1980.
9. Dennis, J., Waller, C., Timpl, R. and Schirrmacher, V. Nature 300:274-276, 1982.
10. Yogeeswaran, G. and Salk, P. L. Science 212: 1514-1516, 1981.
11. Altevogt, P., Fogel, M. Cheingsong-Popov, R., Dennis, J., Robinson, P. and Schirrmacher, V. Cancer Res. 43: 5138-5144, 1983.
12. Klein, G. and Klein, E. Ann. N. Y. Acad. Sci. 63: 650-661, 1956.
13. Schirrmacher, V., Altevogt, P., Fogel, M., Dennis, J., Waller, C. A., Barz, D., Schwartz, R., Cheingsong-Popov, R., Springer, G., Robinson, P. J., Nebe, T., Brossmer, W., Vlodavsky, I., Paweletz, N., Zimmermann, H.-P. and Uhlenbruck, G. Invasion and Metastasis 2:313-360, 1982.
14. Cheingsong-Popov, R., Robinson, P., Altevogt, P., and Schirrmacher, V. Int. J. Cancer 32:359-366, 1983.
15. Ashwell, G., and Harford, J. Ann. Rev. Biochem. 51: 531-554, 1982.
16. Springer, G. F., Cheingsong-Popov, R., Schirrmacher, V., Desai, P. R., and Tegtmeyer, H. J. Biol. Chem. 258: 5702-5706, 1983.

17. Schirrmacher, V., Fogel, M., Russmann, E., Bosslet, K., Altevogt, P., Beck. L. Cancer Met. Rev. 1:241-274, 1982.
18. Schirrmacher, V. and Appelhans, B. Clin. & Exp. Met., in press.
19. Pouyssegur, J., Willingham, M. and Pastan, I. Proc. Natl. Acad. Sci. 74:243-247, 1977.
20. Pena, S. D. J. and Hughes, R. C. Nature 276:80-83, 1978.
21. Butters, T. D., Devalia, V., Alpin, J. A. and Hughes, R. C. J. Cell. Sci. 44: 33-58, 1980.
22. Dennis, J. W., Waller, C. A. and Schirrmacher, V. J. Cell. Biol. 99:1416-1423, 1984.
23. Tao, T.W. and Burger, M.M. Nature (Lond.) 270:437-438, 1977.
24. Reading, C.L., Belloni, P.N. and Nicolson, G.L. J.Natl.Cancer Inst. 64:1241-1249, 1980.
25. Dennis, J., Donaghue, T.P., Florian, M. and Kerbel, R.S. Nature (Lond.) 292:242-245, 1981.

Part Three

Clinical Perspectives and Applications

21

TUMOR HETEROGENEITY AND EMPIRICAL CLINICAL CANCER CHEMOTHERAPY: CURRENT STATUS AND FUTURE PROSPECTS

S.K. CARTER

Pharmaceutical Research and Developmental Division, Bristol-Myers Company, 345 Park Avenue, New York, NY 10009

Cancer chemotherapy is a modality of treatment which has established itself to a greater degree than many early skeptics thought possible, but has failed, as yet, to achieve the full promise of its more ardent supporters. From its earliest days, chemotherapy was seen as heterogeneous in its activity spectrum. Cancer was many different diseases and each individual drug had its own activity profile within these diseases. A drug's activity, or lack thereof, in leukemia predicted nothing about what responsiveness might be observed in lung cancer, and vice-versa. New drugs, therefore, were tested empirically, in a phase 2 fashion against a range of tumor types in the hopes of finding at least one tumor type which would shrink measurably in a reproducible fashion. This has led to a complex armementarium of drugs active against selected histologic types, and an even more complex array of disease-oriented therapeutic strategies involving either drug(s) alone or drug(s) in combination with other modalities[1-3].

Experimentalists involved in the modeling of cancer chemotherapy have long struggled with the observational reality of clinical cancer chemotherapy and the relevance of their models to this observed reality. The large clinical data on cytotoxic drugs can lead to some broad generalizations as follows: 1) some human cancers are very drug responsive, to the point of cure, while others are highly resistant; 2) any given drug will achieve a consistent meaningful response in some tumor types but not in all and each drug has its own spectrum of "sensitive" and "resistant" human tumors; 3) within a tumor sensitive to a given drug, or drug regimen, there will be observed fluctuating degrees of response in similarly treated individuals bearing apparently similar tumor burdens at the initiation of therapy; and 4) with a few exceptions, combinations of drugs are superior to single agents where the tumor is considered highly sensitive to drug treatment.

Cancer chemotherapy is currently at a plateau which has disappointed its enthusiasts and had led to some burn out in its community practioners who must deal with its lack of significant survival prolongation effect in the majority of the common solid tumors they have to treat. Cytotoxic drugs alone can achieve cure in childhood leukemia, Hodgkin's disease, diffuse lymphomas, gestational trophoblastic neoplasms and testicular cancers. In combination with surgery and/or x-ray, it achieves cures in several childhood tumors such as Wilms' tumor, Ewing's sarcoma and embryonal rhabdomyosarcoma. Chemotherapy achieves complete remissions, and survival prolongation, in a wide range of hematologic neoplasms, small cell lung cancer and ovarian cancer. For the great mass of solid tumors, cytotoxic drugs achieve a low to neglible rate of complete remissions and a range of partial response rates which are of more meaning to clinical researchers than to community practioners. Adjuvant chemotherapy, designed to eradicate micrometastases, has been a disappointment. It is questionable, at this time, whether it has meaningfully changed the natural history of long-term overall survival, after local control therapy, in breast cancer(4) and osteosarcoma(5), despite the early enthusiasm and publicity. It has clearly not been effective in lung cancer, colon cancer and many other solid tumors.

Unfortunately, the cytotoxicity of cancer drugs is not restricted to neoplastic cells. All cancer drugs damage normal cells to some degree. The titration of drug dosage to the limits of patients tolerance so as to achieve the maximum tolerated dose (MTD) is one of the prime rules of medical oncology. This means that all drug treatment must be analyzed in terms of a cost vs benefit ratio which is called the therapeutic index. A consistent failing of the chemotherapy literature is that is gives maximal emphasis to efficacy determinants from a research perspective and tends to neglect the therapeutic index analysis from a community patient care vantage point. The Director of the National Cancer Program has often correctly stated that there is no standard treatment and that research is continuous. Community oncologists have all trained in research settings and tend to modulate their therapeutic prescriptions based on the latest research reports. Because of this, the neglect of the therapeutic index analysis in the oncology literature is a major problem. One of the great challenges for cancer control research is to analyze the translation of new treatment research findings into population based community practice from the perspective of the therapeutic index and the quality of life.

Why does cancer chemotherapy fail to eradicate the last neoplastic cell, and thus achieve cure, in most cancer patients? The reason, as eloquently, if simply, stated by Howard Skipper is the overgrowth of resistant cells(6). Tumor cells are either intrinsically resistant, or become resistant, to individual cytotoxic drugs. Unfortunately, the sensitive normal tissues, such as the bone marrow cells, never become resistant. Chemotherapy is administered at the MTD to achieve the highest degree of fractional cell kill of the sensitive tumor cells. A given chemotherapy regimen is utilized until there is clinical evidence of tumor progression indicative of the overgrowth of resistant cells. The study of resistance has been continuous through the modern history of cancer chemotherapy. The evolution of that study has been reflective of the changing view and increased knowledge of the concept of tumor cell heterogeneity.

The first recognized level of heterogeneity by the chemotherapist was histologic. Different tumor types responded differently to individual drugs. Some histologic types were "sensitive" to drugs and others were "resistant". This was an empiric observation which resulted in pragmatic treatment and research strategies. The phase II trial as an empiric clinical screen for efficacy, along disease-oriented lines, was one of the important results. This recognition also led to one of the pivotal dichotomies in clinical cancer research: the dynamic tension between disease-oriented and modality-oriented strategies. The modality-oriented strategy attempts to evaluate new drugs with the goal of elucidating their role in a manner which will allow their regulatory approval for widespread community usage. It is recognized that the final common pathway of the modality-oriented strategy is within different disease-oriented strategies. The disease-oriented strategy involves elucidating the optimal treatment for each stage, and therapeutic situation, within the disease. This may often involve combination regimens whose evaluation precludes any uncovering of what a single drug component may, or may not, be accomplishing.

Relatively early in the history of chemotherapy, tumor cell kinetic studies evolved the concept of kinetic heterogeneity and kinetic resistance. It was observed that tumor cells went through the normal growth sequence of phases such as mitosis(M)$\rightarrow G_1 \rightarrow S \rightarrow G_2 \rightarrow$M. The S-phase was shown through in-vitro studies to be most sensitive to cell kill by many drugs, particularly the anti-metabolites. It was revealed that cells actively moving within the phases of the growth cycle were more sensitive to drug kill than those that were not. The concept of the Go cell was highlighted to describe a cell which was kinetically dormant but still capable of entering the growth cycle and reproducing itself. A dominant idea was that the kinetic resistance of the Go cell was the major cause of drug failure. The cell kill hypothesis supported the view that the larger the tumor the greater the percentage of cells in Go, the lower the growth fraction, the longer the doubling time, and more resistance there was to cytotoxic drug treatment. A variety of clinical research strategies derived from this. These included attempts at the synchronization of tumor cells to put more of them into the most sensitive phase(s) of the growth cycle. These approaches grew out of elegant cell culture studies which were impossible to meaningfully translate into the in vivo reality of the cancer patient. Another approach that was evaluated unsuccessfully was the fixed sequence of two drugs with different kinetic profiles of cell killing in vitro. The first drug would kill in all phases of the cell cycle and be cytoreductive. This would result in a smaller tumor mass, with a higher growth fraction, which would be more sensitive to an anti-metabolite which killed in S-phase. Examples of such approaches included cyclophosphamide $\rightarrow$ methotrexate, nitrosourea $\rightarrow$ cytosine arabinoside and nitrosourea (CCNU) $\rightarrow$ hydroxyurea. It eventually became understood that cell kinetic study methodology was very limited, that deep seated human malignancy could not be meaningfully studied and that cell kinetics was, at best, a unifactorial part of a complex multifactorial problem.

Pharmacology has been an important area of study and can be viewed as a host-drug interaction which can lead to the inability of drugs to kill tumor cells. It is well accepted that a cancer drug to be effective must achieve a minimal concentration over some minimal period of time. The clinical pharmacologist expresses this as concentration x time (C x T).

Pharmacokinetics is known to be extremely variable from one patient to the next and is probably one factor which explains the inconsistent responses of tumors that are "sensitive" in vitro. Critical to the success of a cancer drug is that it must penetrate tumor cell membranes before interacting with intracellular target sites to achieve their pharmacologic effects. The access of drug to tumor tissue is dependent upon the vascular perfusion of the tumor and diffusion within the tumor matrix. When a drug arrives at the tumor cell, a variety of structural, electrical and thermodynamic factors determine the qualitative and quantitative properties of drug transport across the tumor cell membrane. The penetration of the cell membrane by a cancer drug is a highly complex process. The molecular interactions which take place between the drug and the membrane constituents are not well understood at this time and is finally receiving increasing research emphasis. The information which exists has been derived from animal systems and it is questionable as to the level of correlation with the human situation.
For a long period of time the clinical pharmacology of cancer drugs was studied as plasma pharmacokinetics. It was recognized, however, that the critical question was the C x T of the active moiety of the drug at the tumor cell level within the tissue and the resultant levels over time within the cell itself. This recognition has had little practical impact save for the emphasis given to pharmacologic sanctuaries, such as the CNS where it was understood that many drugs could not achieve high concentrations because of the existence of the so-called blood-brain barrier. The capabilities to routinely study the tissue pharmacology of cancer drugs, within deep seated organs, does not currently exist. Because of this, clinical pharmacology has minimal practical impact on cancer chemotherapy despite the elegance of many of the studies. It remains more a talisman of good science,

accruing data for reference usage while rational empiricism totally dominates drug scheduling and combination regimen development.

A "pharmacologic" strategy to overcome resistance has been predicated on the tenets of the cell kill hypothesis. This strategy attempts to increase the C x T of the active moeity of the drug selectively at the tumor cell level. A wide variety of approaches has been, and is still being, attempted to accomplish this. One of the oldest of these, which has now returned again to active investigation, is the use of alternate routes of drug administration to increase the concentration at selective sites. The most successful of these approaches has been the use of intra-thecal methotrexate in childhood acute lymphocytic leukemia. This technique has been utilized to treat both evident meningeal leukemia and as prophylaxis after initial curative intent induction. In both clinical situations, higher concentrations of drug have been achieved, at the selected site, with resultant benefit, and in the case of the adjuvant type prophylactic usage, increased cure.

The use of intra-arterial drug has been repetitively attempted with the goal of bringing higher drug concentrations to selected sites. The cannulation of the hepatic artery to infuse the liver with rapidly metabolized anti-metabolites, such as the fluorinated pyrimidines, has consistently led to high response rates in single arm phase 2 type experiences. It has never been demonstrated in a properly designed randomized comparison of intra-arterial versus intravenous drug that the arterial approach improves the therapeutic index. This leads one to the suspicion that the positive phase 2 studies are due to case selection bias.[7] In essence, however, there is no way to totally reject this approach. Any given drug, at the right C x T for a critical number of cells, might result in the hoped for tumor cell eradication. Since predictive models are yet to be found, the strategy remains mainly empirical and worthy of pursuit.

A new approach to selectively increasing drug concentration through route manipulation utilizes intraperitoneal drug administration.[8] This is conceptually attractive in ovarian cancer where the tumor is often diffusely spread within the regional confines of the abdominal cavity. The data are still of a phase 2 nature but the use of I.P. cisplatin, and other drugs, are giving encouraging results which need to be followed by the appropriate randomized phase 3 type studies. Howell and associates[9] have combined I.P. cisplatin with the systemic administration of sodium thiosulfate which blocks the renal toxicity. This allows a higher dose of drugs to be given and combines two approaches to pharmacologically overcoming resistance through increased concentration.

The strategy of pharmaceutically blocking toxicity is attractive because of its potential ability to increase concentration systemically within the limitations of the existing pharmacologic sanctuaries for the drug(s) in question. The drugs used for toxicity blocking have tended to be linked to the amelioration of the side effects of selected drugs. The example which most readily comes to mind is that of methotrexate and calcium leucovorin. The sequential use of methotrexate followed by calcium leucovorin have enabled the doses of the anti-fole to be increased by a log order magnitude. Unfortunately, it appears that the blocking of toxicity also involves the tumor cells and the resultant selectivity of cell kill is no greater than is achievable with lower doses without pharmacologic "rescue". No prospectively randomized comparison of high dose methotrexate + calcium leucovorin with standard dose drug has ever demonstrated a higher degree of efficacy.

Other pharmacologic toxicity modulators include the unoprotectant Mesna for Ifosfamide and cyclophosphamide as well as the already mentioned sodium thiosulfate for cisplatin. Clinical experiments have been performed with the radioprotecting drug WR-2721 and several drugs but they are still too early to meaningfully evaluate.

Biochemical heterogeneity, as a reason for resistance, has also been studied for nearly the entire modern era of chemotherapy. The anti-metabolities were among the earliest drugs discovered and their mechanism of action was determined to be rational, in a biochemical sense, based on in vitro laboratory studies. Each of these anti-metabolites was postulated to have a specific enzyme target within the cell and the corollary was that resistance occurred because of modulation of the critical enzymatic balance within the cell. The elegant laboratory studies of the biochemical mechanisms of action and resistance of the anti-metabolites provided a patina of rationality to what was an extremely empirical science. The resultant papers could fill multiple volumes but this practical impact has been negligible. The graveyard of drug development is full of compounds "rationally" designed to overcome the resistance to the thiopurines, fluorinated pyrimidines or methotrexate. While 5-Fluorouracil was considered to be an example of the success of "rational" drug development, the subsequent failures of 5-FUDR, F_3TDR and other analogs revealed the translation of postulated biochemical paradigms to practical reality was extremely difficult.

A new concept that is now receiving increased research interest is that drug resistant neoplastic stem cells often arise from mutations and that the drug-resistant phenotype is inherited and propogated. In cell culture studies, three major mechanisms of genetic resistance have been delineated and characterized:

1) overproduction of the drug target (e.g. gene amplification)
2) reduced drug permeability; and
3) altered interaction of the drug with its target.

The concept of the spontaneous mutation of cancer cells to drugs they have never seen has given rise to the so called Goldie-Coldman Hypothesis[(10)] which is now viewed by some as either a revisionist or alternative paradigm to the cell kill hypothesis which has dominated medical oncology since its inception.

The new paradigm to explain chemotherapy's successes and failures focuses on resistance. It draws strong analogy between cancer chemotherapy and the successful use of antibiotics for the treatment of infectious disease. It starts with the observations of Luria and Delbruck in 1943.[(11)] Their observations indicated that bacteria have an inherent capacity to mutate toward resistance to organisms they have never seen. The size of a resistant population in sub-cultures of similar size differed depending upon when this mutation occurred. This type of resistance was a permanent genetic change.

Goldie divides tumor growth history into two discrete time intervals. The first period is that of early growth when no resistant cells will have developed. At some point in time, there will emerge the first stable phenotypically resistant cell and the second critical phase of growth will begin. Therefore, within any tumor, the transition from curability to incurability with drugs will occur as a sudden discontinuous event. The implication of this is that any tumor system with a high mutation rate, which has reached a large size, will have a near certain probability of having at least one resistant cell within it. The mathematical model that is derived from this reveals a steep curve of moving from sensitivity to resistance as a function of tumor growth. The change from a condition where 95% of the tumors can be expected to have zero resistant cells to where only 5% will have no resistant cells occurs over an increase in size range of approximately 1.8 logs. This is consistent with the steepness of the cure rate curve in a system like L1210 leukemia in relation to tumor size.

The Goldie-Coldman model for relating drug sensitivity of tumors to the occurrence of spontaneous mutations makes a variety of assumptions.[10] These assumptions are the following:

1) the mutations rates to resistance, to a variety of anti-neoplastic agents in a variety of mammalian tumor cells, has been estimated to range approximately 10^{-4} to 10^{-7} and that this rate is similar in man;

2) that all cells within the tumor have stem cell capacity which means they are capable of indefinite self-renewal without cell loss due to the generation of progeny with limited proliferative capacity. It is recognized that the human tumor situation is more complex but it is assumed that any more detailed mathematical model would yield "the same general conclusions as are obtained from the minimal model";

3) resistant phenotypes arising within a tumor population are totally resistant to the maximum tolerated doses of drugs that can be employed. Again, the complexity of partial resistance is recognized but is felt not to change the conclusions drawn from the model;

4) the resistant phenotypes will be kinetically identical to that of the parent sensitive strain.

The Goldie-Coldman model inputs a new important variable into the old kinetic cell kill hypothesis model which dominated chemotherapy for so long. It suggests that differentiation and cell death not only contribute to slower tumor growth but also to a marked increase in the proportion of multi drug-resistant stem cells at a given tumor size. Therefore, the overall relationship between tumor size, growth rate and age versus drug responsiveness may be a reflection of the natural selection of permanently drug-resistant mutant tumor stem cells.

The practical implications of the Goldie-Coldman Hypothesis, on the use of chemotherapy, are in two areas, both of which have an empiric basis of study, which predate the first publications on the paradigm. The first area involves the use of preoperative chemotherapy for locally advanced solid tumors at initial diagnosis. This has been called "neoadjuvant therapy" by some and "induction therapy" by others. The goal is: 1) to improve the possibility of local-regional control by surgery and/or irradiation by shrinking the primary tumor and eradicating its marginal metastatic spread; and 2) to improve overall cure by an early attack on disseminated micrometastasis. The Goldie-Coldman Hypothesis supports the view that starting chemotherapy as soon as possible after diagnosis and clinical staging might hit the neoplastic cells before they have mutated to resistance. The use of neoadjuvant chemotherapy in soft tissue and bone sarcomas in the limb has allowed local control to be achieved without the mutilation of radical amputation. Whether it has improved the cure rate is debatable at this time since all the positive data derive from studies with historical controls which are notoriously poor.

Neoadjuvant therapy is currently under the most vigorous investigation in stage 3-4 head and neck cancer. The regimen of cisplatin plus continuous infusion 5-fluorouracil has been shown to induce a high response rate in the primary tumor but its long-term impact on local control and disease-free survival will have to await the randomized trials and a proper follow-up time.

The second area of practical impact involves the use of fixed sequences of non cross-resistant combinations. The most successful example of this is a trial in Hodgkin's disease begun empirically years prior to the Goldie-Coldman papers. This has involved the fixed sequence of the MOPP four drug combination alternated with the four drug ABVD combination(12). In previously untreated patients with advanced stage disease, MOPP alternating with ABVD has been shown to be superior to MOPP alone as would be predicted by the Goldie-Coldman mathematical model. While this is encouraging, similar type trials in advanced breast cancer and small cell lung cancer, again started pragmatically, have failed to give positive results. The jury on this approach must still be viewed as being out.

Despite the current recognition of its oversimplifications and imperfect assumptions, the cell kill hypothesis still dominates the strategic thinking of the clinical chemotherapist. The aspect of the cell kill hypothesis which still dominates is the concept of increasing the drug concentration at the tumor cell level to increase the fractional cell kill and possibly achieve eradication of the last neoplastic cell.[13] If the concentration of drugs could be driven high enough, either existing resistance could be overcome or eradication accomplished before resistance occurs. This approach is pragmatically reasonable with drugs acting cytotoxically, and, in essence, does not need an old hypothesis to support it. The recent approaches to increasing the concentration of drugs at the tumor cell level include supportive care, regional delivery, toxicity blocking and targeting, some of which have already been mentioned.

The strategy behind the use of supportive care is to protect against, or ameliorate, the effects of bone marrow suppression which is the main dose limiting toxicity of chemotherapy. If this toxicity can be ameliorated, then higher doses of drugs can be administered. One approach to this has been the use of protected environments to prevent the lethal infection which often follows severe leukopenia. While the trials in leukemia did demonstrate that somewhat higher doses of anti-leukemic therapy could be given, the resulting efficacy was not meaningfully enhanced. The major thrust of the supportive care strategy has been the utilization of either allogeneic or autologous bone marrow transplantation. The most positive results have been obtained with the use of high dose cyclophosphamide plus total body irradiation in patients with acute leukemia in first remission after standard type induction chemotherapy.[14-15] The support of allogeneic marrow transplantation, from a histocompatible donor, has enabled the apparent eradication of residual leukemia in a significant number of patients surviving the initial therapy and the dangers of graft versus host disease. This is encouraging since it supports the concept that high enough doses of cytotoxic therapy with treatment which, at standard doses, cannot achieve complete remission, can be administered so as to obtain "cure". This interpretation is complicated by the potential role of the immunomodulation caused by graft versus host disease. The possible importance of this is highlighted by the data showing that patients with graft versus host disease (who survive) have a better therapeutic result than those without any evidence of the rejection problem.

The definitive role of allogeneic bone marrow transplantation in the treatment of acute leukemia still awaits the properly designed prospective randomized trial. Its use in other diseases is still in the early phase 2 phase. The application of this methodology to large patient populations will be difficult as long as histocompatible donors must be used and the technique remains as labor intensive and expensive as it currently is.

The use of autologous marrow transplantation has a greater potential for widespread usage if proven successful. This is because it uses the patient's own frozen marrow and does not require the large supportive care apparatus needed for the immunologic imperfections of histocompatible allogeneic marrow. The critical question with autologous marrow transplantation after high dose chemotherapy with today's drugs is whether the normal tissues, apart from the marrow, will tolerate high enough doses to achieve total cell kill. The early results are not encouraging. Drugs such as L-phenylalamine mustard, the nitrosoureas and mitomycin C can be given at higher doses than usual but the order of magnitude of the increase has not been as large as hoped for. More importantly, the phase 2 type therapeutic results have not been dramatically positive. Severe organ toxicity, outside the bone marrow, has proven dose limiting. The ultimate success of this approach will require the discovery and/or the development of drugs with only marrow damage as their major toxicity even as the standard "MTD" is significantly increased.

The idea behind targeting is to specifically deliver the drug to the tumor cell by either attaching it to a carrier, such as an antibody, or to incorporate it within a delivering material such as a liposome. The result would be a higher C x T at the tumor cell level and a lower C x T at the normal cell level. All approaches, to date, must be viewed as highly experimental and fraught with difficulties. The liposomes approach is currently severely limited by the phagocytosis of liposomes by the reticuloendothelial system and the accumulation of liposomes in the liver and spleen. The targeting with monoclonal antibodies must overcome the problem of the failure to discover tumor specific antigens and the challenge of linking drug and antibody together in an appropriate manner to achieve the goal of targeting.

In conclusion, chemotherapy, despite its achievements, remains at a plateau in its ability to deal with the heterogeneity of human malignancy. The search for new chemotypes continues and the discovery of a new active drug will most probably bump the plateau up to a higher level as had been the case in the past. The same is true for analogs of the existing known drugs which could also improve both the spectrum of activity as well as the therapeutic index. It is essential that clinical research allow the drug-oriented strategy on optimal patient population to evaluate new drugs. Where the disease-oriented strategy lacks for innovative ideation, as is the case for many tumors today, it should not be allowed to predominate.

The study of resistance is leading to newer understandings. It is now recognized that kinetic resistance is not exploitable at this time and that biochemical resistance has not yet solved the problem of translating in vitro studies to clinical reality. Classical pharmacology is also severely limited and the need for a tissue pharmacology approach is self evident. The empirical approaches to overcoming pharmacologic resistance through increasing drug concentration have made some progress and are worthy of continued emphasis within the framework of well designed clinical trial strategies. There has been, to date, far too much emphasis on the small phase 2 study as the major generator of literature inspired enthusiasm for this area.

The research into the genetic basis of drug resistance and the integration of cell biology with molecular biology offers great hope for the future although little practical impact at present.

The greatest challenge for the future will be to achieve the optimal research balance between the broad areas of cytotoxic versus non cytotoxic drug therapy. Hopefully, this can be viewed as complimentary and not competitive in this oncopolitical world of diminished funding and continued high visibility for cancer research.

REFERENCES

1. Carter S.K., Bakowski M. and Hellmann K.: Chemotherapy of Cancer, 2nd Edition. John Wiley and Sons, NY, 1981.

2. Carter S.K., Glatstein E., and Livingston R.B.: Principles of Cancer Treatment. McGraw-Hill, NYC, 1982.

3. DeVita V.T., Hellman S. and Rosenberg S.: Cancer Principles and Practice of Oncology. J.B. Lippincott Co. Philadelphia, 1982.

4. Henderson I.C.: J. Clin Oncol 3:140-143, 1985.

5. Carter S.K.: J. Clin Oncol 2:147-148, 1984.

6. Skipper H.E.: Southern Research Institute, Booklet 3, 1982.

7. Stagg R.J., Lewis B.J., Friedman M.A., et al: Ann Int Med 100:736-743, 1984.

8. Myers C.E.: In: DeVita V.T., Hellman S., and Rosenberg S.A.: Important advances in oncology 1985. J.B. Lippincott, London pp. 218-226, 1985.

9. Howell S.B., Pfeifle C.E., Wung W.E.: Ann Int Med 97:845-851, 1982.

10. Goldie J.H., Coldman A.J.: Cancer Treat Rep 63:172, 1979.

11. Luria S.E., Delbruck M.: Genetics 28:491, 1943.

12. Santoro A., Bonadonna G., Bonfante V., et al: N. Engl. J. Med. 306:770-775, 1982.

13. Frei E. III, Canellos G.P.: Am J. Med 69:585-594, 1980.

14. Thomas E.D.: Cancer 49:1963-1969, 1982.

15. Applebaum F.R., Dahlberg S., Thomas E.D., et al: Ann Int Med 101:581-588, 1984.

22

HORMONAL REGULATION OF METASTASES: PROSPECTS FOR PHARMACOLOGICAL MANIPULATION

RUSSELL G. GREIG

Department of Tumor Biology, Smith Kline and French Laboratories, Philadelphia, PA 19101

INTRODUCTION

Many biological processes operate under strict paracrine and endocrine control and aberrations in these regulatory mechanisms can result in the development of disease. For decades this fundamental realization has provided the intellectual infrastructure for the design and development of several classes of pharmacological agents that have subsequently demonstrated efficacy against a spectrum of important human diseases, including diabetes, heart disease, autoimmune disorders and asthma. As our understanding of the molecular mechanisms underlying the complex etiology of each of these diseases becomes more sophisticated, new opportunities for pharmacological intervention steadily emerge. Although investigations into the hormonal[1] regulation of tumorigenesis have been extensive and enjoyed a degree of diagnostic and clinical success (especially for breast carcinoma) (1,2), parallel studies on the metastatic spread of malignant tumors have received much less attention, particularly at the molecular level. If more effective therapeutic strategies are to be developed against disseminated malignant disease, then new biochemical approaches are needed to identify novel pharmacological targets unique to metastatic tumor cells. This, in turn, will provide an essential stepping stone for the discovery of antineoplastic drugs able to occupy the substantial therapeutic void currently unfullfilled by present cytotoxic agents

---

1 The term hormone is employed here to describe auto- para- and endocrinological chemical mediators including neurotransmitters.

that for the most part display minimal activity against solid established metastases.

The purpose of this article is to discuss whether certain of the pharmacological concepts that have emerged from our study of endocrinological mechanisms in human disease can be constructively applied to the study of metastasis, to achieve two goals: first, to determine which steps in the metastatic process are hormone-dependent and second, to exploit this knowledge for therapeutic gain.

Frequently it is stated that the final demise of the cancer problem will be attained only when the elusive secrets of the neoplastic cell are documented in molecular detail. While this may be true, it should not distract from the fact that many incapacitating human diseases whose etiology is poorly understood can be substantially alleviated by drugs whose mechanisms of action are superficially documented but whose discoveries were predicated upon insightful (and in some cases, brilliant) interpretations of precocious working hypotheses (3,4). However, in oncology the sad fact remains that the single most important barrier frustrating the discovery of new antineoplastic agents is the near total absence of any working hypothesis defined in molecular terms to explain the emergence of malignancy and the establishment of metastases.

The depth of this void is especially appreciated when one compares our pharmacological understanding of major human ailments such as heart disease and asthma with malignancy. Although heart disease encompasses several distinct syndromes, all of which display complex and multifactorial etiologies, the fact that cardiovascular performance can be rationalized in terms of several distinct but interacting ligand-receptor systems, including adrenergic, cholinergic, dopaminergic and peptidergic (that display a spectrum of receptor subtypes and complex anatomical distribution), has laid the essential pharmacological groundwork for permitting the development of novel cardiovascular therapeutic strategies (5). In addition, the constant discovery of new endogenous mediators of hemodynamic function (eg atrial natriuretic factor and leukotrienes) (6,7) provides not only new targets for clinical exploitation but also contributes to our understanding of the mosaic of regulatory mechanisms controlling cardiovascular output. The importance of the latter consideration cannot be overemphasized since this fuels the ideas for future generations of drugs. Similarly, in asthma, elucidation of the pharmacology of bronchoconstriction (8) illustrates the quality of information required for rational drug discovery. In this condition, pulmonary mast cells, following antigenic challenge, release a pulse of inflammatory mediators that causes bronchoconstriction, among other effects. With this sequence of events established, several stages can be entertained for possible pharmacological intervention (8). These include: 1) antagonism of the mast cell release reaction; 2) effective antagonists of mast cell release products (leukotrienes, prostaglandins, histamine and platelet-activating factor); and 3) bronchodilators. Some of these strategies have been successfully pursued (especially the use of β-adrenergic agonists and theophylline as bronchodilators) (8,9) and several others (eg the search for leukotriene and platelet-activating factor antagonists) (10) are the subject of intense interest in both industrial and academic laboratories.

The main point here is that the level of pharmacological sophistication which is deemed necessary to embark upon a program of rational drug discovery in either cardiovascular dysfunction, asthma or any other human disease of importance greatly exceeds our still essentially descriptive understanding of metastatic disease. Perhaps more galling is the realization that even equipped with this knowledge, effective drugs for the treatment of heart disease and asthma are proving increasingly difficult to develop and there is no compound for either condition that has been discovered or is likely to be discovered in the near future that cures the disease rather than treating the symptoms.

It is against this backdrop that efforts to design drugs effective against established micrometastases must be evaluated. The first step in this process is an awareness of our current ignorance of the molecular mechanisms responsible for the expression of metastatic disease. We presently lack information on the biochemistry and pharmacology of metastasis and it is unrealistic to set (within the next 10 years) a rigid timetable for the rational discovery of antimetastatic drugs. This does not preclude the possibility that random and semi-empirical screening methods will generate compounds with a desirable profile of action. However, based on historical evidence the probability of success must be viewed as minimal. A more obtainable objective is to survey the quantity and quality of information generally viewed as necessary to launch a new drug discovery program and to apply the same criteria to efforts to study the process of metastasis. From a cursory examination of the literature and the arguments presented here, we lack the information required to begin to speculate on pharmacological targets unique to metastatic tumor cells. If we accept this conclusion, then the goal of contemporary metastatic research should not be predicated upon the continued empirical testing of microbial broths, fungal cultures, plant extracts and miscellaneous chemicals in animal tumor models, but instead should be geared toward generating a pool of molecular information gleaned from relevant tumor systems that will provide a springboard for identifying potential pharmacological targets in metastatic tumor cells. Only when this formidable challenge has been met can a rational process of drug discovery process begin.

HORMONAL REGULATION OF METASTASIS

While there is no convincing evidence to support the contention that metastasis is a hormone-dependent process, this reflects the paucity of studies that have directly addressed this issue rather than the existence of contrary observations. Indeed, based on the all-pervasive control that hormones appear to exert in cell/tissue/organ homeostasis it would be surprising if metastasis were exempt from similar regulation. In this context, however, regulation can have (at least) three different definitions. It can refer to hormones that: 1) are inappropriately expressed by the tumor cell or by host-cell types influenced by the presence of a tumor and which encourage the development of the metastatic phenotype; 2) are produced by host-cells as a defense mechanism to frustrate the expression of metastatic properties; and 3) are produced by host-cells as part of the defense response to the tumor but whose actions inadvertently aid metastatic spread. Each of these classes of regulation operate in addition to the normal hormonal controls that regulate the target tissue in the absence of disease. Possible examples of each classes (designated class I, II and III) will be suggested for each stage in the metastatic process.

The purpose of studying the hormonal regulation of metastasis is to define the hormonal dependency of each step in the metastatic cascade with a view to developing future therapeutic strategies. For descriptive clarity the complex process of metastasis is usually viewed as a series of sequential steps beginning with tumor cell invasion and intravasation and culminating in arrest, extravasation and proliferation (11). While each stage is considered a discrete biological event, a biochemical analysis (which is currently not possible) may reveal that these apparently different stages display considerable overlap in their molecular mechanisms and hormonal control. During invasion, tissue degradative enzymes secreted by the tumor cell are thought to facilitate invasion by degrading components of the extracellular matrix, including basement membrane (12,13). Since tissue invasion is a necessary prerequisite for metastasis, strategies that antagonize this process may have clinical utility (but see below). Hormonal control of tumor cell invasion is poorly understood but several pertinent questions can be raised. Do tumor cells secrete autocrine components that stimulate the synthesis and secretion of their own proteolytic enzymes (class I regulation) (14)? Do macrophages respond to lymphokines (secreted by tumor cell-stimulated lymphocytes) by producing endogenous inhibitors of collagenase and plasminogen activator in an effort to control local damage (class II regulation) (15)? Or, do macrophages belonging to distinct functional subsets, but also activated by lymphokines, produce their own degradative enzymes and thus expedite the invasive process they were supposed to curtail (class III regulation) (16,17)? Identification of these hormonal loops is technically demanding but may offer new insights into the pharmacological control of tumor cell invasion.

A similar series of arguments can also be presented for other steps in the metastatic process including the interaction of tumor cells with the formed elements of blood (to produce emboli) and tumor cell arrest and extravasation. Tumor cell-platelet interactions have received considerable attention (18). Here there is some circumstantial evidence to suggest that embolus formation may be influenced by tumor cell-secreted products (class I regulation) (19). Work by Honn and colleagues suggests that *in vitro* the tumor cell-platelet interaction can be modulated by agents (prostacyclin and theophylline) that affect cAMP metabolism (20,21) but whether the reported antimetastatic effect of these same agents administered *in vivo* can be attributed to a similar mechanism remains unclear (22). The interaction of tumor cells with platelets could also elicit class II and III hormonal mechanisms. Examples of both classes are the platelet-release products some of which (eg histamine, prostaglandins and leukotrienes) may serve as signals for activation of host-defense mechanisms (class II) (23) and others (eg platelet-derived growth factor and transforming growth factor) which may stimulate tumor cell growth (class III) (24,25). The factors regulating the arrest and subsequent extravasation of metastatic tumor cells are poorly understood. However current knowledge of endothelial cell biology and the inflammatory response would suggest that we can reasonably anticipate that these events are responsive to, and probably controlled by, an array of hormonal influences including tumor cell-secreted components that induce endothelial cell retraction (class I), factors of endothelial cell origin that are chemotactic for immune cell types (class II) but which may also serve to recruit additional tumor cell emboli (class III) (26). Collectively, these examples suggest a series of events in which metastatic performance would be influenced by several competing and, in some cases, opposing hormonal actions, which exert their effects at each step in this process. Understanding these interactions at the molecular level (an undertaking that will likely take several years) could lead to pharmacological opportunities able to tilt the sum of the hormonal balances in favor of the host.

While such strategies are critically important in probing the biochemistry of malignant disease their clinical value is debatable. The majority of patients with malignant tumors continue to present with established micrometastases. Consequently, developing therapeutic interventions to antagonize the metastatic process rather than agents to treat established metastases will be of limited value and a classic example of closing the stable door after the metastatic "horse" has bolted. A major challenge then, it to define the hormonal regulation of the ultimate and lethal step in the metastatic cascade, establishment and proliferation of micrometastases.

Apart from its obvious clinical importance, the mechanisms by which cells from one tissue can proliferate in another is intriguing. For example, does the metastatic tumor cell exploit the local hormonal circuitry that normally regulates the proliferative capacity of the colonized tissue to faciliatate its own proliferation? Current concepts of growth regulation hold that the progress of cells through the cell cycle is controlled by both endocrinological and paracrinological events (27). Tumor cells also appear subject to autocrinological control in that they secrete and respond to their own growth factors (27,28). While it might be reasonable to expect that the endocrinological and autocrinological demands of a metastatic tumor cell in a colonized organ can be satisfied by systemic and self-delivery of hormone, respectively, the ability of the host organ to provide appropriate paracrinological signals may be critical in determining the establishment of micrometastases. Thus, do colon carcinoma metastases prosper in the liver because paracrinological hormones secreted by hepatocytes mimic the action of similar molecules produced by colon fibroblasts and/or smooth muscle cells? Conversely, is the brain an extremely infrequent site for colon metastases because it fails to synthesize suitable paracrinological signals? The concept underlying these questions is, of course, not new and simply respresents one of several possible "molecular" interpretations of Paget's "seed and soil" hypothesis (29). However, with progress in the identification of growth factors and elucidation of their mechanisms of action (30), coupled with advances in tissue culture technology that permit the in vitro propagation in defined medium of human epithelial cells (31,32), this hypothesis is now testable. Defining the growth regulatory molecules that inhibit, permit or encourage the proliferation of "foreign" histological cell types in the local milieu of the colonized organ will likely be a key component in elucidating the molecular jigsaw of metastatic spread and, looking to the future, probably represents a potentially productive strategy for developing novel forms of therapeutic intervention.

An important experimental implication of this strategy relate to the selection of appropriate tumor models. While popular metastatic animal tumors such as the murine B16 and K1735 melanomas have been invaluable in describing the cell biology of metastasis (11,33), their usefulness as models for defining the paracrinological hormones that regulate the proliferation of different histological classes of human metastases must be considered dubious. Normal and neoplastic tissue of human origin is clearly the experimental material of choice. At the same time, the failure of human xenografts to metastasize in the nude mouse (unless accompanied by specialized maneuvers like intrasplenic injection) may reflect, at least in part, the inability of the (murine) host to provide the appropriate hormonal (and perhaps species-restricted) signals required for the expression of the (human) metastatic phenotype (34-37). The analogous observation made in both allogeneic and syngeneic animal models that metastatic performance depends on the site of implantation of the primary tumor (subcutaneous versus intramuscular) may also be attributable to the paracrinological environment of the (tumor cell) inoculation site (35,38-40). Clearly, tumor cells injected subcutaneously into the ear or into the footpad would be expected to encounter distinctly different hormonal signals that may affect metastatic spread through selection of tumor cell subpopulations or by turning on (or off) biochemical pathways essential for the expression of metastatic properties. At a practical level this requires, indeed dictates, that efforts to define the hormonal regulation of metastatic spread be conducted on the (human) neoplasms of eventual therapeutic interest or adopting the current vernacular, it must be "disease-driven" (47). Just as pharmacological studies on the vasodilation of the mesenteric artery would never be accepted as a reliable experimental model for designing drugs to vasodilate the coronary artery, we cannot realistically expect investigations of murine (or even human) melanomas to identify pharmacological opportunities for antagonizing, for example, the organ-specific metastatic spread of gastric carcinoma to the ovaries. The anticipation that different histological classes of tumors employ distinct molecular mechanisms of metastatic spread is unsurprising. From a therapeutic standpoint a potentially more troubling complication would be if a single histological type of tumor was capable of forming metastasis by more than one (hormone-dependent) mechanism. If this is so, then this particular aspect of tumor cell heterogeneity (11) will pose formidable problems even for the most sophisticated form of therapeutic intervention.

HORMONES, SECOND MESSENGERS AND GENE REGULATION

Based on the assumption that metastasis is a hormone dependent process, how can we begin to map out appropriate therapeutic strategies to exploit this fact? Those drugs whose mechanisms of action are fairly well understood can be classified into either first or second messenger antagonists (or agonists). A notable exception to this classification are the majority of currently employed antineoplastic agents; while there is general agreement that they appear to act by disrupting chromatin (41), the precise cytotoxic mechanisms of this chemically reactive class of compounds remains poorly documented. Propranolol (β-adrenergic blocker) and cimetidine (H2 histamine blocker) are popular examples of first messenger antagonists currently used for the control of angina and GI uclers, respectively. Second messenger antagonists include the calcium channel blockers, verapamil and nifedipine (used clinically in cardiovascular disorders), the phosphodiesterase inhibitor, theophylline (bronchodilation) and the calmodulin antagonist, thorazine (antipsychotic). While some of these agents (eg thorazine) were discovered, developed and clinically employed long before their mechanism of action was investigated (and in this case, before calmodulin was discovered), others (such as propranolol and cimetidine) were highly successful and purposeful attempts at designing antagonists of known, endogenous mediators of cardiac and gastrointestinal function, respectively.

Ideally, the latter approach is precisely the strategy that is more attractive in the quest for antimetastatic drugs but the problems and weaknesses of embracing such a strategy in 1985 are obvious and simple; the identity of the endogenous mediators which serve either to encourage or discourage the metastatic spread and proliferation of specific human neoplasms are unknown. Redress of our ignorance of the molecular biology of these important processes is a matter of high priority and represents a critical objective in our efforts to seek new pharmacological opportunities to control malignant disease.

In addition to eliciting alterations in second messenger levels (cAMP, cGMP, calcium and inositol phosphate) several hormones and growth factors also induce the expression (and repression) of specific genes, including oncogenes (24,42,43). Precisely what physiological role the products of these genes play in mediating hormone action is not clear. For the oncogene products expressed in response to growth factor challenge, their role appears directly associated with cell cycle control and commitment to cell division. An intriguing question prompted by the rapid progress in molecular biology is whether specific genes are switched on (or off) during the process of metastasis and whether these events are essential for the establishment of micrometastases. Not surprisingly, this question is attracting increasing attention (44-46) and if answered in the affirmative would herald a new (and welcome) aspect of the study of metastasis. The opportunity for posing novel, mechanistic questions will be enormous. What genes and gene products are involved? And directly germane to the current discussion, which mechanisms (first messengers, second messengers) regulate their expression? The power of molecular genetics to provide answers to these questions offers the exciting prospect of transforming this aspect of research from one bereft of mechanism and molecules into one rich with metastasis-associated genes and gene products. The implications for future drug therapies are obvious. The prospect of identifying, cloning, expressing and manipulating (families of) genes that account for a metastatic tumor cell's lethal behavior is an ambitious goal, but one that demands detailed scrutiny.

CONCLUDING REMARKS

Multinational efforts in academia and industry conducted over the last 30 years have failed to generate compounds with effective clinical utility against solid disseminated malignancies. The reasons for this are varied, but a common underlying theme is the lack of a coherent molecular hypothesis to account for the expression of the metastatic phenotype. The void created by the absence of such a hypothesis has resulted in essentially random and semi-empirical screening efforts that have sought and continue to seek antimetastatic drugs. With the advantage of hindsight the reasons for the failure of these efforts is evident. Without any insight into the biochemistry of the metastatic process and the frequent, if not total, reliance on animal (benign) tumor models, the chances of discovering compounds active against established (human) micrometastases by screening, has to be viewed as extremely small. Unfortunately, this has proven to be the reality. As a result, the search for metastatic drugs has now reached a critical watershed. On the one hand we can continue to pursue a semi-empirical approach in which attempts are made to employ more relevant screens including human neoplastic cell lines and metastatic animal tumor models, but which are still devoid of any molecular rationale (47). The alternative approach, and the one favored here, is to recognize that the well of empiricism has run dry and we therefore need to develop new strategies predicated upon a molecular understanding of the metastatic process. In 1985, we do not yet have this information and consequently our major objective should be in designing ways to acquire it. By applying to metastasis the criteria we apply to other drug discovery efforts we can immediately recognize the depth of our ignorance and the magnitude of the challenge. But the only way we can hope to produce new, effective antimetastatic drugs by the 21st century is to invest heavily in understanding the molecular biology and genetics of metastatic spread in specific human neoplasms. To adopt a strategy like this which in essence delays promise of any lead compounds for at least 10 years is, under the full glare of the public eye and the watchful stockholder, difficult to advance compared to the usual statements from government and industrial sources which routinely state how many thousands of compounds yearly have been synthesized, modified or extracted and screened for "anticancer" activity. While this generates good, short-term press, we believe that the long-term approach holds more promise for generating effective drugs for the therapy of metastatic disease.

ACKNOWLEDGEMENTS

I thank Nancy Signora for excellent secretarial assistance and Drs. Deborah Trainer and George Poste for their constructive reviews of the manuscript.

REFERENCES

1. Lippman, M.E. and Allegra, J.C. NEJM, 299: 930-933, 1978.
2. Iacobelli, S., Lippman, M.E. and Robustelli Della Cuna, G. (Eds). The Role of Tamoxifen in Breast Cancer, Raven Press, NY., 1982.
3. Bindra, J.S. and Lednicer, D. (Eds.) Chronicles of Drug Discovery. Vol. 1. John Wiley and Sons, NY., 1982.
4. Gross, R. (Ed). Decision Making in Drug Research. Raven Press, NY., 1983.
5. Gilman, A.G., Goodman, L.S. and Gilman, A. (Eds). The Pharmacological Basis of Therapeutics (Sixth Edition). Macmillan Publishing Co., NY., pp 729-847, 1980.
6. Currie, M.G., Geller, D.M., Cole, B.R., Boyland, J.G., Sheng, W.Y., Holmberg, S.W. and Needleman, P. Science, 221: 71-73, 1983.
7. Murphy, R.C., Hammarstrom, S. and Samuelsson, B. Proc. Natl. Acad. Sci. USA, 76: 4275-4279, 1979.
8. Webb-Johnson, D.C. and Andrews, A.L., Jr. NEJM, 297: 476-482, 1977.
9. Weinberger, M. and Hendeles, L. NEJM, 308: 760-764, 1983.
10. Vargaftig, B.B. and Benveniste, J. Trends Pharmacol. 4: 341-344, 1983.
11. Poste, G. and Fidler, I.J. Nature, 283: 139-146, 1980.
12. Kramer, R.H., Vogel, K.G. and Nicolson, G.L. J. Biol. Chem. 257: 2678-2686.
13. Rifkin, D.B., Moscatelli, D., Gross, J. and Jaffe, E. In: Cancer Invasion and Metastasis: Biological and Therapeutic Aspects (G.L. Nicolson and L. Milas, Eds). Raven Press, NY., pp 187-200, 1984.
14. Liotta, L.A., Thorgeirrson, U.P. and Garbisa, S. Cancer Met. Rev. 1: 277-288.
15. Kuttner, K.E., Hiti, J., Einstein, R. and Harper, E. Biochem. Biophys. Res. Comm., 72: 40-46.
16. Biswas, C. In: Tumor Invasion and Metastasis (Eds. L.A. Liotta and I.R. Hart), Martinus Nijhoff, The Hague, pp 405-425, 1982.
17. Mainardi, C.L. Biochim. Biophys. Acta. 805: 137-142, 1984.
18. Gasic, G.J., Gasic, T.B. and Steward, C.C. Proc. Natl. Acad. Sci. USA., 61: 46-52, 1968.
19. Hara, Y., Steiner, M. and Baldini, M.G. Cancer Res., 40: 1217-1222, 1980.
20. Honn, K.V. Clin. Exp. Met., 1: 103-104, 1981.
21. Honn, K.V., Busse, W.D. and Sloane, B.F. Biochem. Pharm., 32: 1-11, 1983.
22. Hohn, K.V., Cicone, B. and Skoff, A. Science, 212: 1270-1272.
23. Henson, P.M. and Ginsberg, M.H. In: Platelets in Biology and Pathology., Vol. 2 (Ed. J.L. Gordon) Elsevier/North Holland, pp 265-308, 1981.
24. Miller, R, Bravo, R. and Burckhardt, J. Nature, 312: 716-720, 1984.
25. Assosian, R.K., Komoriya, A., Meyers, C.A., Miller, D.M. and Sporn, M.D. J. Biol. Chem., 258: 7155-7160.
26. Nossel, H.L. and Vogel, H.J. (Eds). Pathology of the Endothelial Cell. Academic Press, NY., 1982.

27. Sporn, M.B. and Todaro, G.J. NEJM., 303: 878-880, 1980.
28. Sporn, M.B. and Roberts, A.B. Nature, 313: 745-747, 1985.
29. Paget, S. Lancet, 1: 571-573, 1889.
30. Kris, R.M., Libermann, T.A., Avivi, A. and Schlessinger, J., Biotechnology, 2: 135-140.
31. Moyer, M.P. and Aust, J.B. Science, 224: 1445-1447, 1984.
32. Hammond, S.L., Ham, R.G. and Stampfer, M.R. Proc. Natl. Acad. Sci. USA., 81: 5435-5439.
33. Fidler, I.J., Gruys, E., Cifone, M.A. and Barnes, Z. J. Nat. Cancer Inst., 67: 947-956, 1981.
34. Sharkey, F.E. and Fogh, J. Cancer Met. Rev., 3: 341-360, 1984.
35. Meyvisch, C. Cancer Met. Rev., 2: 295-306, 1983.
36. Kerbel, R.S., Man, M.S. and Dexter, D. J. Nat. Cancer Inst., 72: 93-108, 1984.
37. Sharkey, F.E. and Fogh, J. Int. J. Cancer, 24: 733-738.
38. Auerbach, R., Morrissey, L.W. and Sidky, Y.A. Cancer Res., 38: 1739-1744.
39. Keller, R. Inv. Metast., 1: 136-148, 1981.
40. Kyriazis, A.A. and Kyriazis, A.P., Cancer Res., 40: 4509-4511, 1980.
41. Crooke, S.T., Mirabelli, C.K., Huang, C.H., Mong, S. and Wong, A. In: Development of Target-Oriented Anticancer Drugs, (Eds. Cheng et al) Raven Press, NY, 207-218, 1983.
42. Goyette, M., Petropopulos, C.J., Shank, P.R. and Fausto, N. Science, 219: 510-512, 1983.
43. Kelly, K., Cochran, B.H., Stiles, C.D. and Leder, P. Cell, 35: 603-610, 1983.
44. Thorgeirsson, U.P., Turpeenniemi-Hujanen, Y., Williams, J.E. Westin, E.H., Heilman, C.A., Talmadge, J.E. and Liotta, M.A. Molec. Cell. Biol., 5: 259-262, 1985.
45. Kerr, I.B., Finlay, G., Birnie, G.D. and McArdle, C.J. Brit. J. Cancer, 70: 686, 1983.
46. Greig, R.G., Koestler, T.P., Trainer, D.L., Corwin, S.P., Miles, L., Kline, T., Sweet, R., Yokoyama, S. and Poste, G. Proc. Natl. Sci. USA., In Press, 1985.
47. The Cancer Letter, March, pp 4-8, 1985.

23

USE OF ANTI-TUMOR MABs FOR DIAGNOSIS AND IMMUNOTHERAPY OF HUMAN TUMORS

HILARY KOPROWSKI

The Wistar Institute of Anatomy and Biology, 36th Street at Spruce, Philadelphia, Pennsylvania

> "The discovery of antitoxins by Von Behring, fundamental in itself, has opened to pharmacology and therapeutics this new field in which the principle of distribution is exemplified in an ideal manner, for antitoxins and anti-bacterial substances are, so to speak, Magic Bullets or Zauberkugelin which strike only those objects for whose destruction they have been produced by the organism."

These were the words of Paul Ehrlich spoken at the Royal Institute of Public Health in London on June 5, 1908.

In the last decade, monoclonal antibodies (MABs) became candidates for Ehrlich's Magic Bullets. In principle, they do not qualify for such a category of weapons because in the words of Ehrlich, they should be "produced by the organism" in order to strike "those objects for whose destruction they have been produced." MABs are at present man-made tools which are passively administered to the recipient. It also became clear that their administration produces a cascade of immune responses to evoke in the patients an antibody which may recognize the antigen. Whether the presently available MABs merit the Magic Bullet classification will be discussed.

Diagnosis

A polysaccharide-sialyl-Le^{a}-active pentasaccharide, linked either with ceramide or with mucin, is the determinant of one of the best characterized antigens of human tumors. On the membrane of gastrointestinal cancer cells, the polysaccharide is linked with ceramide to form a glycolipid; in the blood of cancer patients, it is linked with mucin. An MAB reacting with this determinant presents a powerful diagnostic tool in serology, histology and immunoscintigraphy of gastrointestinal cancers. The gastrointestinal cancer antigen (GICA), as detected in the MAB 19-9 antigen assay, circulates in the blood of almost all patients with pancreatic cancer. In one study, 82% of 303 patients with advanced pancreatic cancer and 70% of 30 patients with resectable cancer were found to secrete GICA into the blood. Similar values were obtained for patients with hepatobiliary cancer. These values contrast strikingly with those for healthy subjects and for patients with inflammatory disease. The sensitivity and specificity of the assay is summarized in Table 1 by comparing the GICA in blood of tumor-bearing nude mice and pancreatic cancer patients.

Table 1. Sensitivity and specificity of 19-9 antigen assay (pancreas).

	% in Nu/Nu		% in Patients	
	CEA	19-9	CEA	19-9
Sensitivity	33	86	47	96.1
Specificity	42	86	32	91

In pancreatic carcinoma, simultaneous determination of 19-9, CEA and CA-125 increased to 95% detection rate.

Both from the point of view of specificity and sensitivity, the test with MAB 19-9 is far superior to the test for CEA. The fact that the 19-9 antigen cannot be synthesized by patients who do not express the Le^a or Le^b may account for the small percentage of negative results. Only 60-70% of patients with colorectal cancer secrete the antigen into blood. However even in this disease, it is a very sensitive test for detection of a recurrence of GI cancer prior to the development of symptoms. As shown in Table 2, it is possible to obtain a very accurate prediction of a recurrence of cancer, on some occasions as early as 18 months before symptoms appear.

Radiolabeled MABs play an important role in localizing tumors in the human body. Using MABs reacting with colorectal cancer, Chatal and associates have evaluated the role of a single MAB as a diagnostic tool in comparison to a "cocktail" of MABs. The results indicate that tumor localization is best accomplished by simultaneous administration in 1 mg quantities of two radiolabeled antibodies rather than one.

Table 2. Analysis of recurrence and CA 19-9 level.

		CA 19-9 level (unit/ml)	
		<35	>35
Patient number	95	75	20
Patient with a recurrence	35	20	15
Percent patients with recurrence	37%	26%	75%

These finding and current exploration of the use of MABs in NMR may finally lead to the establishment of anti-tumor MABs as indispensable tools in the precise localization of cancer in the human body.

Oncogenes and growth factors

It is generally agreed that the function of oncogenes is to assure cell proliferation, either physiological during ontogeny, or pathological when the cells from the quiescent state change to rapidly and uncontrollably proliferating cancer cells. Although mechanisms activating oncogenes are now known, particularly in leukemias and lymphomas, chiefly through the studies of my colleague, Carlo Croce, the actual mechanism that triggers cells to proliferate uncontrollably has yet to be discovered. In the last year or two, several scientific groups reported homology between certain oncogenes and portions of receptor genes encoding several growth factors. It is possible that expression of the gene for growth factor receptor is closely linked with amplification of the oncogene function and that this may be the mechanism facilitating abnormal proliferation of the cell. No oncogene has been clearly associated with human melanoma. Curiously enough, however, expression of EGF receptors is limited to cells in the most advanced stages of malignancy. The non-random chromosomal abnormalities characteristic for the melanoma metastases quite often involve trisomy of chromosome 7. The gene for the EGF receptors is located on the short arm of chromosome 7 and only those melanoma cells displaying chromosome 7 trisomy expressed EGF receptors. Thus, an increased dosage of a gene gave the cells an apparent advantage in the greater utilization of a growth factor in the last stages of the malignant process. What does this have to do with MABs? Simply that MABs reacting with the growth factor receptors may block growth of the tumor cells. An epidermoid carcinoma and malignant melanoma were implanted into nude mice which were injected with MABs against the EGF receptor. The epidermoid cancer-implanted mice were treated constantly with MAB, whereas the melanoma WM9-implanted mice were treated for only 5 days after implantation. The results are encouraging since the growth of implanted tumors was arrested. The results also suggest that in order for the antibody to EGF receptor to block binding and utilization of EGF, it has to be present constantly in circulation. It is also possible that in the clinical use of MAB conjugates, it may become necessary to maintain the antibody for longer periods of time in the blood of the patient. Is this feasible? It depends on the initial dose of MAB, on the frequency of subsequent treatments with MAB, and on the onset of anti-murine antibody responses. As shown in Table 3, patients who receive initially 700 mg of MAB followed by biweekly administrations of 100 mg either do not develop anti-murine antibodies or the onset of the immune response is delayed in contrast with those subjects who initially received a 100 mg dose of MAB followed biweekly by the same amount. All of these subjects developed anti-murine antibody responses, with an early average onset of 13.7 days after the first injection of MAB. None of the patients in these two groups showed any severe side reaction. If these results do, in fact, reflect the circulation of MAB in the blood of the patient for 50-60 days, it may be possible to increase the effectiveness of growth factor receptor blocking antibodies in the treatment of cancer.

Table 3. Correlation between development of human anti-17-1A MAB and dose of MAB.

Initial 17-1A MAB dose (mg)	Number of Patients: Total	Not Responding	Responding Days After MAB: 7-8	13-23	35-99
700	10	3	2	2	3
100	9	1	4	4	None

Idiotypes and anti-idiotypes

Following the injection of patients with MAB at therapeutic dosages i.e. from 350-550 mg, mouse globulin circulates in the blood of the patients from day 1 after injection through a maximum of 40-50 days. On the average, the mouse globulin is no longer detected after 10-12 days following treatment. This is usually followed by the appearance of anti-murine antibodies. When one examines carefully the chronology of tumor growth inhibition in patients who respond to MAB therapy, one is struck by the fact that destruction of metastases as shown on X-ray or CT scan can, in several cases, occur long after MAB administration. For example, one patient with metastases of colorectal cancer to the lung and liver was treated with 433 mg of MAB following unsuccessful chemotherapy. After MAB treatment, the lung metastases receded and were undetectable at 4-6 months after administration of MAB. In another patient, metastatic cancer to the liver started to recede 17 months after injection with 500 mg of MAB. These patients received no other therapy after a single administration of MABs. Although there is a very slight possibility of spontaneous tumor regression, more likely we are dealing with the indirect effect of mouse MAB, namely the induction of anti-idiotype antibody bearing the internal image of the antigen. This, in turn, induced an Ab3 which may have reacted with the patient's own tumor. Both patients had an anti-immune and an anti-idiotype response. Obtaining even partial proof for this hypothesis was not an easy task. As the first step, sera of patients who received MAB and developed anti-murine antibody response were assayed for the presence of anti-idiotype antibodies. This is a tricky procedure if one uses non-fractionated human serum to search for an anti-idiotype response. My colleague, Dorothee Herlyn, finally devised a test based on the interaction of idiotype adsorbed on a polystyrene bead, i.e. MAB is incubated with patient's serum for 18 hr at room temperature; following thorough washing of the complex, anti-idiotype produced in goat is radiolabeled and added and the test is read as an inhibition of binding of anti-idiotype to Id by the sera in question. As a control, an unrelated anti-idiotype reactive with a different Id (MAB) is used in the assay. Table 4 shows the results of the assay on several consecutive sera from gastrointestinal cancer patients. In all 5 cases, anti-idiotype antibody was induced after exposure to MAB and was maintained during the entire period that the patients were observed.

Table 4. Development of α-Id 17-1A antibody† in sera of colorectal carcinoma patients.

Patient	Day α-Id Detected after MAB Injection
07	22->475
08	13-668
09	28-250
14	135
23	24-337

†MAB 17-1A (2.0 μg/ml) was coupled to polystyrene beads. Beads were incubated with several serum dilutions at 4°C overnight. Percentage inhibition of binding of ^{125}I-labeled goat α-Id 17-1A to 17-1A was calculated by the following formula: % binding inhibition = 100 - [(100 X cpm bound in presence of serum)/cpm bound without serum]. Serum was considered positive when inhibition rabies were significantly ($p < 0.05$) higher than the values obtained with preinjection sera.

By repeated passage through columns of mouse IgG2a hybridoma antibodies of different idiotypes and final adsorption to and elution from the 17-1A MAB column, it was possible to purify anti-Id from the sera of three patients. The results indicated that the yield may vary from 18 μg/ml to 5 μg/ml of serum. Competitive binding assays with the purified anti-idiotype revealed considerable crossreactivity between the anti-Id of sera from different patients.

Next in line in this study was an attempt to determine whether the human anti-Id carries an internal image of the antigen. In this test binding of the MAB to colorectal cancer cells was inhibited by the anti-Id obtained from two patients. Thus, it seems that the anti-Id induced in patients after exposure to mouse MAB may be of Ab2β class carrying an internal image of the cancer antigen. If Ab2β is present in the blood of the patient, would it induce formation of Ab3? Because of the presence of Ab2β and because of the possibility that Ab3, if induced, would form a complex with Ab2, PBL instead of serum were obtained from blood of patients who showed the presence of anti-idiotype and stimulated in tissue culture by homologous purified anti-Id. The tissue culture medium was then tested for binding to cells of human tumors.

The results in Fig. 1 indicate that the patient's PBL secreted into the medium an antibody that bound preferentially to colorectal cancer target cells but also to melanoma cells. This was not surprising since Ab3 in other systems shows a different reactivity than that of the MAB idiotype.

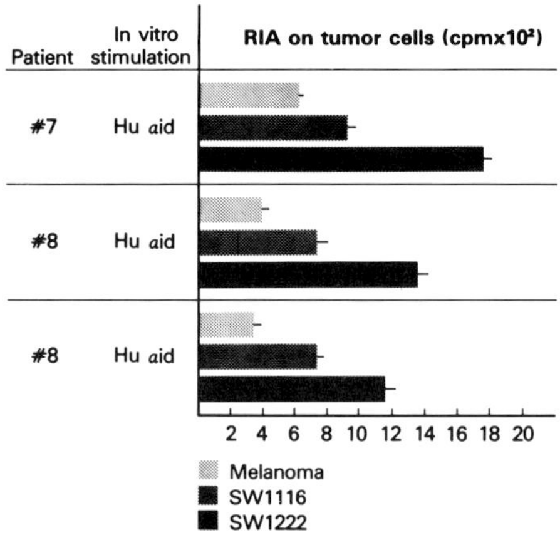

Fig. 1. Induction of Ab3 (anti-Id) which binds to live CRC tumor cells. Experiments were performed using PBMC from two patients stimulated with autologous anti-idiotypic antibody (Ab2). Supernatants were tested in a live cell RIA using CRC and melanoma cells as targets. Stimulation of PBMC of patient 8 was done in two separate experiments from blood obtained 2 weeks apart.

An opportunity arose to administer human purified anti-Id to a patient who had some time received an injection of MAB. The patient received four intradermal injections of purified anti-Id at 0.75 mg per dose in calcium phosphate adjuvant. PBL were then collected after each injection and stimulated in culture either by purified human anti-Id or by the $F(ab')_2$ fragment of the MAB. In this case, the tissue culture medium was tested for binding to goat anti-Id (reacting with the same idiotype as the patient's serum) and to colorectal cancer cells. Figure 2 shows that stimulation with the $F(ab')_2$ fragment of MAB 17-1A did not induce higher levels of anti-idiotype binding to the MAB. Contrariwise, stimulation with either anti-Id or MAB resulted in the secretion of antibody which binds to the goat anti-Id and to colorectal cancer cells.

These are very preliminary results that may also reflect the fact that the donor patient received MAB prior to exposure to anti-Id. It is also quite clear that if one contemplates immunization with anti-Id, a human source of the antibody would be inadequate. We have succeeded in producing anti-Id in goats immunized with $F(ab')_2$ fragments of anti-cancer MABs. These goat anti-Ids are quite specific, reacting only with

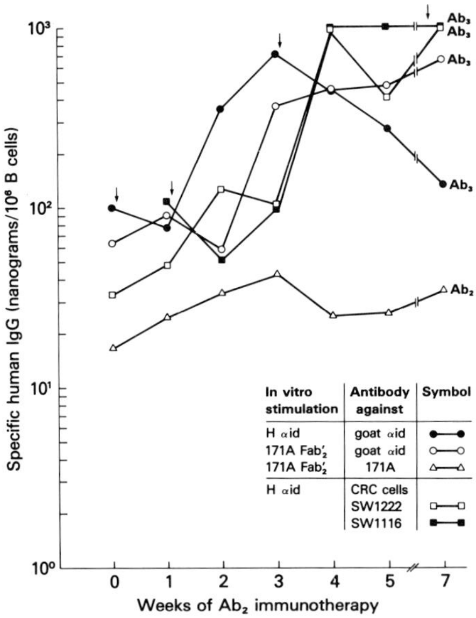

Fig. 2. Induction of Ab2 and Ab3 after in vitro stimulation of PBMC from GIC patient treated with human anti-Id. PBMC taken at various times during therapy were stimulated in vitro with heterologous human anti-Id or 17-1A $F(ab')_2$. Culture supernatants were tested for Ab2 and Ab3 in an ELISA against 17-1A and goat anti-Id, respectively. Ab3, which mimics the binding of Ab1, was detected by a live cell RIA against 17-1A antigen-positive CRC tumor cells lines SW1222 and SW1116.

the MAB used for immunization. The availability of such anti-Id preparations makes it possible to undertake a more precise investigation on the response of the anti-Id immunized host in producing Ab3 and to study the classes of Ab3 produced. If one of the types is Id+Ag+, then this Ab3 antibody may be the source of a thorough study for its specificity and anti-tumor functions. Only time will tell whether this approach will be valuable in the immunotherapy of human cancer.

Future of MAB treatment.

When one tries to analyze data presented by three groups of investigators using MABs for treatment of two types of solid tumors and one type of lymphoma-leukemia, one can see the futility in drawing conclusions from such diverse data. Nevertheless, Table 5 shows the data. Approximately half of the treated patients responded to MAB treatment as evidenced by the objective decrease of the tumor size,

Table 5. Results of MAB immunotherapy in three types of human malignancies.

	Percentage of patients with tumor inhibition		
Disease	Transient	More permanent	Investigator
Melanoma	50	20	Houghton (Sloan-Kettering)
GIC	50	20	Koprowski (Wistar Institute)
B lymphoma	>50	20	Levy (Stanford Univ.)

subjective improvement, and decrease in the level of circulating tumor-associated antigens, such as CEA and 19-9 in GIC. In more than half of these patients, the immunotherapeutic effect was transient and only in about 20% of treated patients was the arrest of tumor growth prolonged. I realize that it is unfair to compare the data obtained by Levy with that of Houghton and ourselves since his mouse monoclonal anti-idiotype antibody used in therapy does not recognize an idiotype shared (if one exists) by lymphomas obtained from different patients, whereas the MABs used to treat solid tumors recognize epitopes of the same tumor species obtained from many different human hosts. In addition, the anti-idiotypes induced in different patients receiving MAB therapy are cross-reactive.

If we accept the figure of 50% for transient immunotherapeutic effect and of 20% for patients who are either in extended remission or cured, we realize that a single administration of one monoclonal antibody may compete with and even successfully replace chemotherapy for a tumor, but it is no yet Ehrlich's magic bullet. What approaches can be developed to improve this situation? One can think about repeated administration of the same antibody. Although this is a safe approach if the patient does not develop IgE anti-immune antibodies, success is minimal as far as tumor destruction is concerned. This lack of success may be attributed either to the emergence, following the first injection of MAB, of a cancer cell population which no longer expresses antigen reactive with the MAB in question; this has been observed in at least two cases of colorectal cancer and is almost the rule in B cell lymphomas treated by anti-idiotype MAB. Alternatively, the relative ineffectiveness of multiple administration of murine MAB might rest in the induction of anti-mouse antibody in patients receiving the first injection of MAB and interference by the circulating antibody to allow the second MAB to reach the tumor tissue. It is because of these two problems that I believe one should treat cancer patients with MABs that react with different antigens of the same tumor species but that are administered simultaneously to avoid interference by circulating anti-tumor antibody. Until recently the availability of different MABs of IgG2a isotype, which is the class of MABs which mediate the tumoricidal effect of the effector cells, was limited. However, more recently it has been possible to generate mutants of murine hybridomas that had switched the class of selected IgG to IgG2a. An example is shown in Fig. 3. We have used a hybridoma secreting MAB reactive with the 19-9 antigen which I have discussed before. This hybridoma secreted an IgG1 antibody, which showed a great affinity for gastrointestinal cancer cells, but did not mediate the tumoricidal effect of macrophages in vitro or in vivo. When we succeeded in producing a variant of this hybridoma that secreted IgG2a globulin, the MAB did mediate the tumoricidal effect of the macrophages. Since both classes of globulin recognize the same chemically defined antigen, the tumoricidal effect of mouse MAB relates to its interaction with macrophages and not with

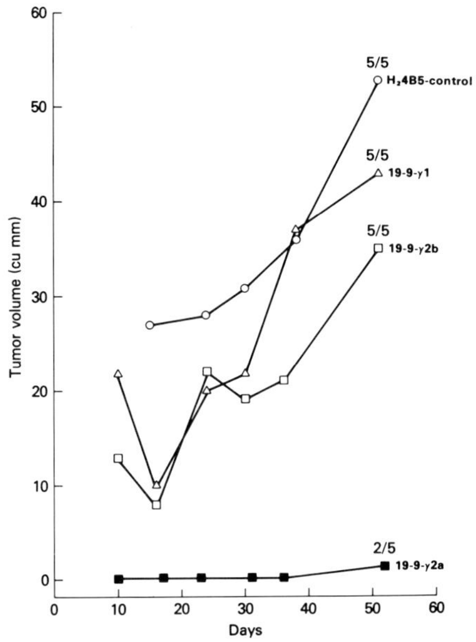

Fig. 3. Summary of the growth inhibition of SW 948 colon carcinoma cells in nude mice. The antibodies are switch variant isolates of CO 19-9 hybridoma (γ1, γ2a and γ2b). Only γ2a inhibits tumor growth.

different antigen(s) on the surface of the target cells. Mutation of hybridomas to secrete "tumoricidal" IgG2a MABs opens a "Pandora's Box," so to speak, for a multitude of mouse MABs capable of mediating the destruction of tumor cells. Their simultaneous administration in a patient for immunotherapeutic purposes should be given a high priority in the near future unless a single MAB is discovered that, by itself, shows the properties of the Magic Bullet; I doubt this very much (Fig. 4).

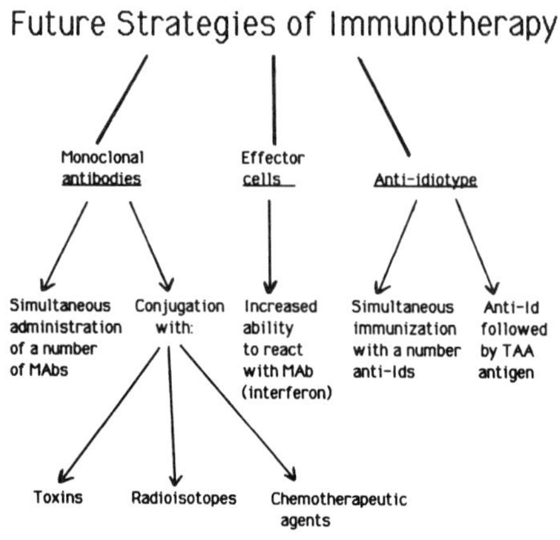

Fig. 4. Future strategies of immunotherapy.

The interaction of MAB with effector cells which become tumorigenic takes place through an Fc receptor for IgG2a mouse antibody. Monocytes obtained directly from human blood do not express such a receptor. When kept in tissue culture, they develop the ability to express the receptor. This can be mimicked by treatment of human cells with γ-interferon and by injection of man with three 500,000 doses of interferon prior to administration of MAB (Table 6).

Table 6. ADCC of monocytes obtained from cancer patients receiving three injections of γIFN.

Percentage Killing			
Day 0		Day 2/3	
NK[a]	ADCC[b]	NK	ADCC
11.3	6.1	15.7	31.7

[a]Natural killing (NK) represents lysis of ^{111}In-labeled SW1116 cells by monocytes armed with anti-influenza virus monoclonal antibody (IgG2a)

[b]ADCC represents lysis of ^{111}In-labeled SW1116 by monocytes armed with antibody 17-1A (IgG2a).

The use of anti-idiotype antibodies for immunotherapy enlarges the scope of anti-tumor MABs available for the production of anti-idiotypes since it is possible to use MABs of various classes to produce anti-idiotypes. Experiments with a rabbit injected simultaneously with a variety of 5 or 6 antibodies representing different idiotypes show that one animal will produce 5 or 6 anti-idiotypes specific for a given idiotype. Thus, in the case of anti-tumor MABs, it may be possible to immunize an animal, a goat for instance, with 5 or 6 different MABs. Serum obtained from the animal can be fractionated to yield a relatively pure preparation of 5 or 6 anti-idiotypes. These can be used for immunization of a patient whose serum and PBL will be processed for the presence of Ab3 reactive with the same species of cancer as that of the patient. If Ab3 is present, it can be further characterized for its immunoglobulin and function, such as binding of complement, and tumor destruction mediated either by complement or by effector cells. The dosage of anti-idiotype for the induction of Ab3 in humans will have to be carefully investigated since there are no data whatsoever on antibody suppression vs. antibody-stimulating effect. In the case of immunization of mice against rabies, the 2.5 μg of anti-idiotype antibody will immunize mice, whereas larger or smaller doses will not.

I personally believe that the future of immunodiagnosis and immunotherapy of human cancer is bright. However, as in all human endeavor, there is a certain element of unpredictability which will be the ultimate and optimal choice of tools.

24

NEW APPROACHES TO THE ADOPTIVE IMMUNOTHERAPY OF ESTABLISHED METASTATIC CANCER USING LYMPHOKINE-ACTIVATED KILLER CELLS AND RECOMBINANT INTERLEUKIN-2

S.A. ROSENBERG AND J.J. MULÉ

Surgery Branch, Division of Cancer Treatment, National Cancer Institute, National Institutes of Health, Bethesda, Maryland 20205

INTRODUCTION

The passive transfer of immune lymphocytes is capable of mediating the regression of established tumor in a variety of animal tumor models (1,2). This approach to cancer therapy, commonly referred to as adoptive immunotherapy, has several advantages compared with more common methods of immunotherapy, which have been based on the stimulation of the host immune system by nonspecific adjuvants. Because patients with cancer are often immunosuppressed, it can be difficult to stimulate immune responses in these patients. The adoptive transfer of previously sensitized cells thus may provide the cancer-bearing host with needed components to mediate effective antitumor responses.

Although successful adoptive immunotherapy of cancer in animals has been achieved using specifically sensitized syngeneic lymphocytes, such an approach has received very little attention in humans (1,2). This is largely because of the difficulty associated with generating large numbers of specifically reactive cells for adoptive transfer (3,4).

We have recently introduced a novel concept of adoptive immunotherapy that has served as the rationale for our efforts in the treatment of weakly- or non-immunogenic tumors. Our laboratory first reported that, in both murine and human systems, fresh lymphoid cells incubated in Interleukin-2 (IL-2) containing supernatants, develop the capacity to lyse fresh tumor cells but not fresh normal lymphoid cells (5-7). IL-2 is the lymphokine solely responsible for the generation of these "activated" killer cells (8,9). Based on kinetics, surface markers and function, these cytotoxic cells appear distinct from both the classical cytotoxic T-lymphocytes and natural killer (NK) cells (10). The term lymphokine-activated killer (LAK) cell has been used to identify this population. The LAK cell phenomenon has been extensively characterized (10), and our recent studies have shown that it has pronouned therapeutic benefit in the treatment of established metastatic cancer (11-18). These studies will now be briefly reviewed.

Generation of Lymphokine-Activated Killer Cells

An example of the ability of IL-2 to generate LAK cells in the mouse is shown in Table 1 (7). In this experiment, splenocytes from

TABLE 1 - Generation of murine LAK cells

Target	C57BL/6 Splenocyte Effector Cells		
	% lysis (± SEM)[a]		
	Fresh (uncultured)	Cultured in Medium	Cultured in IL-2[b]
MCA-102 sarcoma	1 ± 1	2 ± 2	53 ± 2
MCA-106 sarcoma	3 ± 1	4 ± 1	53 ± 1
EL-4 lymphoma	-20 ± 2	-18 ± 1	43 ± 2
Normal splenocyte	-12 ± 3	-20 ± 5	8 ± 2
YAC	20 ± 1	8 ± 1	72 ± 1

[a]Mean ± SEM of triplicate wells at 100:1 effector target ratio in a 4 hr ^{51}Cr-release assay.

[b]LAK cells generated by the 5-day incubation of fresh splenocytes in lectin-free IL-2 containing medium.

(From Reference 7)

C57BL/6 mice were tested against a variety of target cells for their ability to lyse tumor and normal syngeneic cells. The lytic activity of the splenocytes was tested before and after incubation in medium containing IL-2. Fresh uncultured splenocytes were capable of lysing the cultured YAC target, indicating that they had NK cell activity. Fresh methylcholanthrene-induced sarcomas (MCA-102 and MCA-106) and the syngenic EL-4 lymphoma were not lysed by these fresh splenocytes, indicating that they were resistant to lysis by NK cells. However, culture of these fresh splenocytes in IL-2 for 5 days, generated LAK cells that were capable of killing the MCA-102 and MCA-106 sarcoma, the EL-4 lymphoma and YAC cells, but not normal lymphocytes. None of these targets were lysed by normal C57BL/6 splenocytes cultured in complete medium (without IL-2) for 5 days.

Successful Reduction of Established Metastases with LAK-Cells Plus Recombinant IL-2.

Because LAK cells are easily generated and show broad tumor specificity, these cells are logical candidates for use in adoptive immunotherapy. The ability to use LAK cells *in vivo*, however, depended on the availability of large amounts of IL-2 (2,17). In addition to the IL-2 needed to activate LAK cells, the exquisite dependence of these cells on the continued presence of IL-2 for their survival suggested that IL-2 administration to mice receiving LAK cells would also be required. Moreover, since IL-2 was the sole activating agent responsible for the generation of LAK cells, it was further hypothesized that the systemic administration of this lymphokine may induce these cells directly in vivo. Donohue and Rosenberg (19) had shown that the serum half-life of IL-2 was short (2-3 minutes) and thus large amounts of IL-2 were required to obtain sustained bioavailability for the in vivo experiments.

Large quantities of IL-2 have now become available through recombinant DNA technology (9,20). We have shown that this recombinant IL-2 (RIL-2) is effective in the generation of LAK cells from murine and human lymphocytes in vitro and is well tolerated when injected intraperitoneally into mice.

To study the effect of LAK cell therapy on the established pulmonary metastases, tumor was injected intravenously and from 3 to 10 days later, therapy was instituted. An illustration of the phenomenon is presented in Table 2 using the NK-resistant, B16 melanoma.

TABLE 2 - LAK cells plus recombinant IL-2 markedly reduce the number of artificially-induced, established pulmonary metastases of B16 melanoma

Treatment Group[a]	No. of metastases per mouse[b]	Mean	P[c]
HBSS	250, 250, 151, 129, 114	179	--
RIL-2	250, 250, 57, 26, 23	121	0.34
LAK cells + HBSS	250, 250, 53, 42, 12	126	0.34
LAK cells + RIL-2	22, 22, 7, 7, 4	12	0.01

[a]The number of B16 melanoma cells injected i.v. was 2.5×10^5. RIL-2 (6,000 U/injection) or HBSS was injected i.p. approximately every 8 hr. from days 3 through 8 after tumor injection. 1×10^8 and 4×10^7 LAK cells were injected i.v. on days 3 and 6, respectively, after tumor injection. LAK cells were generated by the incubation of fresh splenocytes in medium containing pure recombinant IL-2 for 3 days in vitro.
[b]No. of metastases at day 14.
[c]Wilcoxon rank sum test of treated groups compared to group receiving HBSS alone (two-sided).
(From Reference 18)

In this particular experiment, mice bearing established pulmonary micrometastases three days after the intravenous injection of B16 melanoma cells received the adoptive transfer of LAK cells on days 3 and 6. Mice also received intraperitoneal injections of Hank's Balanced Salt Solution (HBSS) or recombinant IL-2 (RIL-2) at a dose of 6,000 U/injection three times a day from days 3 through 8. Numerous micrometastatic foci were evident histologically when the immunotherapy was begun. Fourteen days after tumor injection, the mice were eartagged, randomized and the number of pulmonary nodules evaluated in a blinded fashion. Administration of LAK cells alone or RIL-2 alone slightly reduced the number of metastases (mean = 126 and 121, respectively). These slight reductions are often seen in our experiments, but they are rarely statistically significant due to variation within each group. However, mice given LAK cells plus RIL-2 had an average of 12 metastases, a reduction in pulmonary B16 melanoma metastases of 93%. A similar therapy experiment using the NK-resistant, weakly-immunogenic MCA-105 sarcoma is shown in Table 3.

TABLE 3 - LAK cells plus recombinant IL-2 markedly reduce the number of artificially induced, established pulmonary metastases of MCA-105 sarcoma

Treatment Group[a]	No. of Metastases per mouse[b]	Mean	P[c]
HBSS	250,250,178,158,141,134,114,98,77,0	140	--
RIL-2	250,250,250,222,163,68	201	0.16
LAK + HBSS	202,115,102,84,21,2	88	0.26
LAK + RIL-2	31,6,1,0,0,0	6	0.01

[a]The number of MCA-105 sarcoma cells injected i.v. was 3 x 10^5. RIL-2 (7,500 U/injection) or HBSS was injected i.p. approximately every 8 hr. from days 3 through 5 after tumor injection; 1 x 10^8 LAK cells were injected i.v. on day 3. LAK cells were generated by the incubation of fresh splenocytes in medium containing pure recombinant IL-2 for 3 days *in vitro*.
[b]No. of metastases at day 13.
[c]Wilcoxon rank sum test of treated groups compared to group receiving HBSS alone (two-sided).
(From Reference 18)

The combination of LAK cells plus RIL-2 effectively reduced the number of established metastatic foci by 96%. As shown in Table 4, reduction of metastases was dependent on the transfer of LAK cells since fresh, normal splenocytes, splenocytes cultured for three days without RIL-2 or either combined with RIL-2 failed to mediate any significant anti-tumor effect upon adoptive transfer (12). Thus, the adoptive transfer of LAK cells plus repetitive injections of recombinant IL-2 was necessary in these models to mediate antitumor effects. The successful treatment of 3-day established pulmonary metastases is highly reproducible and has been seen in over 40 consecutive experiments (21). Regression of pulmonary metastases has been seen not only when using the weakly-immunogenic MCA-105 sarcoma and the B16 melanoma but for the treatment of two

TABLE 4 - Immunotherapeutic effect of LAK cells and recombinant interleukin-2 on established pulmonary sarcoma metastases

Cells[b]	No. of Metastases (Mean) at day 13[a] HBSS	RIL-2
None	228	152
Cultured	183	156
Fresh splenocytes	214	191
LAK	200	20[c]

[a]The number of MCA sarcoma cells injected i.v. was 3×10^5; each group contained 6 mice. RIL-2 (15,000 U/injection) or HBSS was injected i.p. approximately every 8 hr. from days 3 through 8 after tumor injection.
[b]1×10^8 LAK cells, fresh normal splenocytes, or normal splenocytes cultured for 3 days without RIL-2 were injected i.v. on days 3 and 6 after tumor injection.
[c]LAK + RIL-2 vs. RIL-2 alone; $p<.01$.
(From Reference 12)

non-immunogenic MCA-induced sarcomas, MCA-101 and MCA-102, as well as in the treatment of pulmonary metastases from the MCA-38 adenocarcinoma, and the M-3 melanoma (22).

We have also tested the efficacy of the combined therapy of LAK cells and RIL-2 on pulmonary macro-metastases established in the lung ten days after the intravenous injection of MCA-105 sarcoma cells. A representative experiment illustrating the effectiveness of LAK cells plus RIL-2 therapy in the treatment of these larger metastases is seen in Table 5.

TABLE 5 - Treatment of established (day 10) pulmonary metastases with LAK cells plus RIL-2

Recombinant IL-2 (units)[a]	No LAK Cells (mean number of metastases)	Plus LAK Cells[b]
0	227	210
1,000	208	192
6,000	171	61 ($p<.05$)
20,000	56 ($p<.05$)	6 ($p<.02$)

[a]IL-2 injected intraperitoneally every eight hours from days 10 through 17.
[b]10^8 LAK cells injected i.v. on days 10 and 13 after tumor injection.
(From Reference 17)

Note that there was a reduction in pulmonary metastases when RIL-2 was administered alone, which is commonly seen when higher doses of RIL-2 are used in this day 10 treatment model (23). However, when LAK cells were given in conjunction with RIL-2 a more marked reduction in the number of pulmonary metastases was obtained (from 227 metastases in control mice to an average of 6 metastases in animals treated with LAK cells plus 20,000 units of recombinant IL-2 three times a day).

In a new experimental model designed to produce metastases solely in the liver, Lafreniere and Rosenberg undertook studies to assess the effectiveness of adoptive therapy with RIL-2 alone and LAK cells plus RIL-2 on metastatic foci in this location. This model also tested the ability of intravenously injected LAK cells to mediate antitumor effects in an organ that did not represent the first capillary bed (i.e., the lung) encountered by LAK cells after intravenous injection (15,16). The selective induction of hepatic metastases is based on the injection of tumor cells into the spleen followed by immediate splenectomy. Tumor cells flow out of the splenic vein into the portal vein and lodge in the liver where established metastases are formed. A typical experiment in which three day established hepatic metastases were treated with increasing doses of RIL-2 with or without LAK cells is shown in Table 6.

TABLE 6 - LAK cells plus recombinant IL-2 markedly reduce the number of artificially induced, established hepatic metastases of MCA-38 adenocarcinoma

Treatment Group[a]	No. of Metastases per mouse[b]	Mean	P[c]
HBSS	250,250,250,250,250	250	-
RIL-2	230,220,148,143,98,70	151	<0.01
LAK + HBSS	250,250,250,250,247,236	247	>0.05
LAK + RIL-2	5,1,1,0,0,0	1	<0.01

[a]The number of MCA-38 tumor cells injected intrasplenically was 3×10^5. RIL-2 (100,000 U/injection) or HBSS was injected i.p. approximately every 8 hr. from days 3 through 7 after tumor injection; 1×10^8 LAK cells were injected i.v. on days 3 and 6. LAK cells were generated by the incubation of fresh splenocytes in medium containing pure recombinant IL-2 for 3 days *in vitro*.
[b]No. of metastases at day 14.
[c]See corresponding footnote of Table 3.
(From Reference 16)

In this experiment, hepatic metastases were established using MCA-38 adenocarcinoma. Note that the combination of LAK cells plus RIL-2 dramatically reduced the number of hepatic metastases (mean = 1, p = 0.01) when compared to mice treated with HBSS alone (mean = 250) or LAK cells alone (mean = 247, p = NS). RIL-2 alone slightly reduced the number of hepatic metastases to 151 (p = 0.01). The successful therapy of established hepatic metastases with LAK cells plus RIL-2 is highly reproducible and has been seen in 16 consecutive experiments with the weakly-immunogenic MCA-105 sarcoma as well as in experiments with the non-immunogenic MCA-102 sarcoma (15,16).

Regression of Established Metastases Mediated by the Systemic Administration of High-Dose Recombinant IL-2 Alone.

Because recombinant IL-2 is capable of efficiently generating LAK cells *in vitro*, and because the systemic administration of LAK cells has potent antitumor activities, we studied the possible use of high-dose recombinant IL-2 administered directly to tumor-bearing mice, aiming to generate LAK cells *in vivo* and mediate tumor regression (23). To further test the ability of RIL-2 to induce LAK cell activity *in vivo*, mice were injected intraperitoneally with varying doses of RIL-2 approximately every 8 hours for 5 days. On the morning of the sixth day, mice were killed and LAK activity in spleens was assessed using fresh MCA sarcoma target cells. As shown in Table 7, increasing doses of RIL-2 resulted in a successive increase in the lytic activity in the spleen against fresh sarcoma cells.

TABLE 7 - Induction of LAK activity in spleens of mice treated with recombinant IL-2

IL-2[a] (units)	Lytic activity vs MCA-102 E:T[b] %[c] 100:1	20:1
0	-1 ± 2	-3 ± 3
75,000	14 ± 1	-2 ± 1
100,000	17 ± 2	1 ± 2
150,000	19 ± 1	9 ± 1
200,000	21 ± 4	11 ± 1
400,000	32 ± 2	16 ± 4

[a]IL-2 was administered for 5 days i.p. in 0.5 ml, approximately every eight hours.
[b]Effector to target ratio in four-hour, chromium-51 release assay.
[c]Percent lysis ± SEM.
(From Reference 23)

We then undertook a series of immunotherapy experiments in which RIL-2 was administed alone at high doses (23). The results of 10 consecutive experiments in which mice received 6-7 days of treatment with RIL-2 beginning three days after the intravenous injection of tumor cells are shown in Table 8. In 7 of these 10 experiments, a statistically significant inhibition of pulmonary metastases was seen. The mean reduction in pulmonary metastases was 79.5 ± 7.3%. It should be mentioned that lower doses of RIL-2 were ineffective in reducing pulmonary metastases. For example, in 15 consecutive experiments mice receiving between

3,000 and 8,000 units of RIL-2 three times a day experienced a mean reduction in pulmonary metastases of 5.7 ± 5% and in 19 consecutive experiments mice receiving 20,000 to 50,000 units of IL-2 had a 22.2 ± 9.2% reduction in pulmonary metastases. Thus, a significant dose response relationship was seen and very high doses of recombinant IL-2 were necessary in mice in order to achieve these antitumor effects.

TABLE 8 - Reduction of established day 3 pulmonary metastases in mice by treatment with recombinant IL-2

Experiment	Dose of IL-2[a] (units)	Number of metastases[b] (mean)		P[c]
		HBSS	IL-2	
1	170,000	74	7	<.01
2	170,000	216	32	<.05
3	100,000	250	61	<.005
4	170,000	82	4	>.05
5	170,000	37	9	>.05
6	170,000	225	28	<.05
7	170,000	193	32	<.01
8	200,000	230	21	<.05
9	150,000	192	11	<.01
10	100,000	191	161	>.05

[a]Recombinant IL-2 at this dose was administered intraperitoneally in 0.5 to 2 ml every eight to 12 hours for 6 to 7 days starting 3 days after tumor injection.
[b]Mice were killed 14 to 18 days after tumor injection and the number of pulmonary metastases were evaluated in a blinded fashion.
[c]Two sided p values using a Wilcoxon rank sum test.
(From Reference 13)

It should be mentioned that doses in the range of 100,000 to 200,000 units of IL-2 given three times a day can be toxic to mice and therapy is not tolerated for more than seven days. Mice experience weight loss and develop ascites and hepatic dysfunction when these high doses of RIL-2 are administered. The systemic administration of high-dose RIL-2 was also effective against established 3-day pulmonary metastases of the MCA-101, MCA-102, MCA-106 sarcomas, the M-3 and B16 melanomas, as well as the MCA-38 adenocarcinoma (22,23). Further, therapy with high-dose RIL-2 alone is also effective against established 3-day hepatic metastases (15,16).

We have also performed experiments exploring the ability of RIL-2 therapy to treat larger pulmonary metastases (13,23). In a series of experiments, therapy with varying doses of RIL-2 was initiated 10 days after tumor injection. The results of 13 consecutive experiments in which mice received 6-7 days of treatment with RIL-2 beginning 10 days after tumor injection are shown in Table 9.

TABLE 9 - Reduction of established day 10 pulmonary metastases in mice by treatment with recombinant IL-2

Experiment	Dose of IL-2[a] (units)	Number of metastases[b] (mean)		P[c]
		HBSS	IL-2	
1	100,000	>250	58	<.02
2	100,000	>250	16	<.008
3	170,000	71	1	<.008
4	170,000	227	50	<.03
5	100,000	273	74	<.04
6	100,000	212	113	<.03
7	170,000	172	7	<.02
8	170,000	19	3	>.05
9	200,000	105	21	>.05
10	150,000	169	8	<.02
11	100,000	50	24	<.08
12	100,000	53	18	<.04
13	100,000	227	9	<.02

[a]Recombinant IL-2 at this dose was administered intraperitoneally in 0.5 to 1.0 ml every 8 hours for six to seven days (experiments 1 to 4 and 9 to 13) or in 2 mls injected every 12 hours (experiments 5 to 8). Treatment began 10 days after tumor injection in all experiments except for experiment 9 when treatment began on day 8.

[b]Mice were killed 14 to 18 days after tumor injection and the number of pulmonary metastases were evaluated in a blinded fashion.

[c]Two sided P values using a Wilcoxon rank sum test.

(From Reference 13)

A statistically significant decrease in pulmonary metastases was seen in 11 of 13 experiments.

Thus, it appears that the systemic administration of high-dose RIL-2 is an effective approach to the treatment of established metastatic cancer. Tumor regression appears to be mediated by endogenously activated lymphocytes, presumably LAK cells, present at the tumor site and in the organs of mice receiving this lymphokine.

Possible Mechanisms of Adoptive Immunotherapy with Lymphokine-Activated Killer Cells and Recombinant IL-2.

We have been investigating the mechanisms of action of tumor destruction *in vivo* by LAK cells plus RIL-2. The infiltration of activated lymphocytes into the tumor stroma with subsequent tumor destruction by histologic analysis argues for a direct role of these activated lymphocytes

(LAK cells?) in tumor cell killing in vivo. Ettinghausen et al. (24) initiated a series of experiments to test directly for the presence of LAK cells in lungs of irradiated mice receiving adoptively transferred LAK cells i.v. plus RIL-2. Total body irradiation of the host was necessary in order to eliminate background endogenous lymphoid proliferation by RIL-2 (24). Lymphoid cells were recovered directly from lungs of treated mice and used as effectors in an 18 hr. ^{51}Cr-release assay against the fresh tumor target, MCA-102.

As shown in Table 10, the transfer of LAK cells to irradiated mice followed by repeated injections of IL-2 lead to a substantial increase in the number of recovered lymphocytes when compared to mice receiving LAK cells and HBSS, RIL-2 alone or HBSS alone.

TABLE 10 - In vitro cytotoxicity of adoptively transferred LAK cells recovered from recipient lungs[a]

Treatment	Cell Recovery ($x\ 10^{-6}$)	% Lysis[c]	L.U.[d]	L.U. Per Lungs
HBSS[b]	0.6	5.4	<1	<1
IL-2[b]	0.7	22.5	<1	<1
LAK + HBSS	1.6	78.7	8.8	14.4
LAK + IL-2	52.8	86.1	8.8	463.4

[a]Whole body irradiated (500 rads) mice received 10^8 LAK cells i.v. on day 0 or no cells.
[b]IL-2 200,000 units (or HBSS) i.p. 3 times a day from day 0 through 3.
[c]Lymphocytes isolated from lungs of treated mice were used as effectors in an 18 hour ^{51}Cr release assay against a fresh murine sarcoma target. Effector:target ratio was 100:1.
[d]One lytic unit defined as number of effectors to produce 50% lysis of 3×10^3 tumor target cells; L.U. per 10^6 cells.
(From Reference 24)

Highly lytic lymphoid cells were obtained from the lungs of mice receiving LAK cells with or without RIL-2 [8.8 lytic units (L.U.) per 10^6 cells compared to mice treated with HBSS or RIL-2 alone (<1 L.U.)]. However, the higher recovery of lymphocytes from the LAK cell and RIL-2 treated animals yielded a significantly greater number of L.U. per mouse when compared with the LAK cell and HBSS treated group (463.4 and 14.4 L.U., respectively). Thus, the expanding cells seen in mice receiving LAK cells and RIL-2 have the capability to lyse tumor target cells and are consistent with our functional definition of LAK cells. Further, as shown in Table 11, these lytic cells can also be recovered in large numbers from the lungs of mice receiving LAK cells plus RIL-2, but not low-dose RIL-2 alone, during the regression of established pulmonary metastases (21).

TABLE 11 - Adoptively transferred LAK cells can be recovered from lungs of mice during regression of established 3 day pulmonary metastases

Days After Tumor Injection	Group[b]	Cell Recovery ($x 10^{-6}$)[c]	% Specific Lysis of MCA-102 at effector:target ratio of 100:1 — 4 Hour	18 Hour
5	RIL-2	3.5	-0.2 ± 1.7[a]	-16.5 ± 5.8
	LAK Cells + RIL-2	16.5	18.0 ± 2.8	85.3 ± 4.4
6	RIL-2	11.0	-4.0 ± 0.0	-3.5 ± 4.9
	LAK Cells + RIL-2	23.0	34.2 ± 5.9	70.6 ± 4.4

Days After Tumor Injection	Group	Number of Metastases	Mean	P
16	HBSS	250,250,250,250,250	250	--
	RIL-2	250,250,250,250	250	1.00
	LAK Cells + RIL-2	137,122,117,32,23	86	0.01

[a]Mean ± SEM of triplicate wells.
[b]The number of MCA-105 sarcoma cells injected i.v. was 5×10^5. RIL-2 (7,500 U/injection) was injected i.p. approximately every 8 hrs. from days 3 through 8. 1×10^8 LAK were infused i.v. on days 3 and 6.
[c]Number of viable cells recovered per lungs per mouse; these cells were used as effectors in the ^{51}Cr-release assay.

CONCLUSIONS

In this brief review we have presented our approach to developing an adoptive immunotherapy that may be applicable to the treatment of human cancer. Extensive studies of the successful adoptive immunotherapy of established lung and liver metastases from tumors that are weakly- or non-immunogenic with the combination of LAK cells plus RIL-2 and of the systemic administration of high-dose RIL-2 alone have been performed (11-18,21-23). It appears that successful therapy depends on the expansion of highly lytic LAK cells at the site of tumor by RIL-2. The adoptive transfer of LAK cells now appears to be an effective alternative approach to the use of specifically immune cells in the therapy of established metastases of a variety of cancers, regardless of the immunogenicity of the tumor _in vivo_. By using IL-2, human LAK cells with _in vitro_ properties identical to those of murine LAK cells can readily be generated _in vitro_. Further, the availability of large amounts of recombinant IL-2 with known biologic activity, has now made it feasible to evaluate this immunotherapeutic approach in humans. Trials exploring the efficacy in humans with metastatic disease of combined therapy with LAK cells and RIL-2 are now in progress.

REFERENCES

1. Rosenberg, S.A., and Terry, W. Adv. Cancer Res. 25:323, 1977.
2. Rosenberg, S.A. Cancer Treat. Reports 68:233, 1984.
3. Hellstrom, I., and Hellstrom, K.E. J. Biol. Response Mod. 2: 310, 1983.
4. Grimm, E.A., Vose, B.M., Chu, E.W., Wilson, D.J., Lotze, M.T., Rayner, A.A., and Rosenberg, S.A. Cancer Immunol. and Immunother. 17:83, 1984.
5. Yron, I., Wood, T.A., Spiess, P., and Rosenberg, S.A. J. Immunol. 125:238, 1980.
6. Lotze, M.T., Grimm, E.A., Mazumder, A., Strausser, J.L., and Rosenberg, S.A. Cancer Res. 41:4420, 1981.
7. Rosenstein, M., Yron, I., Kaufmann, Y., and Rosenberg, S.A. Cancer Res. 44:1946, 1984.
8. Grimm, E.A., Robb, R.J., Roth, J.A., Neckers, L.M., Lachman, L.B., Wilson, D.J., and Rosenberg, S.A. J. Exp. Med. 158:1356, 1983.
9. Rosenberg, S.A., Grimm, E.A., McGrogan, M., Doyle, M., Kawasaki, E., Koths, K., and Mark, D.F. Science 223:1412, 1984.
10. Grimm, E.A., and Rosenberg, S.A. In:Lymphokines, Vol. 9, p. 279, 1984.
11. Mulé, J.J., Shu, S., Schwarz, S.L., and Rosenberg, S.A. Science 225:1487, 1984.
12. Mulé, J.J., Shu, S., and Rosenberg, S.A. J. Immunol. 135:646, 1985
13. Rosenberg, S.A., and Mulé, J.J. Surgery (in press).
14. Mazumder, A., and Rosenberg, S.A. J. Exp. Med. 159:495, 1985.
15. Lafreniere, R., and Rosenberg, S.A. Cancer Res. (in press).
16. Lafreniere, R., and Rosenberg, S.A. J. Immunol. (in press).
17. Rosenberg, S.A. J. Natl. Cancer Inst. (in press).
18. Mulé, J.J., Ettinghause, S.E., Spiess, P.J., Shu, S., and Rosenberg, S.A. Cancer Res. (in press).
19. Donohue, J.H., and Rosenberg, S.A. J. Immunol. 130: 2203, 1983.

20. Taniguchi, T., Matsui, H., Fajita, T., Takaoki, C., Kashina, N., Yoshimoto, R., and Hamuro, J. Nature (Lond.) 302: 305, 1983.

21. Mulé, J.J., Yang, J., Shu, S., and Rosenberg, S.A. J. Exp. Med. (submitted).

22. Papa, M., Mulé, J.J., and Rosenberg, S.A. J. Exp. Med. (submitted).

23. Rosenberg, S.A., Mulé, J.J., Spiess, P.J., Reichert, C.M., and Schwarz, S.L. J. Exp. Med. 161:1169, 1985.

24. Ettinghausen, S.E., Lipford, E.H., Mulé, J.J., and Rosenberg, S.A. J. Immunol. (in press).

25

PRECLINICAL SCREENING OF BIOLOGICAL RESPONSE MODIFIERS: APPLICATION TO THE TREATMENT OF METASTATIC DISEASE

James E. Talmadge

Preclinical Screening Laboratory, Program Resources, Inc., National Cancer Institute-Frederick Cancer Research Facility, Frederick, MD 21701

Biological Response Modifiers (BRMs) are those agents or approaches that alter the relationship between the tumor and host by modifying the host's biological response to the tumor cells with resultant therapeutic effects (1). BRMs may modify the host's responses in several ways by:

a) increasing the host's antitumor responses through augmentation and/or restoration of effector mechanisms or mediators of the host's defense or decreasing the component of the host's reaction that may be deleterious;
b) increasing the host's defenses by the administration of natural biologics (or the synthetic-recombinant derivatives thereof) as effectors or mediators of an antitumor response;
c) augmenting the sensitivity of the host's tumor cells to endogenous mechanisms for the control of tumor growth;
d) decreasing the transformation and/or increasing differentiation (maturation) of the tumor cells; and
e) increasing the ability of the host to tolerate damage by cytotoxic modalities of cancer treatment.

The in-depth evaluation of even a moderate number of BRMs for their immunomodulatory and therapeutic potential is a formidable task (2-4). The Preclinical Screening Laboratory (PSL) utilizes a series of *in vitro* and *in vivo* models that were selected for their expected relevance to the immunotherapy of neoplastic diseases (5-7). Suitable positive and negative control agents are included in each assay in order to determine the technical adequacy of each experiment or therapy and to allow the comparison of results on different days (4).

The PSL uses a sequential and progressively more demanding series of assays, designed to determine the range of immunomodulatory effects of a BRM as well as to evaluate its therapeutic potential. The initial emphasis is on defining the effects of BRMs on T cells, B cells, NK cells, and macrophages. In addition to this assessment of a BRM's effects on the overall function of the components of the immune system, particular emphasis is placed on determining its ability to augment effector functions *in vivo*, which have been shown to have some relevance to host resistance against tumor growth. Thus, evaluation of the induction of tumoricidal activity by T cells, NK cells, and macrophages is a major component of the screen (5). In addition, the adjuvant activities of each agent for T cell and B cell responses are assessed. The sequence and progression of assays are as follows:

a. *in vitro* activation, followed by *in vitro* testing;
b. *in vivo* activation, followed by *in vitro* testing;
c. *in vivo* activation, followed by *in vivo* testing;
d. therapeutic activity.

This sequence of studies allows the determination of the effective dose range and any concomitant toxicity. Information is also obtained regarding the kinetics and duration of the effects of an agent, which can guide the planning of therapeutic protocols.

Because metastasis is a major threat to cancer patients, the emphasis in the therapeutic evaluation of BRMs is on the assessment of therapeutic efficacy against systemic lesions (Table 1).

Table 1

PRECLINICAL SCREENING OF BRMs FOR IMMUNOTHERAPEUTIC EFFECTS

Treatment of Metastases	Treatment of Autochthonous Tumors
1. Experimental metastases 2. Spontaneous metastases Efficacy indicated by an increase in survival and decrease in metastasis	1. UV-induced skin tumors, BRM given: a. i.v., administered chronically b. Intralesionally 2. NMU-induced mammary tumors in rats, BRM given: a. i.v., administered chronically Efficacy indicated by increase in survival, reduction in growth rate, and decrease in metastasis

Two models of immunotherapy are utilized (7,8), one directed against tumor cells introduced in the host systemically by i.v. injection (experimental metastasis) and the other (a more clinically relevant assessment of therapeutic activity) is directed against metastatic foci that develop after the resection of a primary tumor (spontaneous metastasis). In the spontaneous metastasis model, the host has been "conditioned" by tumor growth prior to the therapy that is initiated following surgical resection of the primary tumor. This is in contrast to therapeutic models directed against metastatic disease, in which therapy is initiated one or two days following primary tumor transplantation. This approach represents a form of metastatic prophylaxis concomitant with the therapy of the primary tumor, since therapy is started prior to the tumor cell dissemination to form metastases. Furthermore, with the latter type of therapy model, the primary tumor burden in the treated animals is often smaller than in saline-treated hosts, thereby reducing metastatic disease and preventing a valid comparison of metastatic burden with the control animals, which have a more extensive primary and secondary tumor burden. Therefore, it is expected that the post-resection experimental design will be more predictive of systemic therapeutic efficacy in the clinical setting, while the latter tumor model might give an overly optimistic assessment of the efficacy of an agent.

It is likely that, as one progresses from models of immunoprophylaxis to more stringent and rigorous models of immunotherapy, fewer positive results will be obtained (7,8). Thus, a BRM with prophylactic activity may not have striking therapeutic activity against established systemic disease in a relatively normal host (experimental metastasis) or in a host "conditioned" by the growth of a primary tumor and bearing preexistent spontaneous metastases. However, even the spontaneous metastasis model utilizes young, relatively healthy hosts bearing transplantable tumors, a situation that may be quite disparate from that of cancer patients who may have been exposed to carcinogenic insult many years before and who have developed progressive primary and metastatic tumor lesions over a very prolonged period. Thus, one might expect that the most clinically relevant tumor model would be primary, autochthonous tumors that metastasize.

To this end, we have developed two metastatic, autochthonous tumor models. The first utilizes ultraviolet (UV) light-induced skin tumors, which include a variety of histologic types, i.e., squamous cell carcinomas, fibrosarcomas, hemangiosarcomas, rhabdomyosarcomas, and other less well differentiated sarcomas. The tumors develop 20 to 50 weeks following the onset of UV irradiation (2.8 Joules/M^2/min for 2 hours, 3 times per week, for 8 months). The second model utilizes the induction of mammary tumors in Sprague Dawley rats by N-nitroso-methyl urea (NMU). The NMU is injected intravenously (i.v.) into 50-day-old female rats at 35 mg/kg. This dose provides an acceptable balance between the tumor induction time (90 to 120 days) and the induction of a limited number of primary tumors. Higher doses of NMU result in the more rapid development of an initial lesion in 100% of animals but also results in multiple primary neoplasms, such that few animals survive to form apparent metastases (7,9). In both therapeutic models, when palpable tumors are detected, the animals are randomly assigned to one of several arms of a protocol and treatment is initiated. These studies are more difficult, tedious, and time consuming than experiments with transplantable tumors, but they provide a rigorous and hopefully more predictive test of therapeutic efficacy. The difficulties associated with such studies are similar to those routinely encountered in clinical trials, including variations and delays in tumor accrual, the need for randomization between treatment and control groups, variability of tumor growth and behavior, the need for extensive and relevant statistical evaluation, and a prolonged study duration.

Obviously, a tumor model must provide an acceptable compromise between the requirement for clinically predictive therapy models and the fiscal and time constraints that bias toward the use of experimental or spontaneous tumor metastasis models. To achieve the parallel goals of predictive and realistic therapy models, studies to determine optimal doses and schedules of BRM administration are performed with the experimental and spontaneous metastasis models. Based on the results of these studies, more limited studies are continued utilizing autochthonous tumor models (Table 1). Thus, the initial assessment of therapeutic activity, using a broad range of doses and therapeutic schedules, can be undertaken for the treatment of experimental metastases such that data are obtained within 6 to 8 weeks for each assay. Selected schedules and doses (3 to 5 doses, 1 to 3 schedules) are then examined in therapeutic models utilizing spontaneous metastases, and data are obtained in 12 to

16 weeks. Then, using the optimized schedule and 1 to 3 doses of a BRM, the BRM is evaluated for its ability to treat autochthonous tumors, and survival data are obtained approximately 1 year following the development of a palpable tumor. It has become apparent from these studies that the use of autochthonous tumors provides a logistical and technical challenge to the evaluation of the therapeutic potential of a BRM. However, because of the closer analogy of the autochthonous tumor models to the clinical situation, we anticipate that positive therapeutic efficacy in the autochthonous tumor models will be more predictive of clinical efficacy than the simpler and less demanding metastasis models.

During the development of the therapy models for the PSL, it was realized that no one animal tumor model is optimal and that all suffer from various limitations and discrepancies from the clinical situation. In assessing the therapeutic efficacy of a BRM, the rational development of therapeutic models and interpretation of results from the tumor model that are utilized is of greater importance than the tumor type. Thus, the aim has been to develop reproducible, standardized, and interpretable models of immunotherapy that may have some predictive value.

The development of successful immunotherapeutic protocols requires new strategies as well as the development of clinical predictive tumor models and protocols (7,8). Beyond the determination of a maximum tolerated dose (MTD), it is also important to determine the optimal immunomodulatory dose (OID) in animals as well as in clinical models. Furthermore, it should be determined whether either the MTD or OID represent the optimal therapeutic dose (OTD). The OTD appears to be dependent not only on the augmentation of effector cell activity, but also on the number, function, and tumor infiltration of effector cells. The determination of an optimal therapeutic protocol (OTP) requires additional information on an optimal therapeutic schedule and duration of treatment. Studies of scheduling parameters have revealed that many BRMs induce host toxicity when delivered frequently at high doses. However, if the BRM is administered at either a lower level or less frequently, there may be a decrease in toxicity that may be paralleled by greater therapeutic activity. Thus, in agreement with studies of _in vitro_ and _in vivo_ immunomodulation (10), a bell-shaped dose response curve for therapeutic activity (11) may be observed. This may be due to decreased immunomodulatory potential at higher BRM doses or it may be associated with host toxicity which, in combination with the tumor burden may result in morbidity or decreased survival as compared to an OTP.

Initial therapy studies of MBL-2 (a Moloney virus-induced lymphoma) tumor ascites revealed a dose-dependent prolongation of survival, in mice receiving rHu IL-2 (Table 2).

Table 2

IMMUNOTHERAPY OF MBL-2 TUMOR ASCITES WITH rHu IL-2[a]

BRM	Dose U/A	Median Survival Time	Percent "cured"
saline	-	36	0
excipient	-	36	0
rHu IL-2	10,000	35	20
rHu IL-2	1,000	41[b]	20
rHu IL-2	100	>55[b]	70
rHu IL-2	25	39[b]	30

[a]Syngeneic C57BL/6N mice were injected with 5000 MBL-2 ascites cells and 48-hr later were placed on a therapeutic protocol of triweekly administration of rHu IL-2, saline or excipient.
[b]Significant prolongation of survival (Kruskal Wallis Test).

The optimal therapeutic activity with rHu IL-2 was observed at 100 units/animal delivered either 3 times per week or daily by intraperitoneal (i.p.) administration. Less activity was observed with higher levels of the rHu IL-2. The animals that had no apparent tumor burden on day 55 rejected a second tumor challenge, which grew progressively and killed control animals.

The schedule and dosage dependency of the therapeutic activity of rHu IL-2 was also observed for the treatment of experimental M109 metastases (Table 3) and B16-BL6 experimental and spontaneous metastases (results not shown).

Table 3

THERAPEUTIC ACTIVITY OF rHu IL-2 FOR THE TREATMENT OF M109 LUNG NODULES[a]

BRM	Dose/Animal	Schedule	MST (range)	Med No. Metastases
HBSS	-	tiw[b]	40.5 (33-56)	181
pICLC	10 μg	tiw	46.5[c] (38-56)	23[d]
rHu IL-2	10,000 U	tiw	43[c] (33-56)	60[d]
rHu IL-2	1,000 U	tiw	45[c] (33-56)	50[d]
rHu IL-2	100 U	tiw	49[c] (39-56)	61[d]
rHu IL-2	10 U	tiw	40 (38-56)	111[d]
rHu IL-2	100 U	qd	50[c] (37-56)	39[d]
rHu IL-2	1,000 U	qd	42.5[c] (33-56)	24[d]

[a]Syngeneic C57BL/6N mice were injected i.v. with 50,000 M109 tumor cells and i.v. therapy was initiated 48 hr later. Therapy was continued for 4 weeks and the mice were necropsied 6 weeks after tumor challenge.
[b]rHu IL-2 was administered either tiw (triweekly) or qd (administered daily).
[c]Significant prolongation of survival (Kruskal Wallis Test).
[d]Significant reduction in experimental metastases (Mann Whitney U Test).

In this study, the optimal prolongation of survival was observed following the administration of 100 units of rHu IL-2, 3 times per week (49 days) or 100 units of rHu IL-2 administered daily (50 days). In contrast, the administration of lower or higher doses of rHu IL-2 had less therapeutic activity. The greatest reduction in the median number of metastases was observed in animals that received 100 or 1000 units of rHu IL-2 administered on a daily basis (39 and 24 median number of metastases, respectively). Although the triweekly administration of 10,000, 1,000, or 100 units of rHu IL-2 had significant therapeutic activity as compared to saline, this reduction in metastases was significantly less than that observed following the daily IL-2 administration. While an MTD of rHu IL-2 was not administered, it is apparent that low levels of this agent delivered on an optimized therapeutic schedule can have greater therapeutic activity than higher doses (10,000 U/A).

During initial therapeutic studies, the BRM is administered twice a week for 4 weeks. This schedule is utilized with several different doses of the BRM and if no therapeutic or toxic activity is observed, additional studies are undertaken until an MTD is determined. If no therapeutic activity is observed with a schedule that includes an MTD administered daily, we conclude that the agent has no therapeutic activity on the basis of either immunomodulation or cytoreductive activity. It should be noted, however, that therapeutic activity may not be obtained in all models of immunotherapy. For example, some BRMs, e.g., and lentinan, have had therapeutic activity only in specific strains of mice (12). In the PSL, lentinan had no detectable therapeutic activity against B16-BL6 in C57BL/6 mice but was active against the Madison 109 tumor in Balb/c mice.

The observation of therapeutic activity in the initial dosage studies stimulates additional studies to determine the optimal therapeutic schedule using a limited number of active doses. The BRM, at 1 to 3 doses, is administered once a week, twice a week, 3 times a week, and daily for a 4-week period. This second phase, to determine an OTP, utilizes an experimental design such that the daily administration is biased towards lower, more easily tolerated doses, whereas a broader range of doses are tested using weekly or biweekly administration. Following these studies, a third therapeutic phase may then be designed to determine the duration of administration required for optimal therapeutic activity. Studies are undertaken in which the agent is delivered by an optimal schedule for 1 to 4 weeks or continuously (as long as the animals remain alive). Studies using the autochthonous tumor models, particularly the NMU-induced tumor model in which multiple primary tumors often develop, have demonstrated that maximum therapeutic activity (long-term survival) is dependent upon continuous administration of the BRM. Therefore, we suggest that the demonstration of clinical immunotherapeutic activity, as well as maximal therapeutic activity in animal models, may also require the prolonged administration of a BRM.

The OTP appears, at present, to be dependent on a complex combination of parameters, including dose, frequency of administration, duration of administration, and route of administration. The route of injection has been shown to be important for ascites tumors, which respond most favorably to the i.p. administration of an agent, in contrast to systemic metastatic disease, which generally is most responsive to the i.v. injection of an agent. Normally, systemic disease is treated by i.v. injection; however, consideration must also be given to intramuscular (i.m.) administration, which in addition to being a more convenient injection site, may prolong the serum half-life of an agent. Indeed, in our studies with chronic administration, particularly in animals receiving daily injections of a BRM, it becomes impossible to continue i.v. lateral tail vein injection and one must resort to alternative injection sites (i.m.). For agents that can be absorbed through the gastrointestinal tract, oral administration also has considerable clinical practicality, since it may allow ongoing treatment on an outpatient basis. Thus, administration of BRMs via gavage should also be considered for immunotherapeutic studies.

In one study in which we examined the treatment of experimental metastases (Table 4), we compared the therapeutic efficacy of 8 BRMs administered at their OTDs.

Table 4

IMMUNOTHERAPY OF EXPERIMENTAL METASTASES[a]

Drug	Dose	No of Metastases Med.	No of Metastases Range	P[b]
HBSS	-	210	26->300	-
Nor-MDP	75 mg/kg	147	0->300	0.05
FK-565	5 mg/kg	0	0-0	0.001
Poly AU	5 mg/kg	124	0->300	0.05
Poly ICLC	0.5 mg/kg	1	0-13	0.001
Lentinan	50 mg/kg	>300	0->300	NS
OK-432	1 KE	60	0->300	0.01
MVE-2	25 mg/kg	69	0->300	0.01
rHu IFN-α A/D	5 x 10^4 U	7	0->300	0.005

[a]B16-BL6 tumor cells were injected i.v. into the lateral tail vein of syngeneic C57BL/6N mice. Two days following tumor challenge, therapy was initiated on a triweekly basis and was continued for 4 weeks. The mice were necropsied 6 weeks post tumor inoculation, and the number of lung colonies were determined.

[b]Significant difference from the HBSS control as determined by the Mann Whitney U Test.

The most active BRMs in this therapy of experimental metastases were FK-565, poly ICLC, and rHu IFN-α A/D. FK-565 is a synthetic derivative of FK-156, which is a dipeptide isolated from the fermentation broth of several Streptomyces species. Poly ICLC is a synthetic double-stranded ribonucleotide that has been complexed with poly L-lysine and carboxymethyl cellulose to protect it from serum ribonucleases. The rHu IFN-α A/D BgL is a hybrid, recombinant molecule from clone-α A and clone-α D human interferon, which has murine activity. In this study, OK-432, a lyophilized preparation of Streptococcus, and MVE-2, a synthetic polymer of maleic anhydride divinyl ether, also had therapeutic activity but were less efficacious than FK-565, poly ICLC, and rHu IFN-α A/D. Lentinan, a hot water extract from an edible mushroom, had no therapeutic activity against the B16-BL6 experimental metastases. This is not surprising, since lentinan has species-specific therapeutic activity (12,13). In the PSL, no activity has been observed against B16-BL6 in C57BL/6 mice, although therapeutic activity has been demonstrated with lentinan and the M109 tumor in Balb/c mice. Minor therapeutic activity was seen with Poly AU, a synthetic polyribonucleotide that is more sensitive to ribonuclease activity that poly ICLC, and Nor MDP, an analog of muramyl dipeptide that is the minimal subunit capable of replacing Mycobacterium tuberculosis in Freund's complete adjuvant. MDP and its analogs are very rapidly excreted and, unless admixed as an emulsion or administered at extremely high levels, have minimal activity.

In another study (Table 5), we compared the therapeutic efficacy of 4 BRMs administered at their OIDs, against spontaneous metastases.

Table 5

IMMUNOTHERAPY OF SPONTANEOUS METASTASES[a]

BRM	Dose	Median no. of Lung Colonies (range)	TB[c]/Total	p[d]
HBSS	-	86 (57->250)	10/10	-
Liposomes[b]	2.5 μmole	34 (0->250)	6/10	0.05
Azimexon	100 mg/kg	16 (6- 111)	10/10	0.01
MDP	50 mg/kg	32 (0- 82)	6/10	0.05
MVE-2	25 mg/kg	92 (0->250)	9/10	NS

[a]B16-BL6 tumor cells were injected into a posterior footpad of each mouse. When the tumors reached a 0.8 to 1.0 cm diameter, the tumor-bearing limb was resected at midfemur to include the popliteal lymph node. Therapy was initiated 3 days later and consisted of the twice weekly i.v. injection of the BRM for 3 weeks. Necropsies were performed 3 weeks after the last injection.

[b]Liposomes: multilamellar vesicles (2.5 μmole/A) incorporating 10 μg of muramyltripeptide-phosphatidyl ethanolamine.

[c]Tumor bearer.

[d]Significant difference compared to the HBSS control as determined by the Mann Whitney U Test.

In this study, both Nor-MDP and a lipophilic variant of MDP (MTP-PE) in liposomes "cured" 40% of the treated mice, while the other 2 BRMs examined

(MVE-2 and azimexon) "cured" 0 and 10% of the mice treated, respectively. Although azimexon did not cure any of the mice, it did significantly reduce ($p < 0.01$ by the Mann Whitney U test) the number of metastatic foci, as did both Nor-MDP and liposomes incorporating MTP-PE (both $p \leq 0.05$). The treatment of spontaneous metastases represents a therapeutic approach against established systemic disease in a "tumor-conditioned" host that had an established primary tumor, has been traumatized by surgery, and has an established extensive secondary disease at the initiation of therapy. Nonetheless, even this model of spontaneous metastasis utilizes relatively young, healthy hosts bearing transplantable tumors and is disparate from a patient which may have been exposed to carcinogenic insult, and who develops a tumor that progresses over a period of many years until it becomes detectable. The therapeutic efficacy of BRMs in more clinically similar autochthonous tumor models (UV-induced tumors) are monitored by the growth kinetics of the primary tumor, prolongation of host survival, and ultimately on the development of metastases. Initial studies demonstrated that liposomes incorporating MTP-PE significantly delayed the growth of UV-induced tumors, prolonged the host's survival, and reduced the incidence of spontaneous metastases from 30% to approximately 10% (9). Recent studies examined OK-432, which was injected i.v. at 1 KE/animal administered biweekly for 3 weeks, followed by 2 weeks of intralesional administration on a weekly basis. The second immunomodulator studied was poly ICLC, which was administered biweekly at 1 μg/animal for 12 weeks and, predicated on the initial experiments, liposomes incorporating MTP-PE were administered on a biweekly basis for 4 weeks. In these studies, as shown in Table 6, it is apparent that all 3 of these agents at a nontoxic OTD significantly prolonged the survival of these animals.

Table 6

MEDIAN SURVIVAL OF UV-INDUCED AUTOCHTHONOUS TUMORS[a]

BRM	Dose	Schedule	Duration	Median Survival Time (Range)
HBSS	-	biw	4 wk	13.5 (5-33)
OK-432	1 KE	biw	5 wk	23.5 (3-33)[b]
pICLC	1 μg	biw	12 wk	20.0 (6-35)[b]
MTP-PE, MLV	2.5 μmole	biw	4 wk	21.5 (7-40)[b]

[a]Tumors were induced by chronic exposure to UV radiation (9). Following the development of a palpable tumor, mice were randomly entered into a therapeutic regimen (N=20). OK-432 was administered i.v. on a biweekly (biw) basis for 3 weeks and then intralesionally weekly for 2 weeks.
[b]Significant prolongation of survival as measured by the Kruskal Wallis analysis.

One question that is still unresolved is whether immunomonitoring studies can be utilized to determine (individually, if necessary) the OTP (dose and schedule) of administration for an immunomodulator. Studies were undertaken in collaboration with Dr. Diethelm Hartmann (Hoffmann-La Roche, Basel, Switzerland) to determine whether there is a correlation between the survival of tumor-bearing animals and the augmentation of specific effector cell activities. As shown in Table 7, we examined the therapeutic activity of poly ICLC using an OID in normal mice (10 μg/animal administered daily), an MTD (50 μg/animal administered daily), and the OTP (100 μg/animal administered biweekly).

Table 7

CORRELATION BETWEEN HOST SURVIVAL AND FUNCTION OF THE EFFECTOR CELLS IN THE TREATMENT OF MBL-2 ASCITES TUMOR-BEARING MICE[a]

Treatment		MST	%Mac[c] in the Ascites	Mac[c] Cytotox.	CTL[c] MBL-2	NK[c] YAC
PBS	qd[b]	20	1	0	3	0
PICLC 10 μg/A	qd	36	65	10	10	10
PICLC 50 μg/A	qd	34	74	11	13	16
PICLC 100 μg/A	biw	43	75	16	17	25
Statistical Analysis[d]	p=	-	0.05	0.01	0.05	0.08

[a]Syngeneic mice were injected i.p. with 5,000 MBL-2 ascites cells and 2 days later therapy was initiated by i.p. administration.
[b]qd = daily administration, biw = biweekly.
[c]Mac.: Macrophages; CTL: Cytotoxic T Lymphocyte activity; and NK: Natural killer cell activity.
[d]Pearson's correlation of medians.

The therapeutic model utilized MBL-2 tumor ascites, and therapy was initiated 48 hours following the i.p. administration of 5,000 MBL-2 ascites cells. Twelve days after the initiation of therapy, we used B16-BL6 target cells to determine the functional activity of macrophages from the ascites,

MBL-2 tumor cells to determine specific CTL activity of spleen cells, or YAC tumor cells to determine the NK cell activity of the spleen cells. The effector cell assays included the 72-hr macrophage assay using ^{125}I-IUdR labeled B16-BL6 target cells, the 4-hr ^{51}Cr release assay using the MBL-2 targets for the CTL assay, and the 4-hr ^{51}Cr release assay for NK cell activity (5). Using the statistical analysis of Pearson's correlation for medians we determined the correlation of host survival with the percentage of macrophages in the tumor ascites (0.05), macrophage effector cell function (0.01), CTL activity against specific (0.05) or nonspecific P815 targets (0.04), and NK cell activity (0.08). In this tumor model, macrophage activation strongly correlated with therapeutic response, suggesting that tumoricidal macrophages represent an important effector cell. In addition, cytotoxic T cell and natural killer cell augmentation also correlated with host survival although at a less significant level (p = 0.05 and 0.08 respectively). These results encourage immunomonitoring studies during clinical trials and suggest that they may be useful in optimizing therapeutic protocols for subsequent clinical trials.

In summary, these results demonstrate that it is possible to efficiently generate a body of data on a series of widely disparate BRMs, which can then be examined for possible correlations, retrospective or prospective, with clinical immunomodulatory and therapeutic effects. Initial studies of clinical correlations have demonstrated the need for a clinical assessment of immunological parameters as similar as possible to those used in preclinical studies. These studies also suggest that such immunomonitoring may be useful in developing effective therapeutic protocols. Ultimately, we hope to determine the usefulness of the PSL results in designing clinical treatment schedules, determining appropriate monitoring assays, and predicting therapeutic efficacy.

REFERENCES

1. Mihich, E. and Fefer, A. In: National Cancer Institute Monograph 63, U.S. Government Printing Office, Washington, D.C., 1983.
2. Oldham, R.K. J. Biol. Resp. Modif. 1:81-100, 1982.
3. Fidler, I.J., Berendt, M. and Oldham, R.K. J. Biol. Resp. Modif. 1:15-26, 1982.
4. Talmadge, J.E., Oldham, R.K. and Fidler, I.J. J. Biol. Resp. Modif. 3:33-109, 1984.
5. Talmadge, J.E., Fidler, I.J. and Oldham, R.K. Screening For Biological Response Modifiers: Methods and Rationale. Martinus Nijhoff Publishing, Boston, 1985.

6. Talmadge, J.E., Fidler, I.J. and Oldham, R.K. Behring Inst. Mitt. 74:189-194, 1984.
7. Talmadge, J.E. and Herberman, R. Cancer Treat. Rev. 1985.
8. Talmadge, J.E., Lenz, B.F., Collins, M.S., Uithoven, K.A., Schneider, M.A., Adams, J.S., Pearson, J.W., Agee, W.J., Fox, R.E. and Oldham, R.K. Behring Inst. Mitt. 74:219-229, 1984.
9. Talmadge, J.E., Lenz, B.F., Klabansky, R., Simon, R., Riggs, C., Guo, S., Oldham, R.K. and Fidler, I.J. Science, Washington, D.C., 1985.
10. Talmadge, J.E., Adams, J. and Phillips, H. Cancer Res. 45:1058, 1985.
11. Talmadge, J.E., Adams, J. and Phillips, H. Cancer Res. 45:1066, 1985.
12. Suga, T., Shiio, T., Maeda, Y.Y. and Chihara, G. Cancer Res. 44:5132-5137, 1984.
13. Rose, W.C., Reed, F.C., Siminoff, P. and Bradner, W.T. Cancer Res. 44:1368-1373, 1984.